Studies in Systems, Decision and Control

Volume 151

Series editor

Janusz Kacprzyk, Polish Academy of Sciences, Warsaw, Poland
e-mail: kacprzyk@ibspan.waw.pl

The series "Studies in Systems, Decision and Control" (SSDC) covers both new developments and advances, as well as the state of the art, in the various areas of broadly perceived systems, decision making and control- quickly, up to date and with a high quality. The intent is to cover the theory, applications, and perspectives on the state of the art and future developments relevant to systems, decision making, control, complex processes and related areas, as embedded in the fields of engineering, computer science, physics, economics, social and life sciences, as well as the paradigms and methodologies behind them. The series contains monographs, textbooks, lecture notes and edited volumes in systems, decision making and control spanning the areas of Cyber-Physical Systems, Autonomous Systems, Sensor Networks, Control Systems, Energy Systems, Automotive Systems, Biological Systems, Vehicular Networking and Connected Vehicles, Aerospace Systems, Automation, Manufacturing, Smart Grids, Nonlinear Systems, Power Systems, Robotics, Social Systems, Economic Systems and other. Of particular value to both the contributors and the readership are the short publication timeframe and the world-wide distribution and exposure which enable both a wide and rapid dissemination of research output.

More information about this series at http://www.springer.com/series/13304

Halit Ünver

Global Networking, Communication and Culture: Conflict or Convergence?

Spread of ICT, Internet Governance, Superorganism Humanity and Global Culture

Foreword by Vinton G. Cerf

 Springer

Halit Ünver
Research Institute for Applied Knowledge
Ulm
Germany

ISSN 2198-4182 ISSN 2198-4190 (electronic)
Studies in Systems, Decision and Control
ISBN 978-3-030-09493-5 ISBN 978-3-319-76448-1 (eBook)
https://doi.org/10.1007/978-3-319-76448-1

Printed on acid-free paper

This Springer imprint is published by Springer Nature
The registered company is Springer International Publishing AG
The registered company address is: Gewerbestrasse 11, 6330 Cham, Switzerland

"I have no special talent. I am only passionately curious."

Albert Einstein
(1879–1955)

"Out beyond ideas of wrongdoing and rightdoing, there is a field. I will meet you there."

Mevlânâ Dschalâl-ed-dîn Rumî
(1207–1273)

Foreword by Vinton G. Cerf

Vinton G. Cerf is Vice President and Chief Internet Evangelist for Google. He contributes to global policy development and continued spread of the Internet. He is the co-designer of the TCP/IP protocols and the architecture of the Internet and widely known as one of the "Fathers of the Internet". He is Fellow of the Royal Society, and has served in executive positions at MCI, as President of ACM, the Corporation for National Research Initiatives and the Defense Advanced Research Projects Agency and on the faculty of Stanford University. His contributions have been acknowledged and lauded, repeatedly, with honorary degrees and awards that include the US National Medal of Technology, the Presidential Medal of Freedom, the Turing Award 2004, the Marconi Prize and membership in the National Academy of Engineering.

The book you are reading is an important contribution toward understanding the nature of global communication, the growth of the Internet and the World Wide Web, the role of smart phones and mobile telephony and the way in which these technologies are applied in different cultural, economic, and educational contexts. It is not an overstatement to say that the Internet with its diverse applications has become a significant and even crucial part of lives of about half of the world's population today. Efforts to develop the Internet protocols were international from the beginning. The Internet has significantly contributed to globalization processes and transcultural interactions which will continue into the future. The formation of the Internet Society in 1992 was partly motivated by my belief that a society would in fact emerge from the presence and spread of the Internet.

The main subject of this empirical research is to identify convergence/divergence trends in the use of ICT in order to assess the existence of digital divides at a country, culture, and worldwide levels. These investigations conclude that the convergence regarding ICT usage proceeds faster worldwide than convergence in terms of wealth and education in general. Multivariate analysis shows that ICT use is most closely related to economic performance, followed by the Education Index, which contributes only slightly less to ICT use than economic performance. However, the adult literacy rate hardly has any influence on ICT use today. These results can be explained by the fact that the (mobile) use of the modern-day Internet is growing less expensive (devices and access) and can also be used by illiterate as well as physically or cognitively disabled people. In the future, Internet-capable devices and access will become less costly and thus facilitate ICT usage worldwide bridging the technical digital divide. But there is still a way to go. As a result, we need to consider means to minimize digital inequality in terms of information asymmetry and to maintain concepts such as the open Internet or net neutrality. In some sense, however, this book is unique in looking at the digital divide from the perspective of different religions and cultures showing hardly any empirical correlation between ICT access rates and religious affiliation, but a positive correlation with religious diversity on ICT access rates in a country. It can be stated that the social participatory effect of ICT usage is accepted by all worldviews and cultures.

This research is highly interdisciplinary and describes in the area of power politics that similar bargaining positions can cross-cultural boundaries, and thus convergence patterns can be observed in the global multi-stakeholder community (technical community, civil society, private sector, and governments) within the Internet governance ecosystem. One key milestone for multi-stakeholder convergence in the field of Internet infrastructure governance was the transfer of unique Internet parameter administration, the so-called IANA transition, by the US to a broader international structure of responsibility. The Internet Assigned Numbers Authority (IANA) stewardship transition is largely complete. On October 1, 2016, the telecommunication authority of the US Department of Commerce (NTIA) allowed the contract with ICANN for the performance of the IANA functions to lapse. For the first time, a multi-stakeholder community controlled and managed the root zone (top-level domains) of the Domain Name System and the global allocation of Internet address space. This is a positive sign, bringing the diversity of cultures worldwide into cooperative relationships and stimulating convergence of power structures and processes in the Internet governance ecosystem. Bridging the cultural gap between the technical and political communities should not be underestimated as cooperation between these two stakeholders plays a central role, especially in the digital transformation of society.

Overall, the author of this book claims that *"it is not clear whether we as humanity will succeed in the foreseeable future to develop a global culture that is more than the sum of all cultures. It is open to the question of whether a largely (digitally) networked world society will form a community, constituting a world culture. However, one thing is true: There will be no global culture without a digitally networked world society."*

Two things are evident as we look at the state of the Internet and the World Wide Web today. The first is that the system is capable of supporting an extraordinary degree of diversity. For proof, one has only to look at the endless collection of Web sites and the hugely diverse platforms such as Facebook, Twitter, WeChat, Amazon, Alibaba, YouTube, and soon the so-called Internet of Things. The second is these platforms are put to use by highly diverse populations that may effectively form disconnected islands by self-selection. This trend is sometimes called a *bubble chamber* effect that reinforces boundaries between groups creating a digital divide of a different kind. We can see both trends at work in today's Internet. At the time the Internet was designed, its inventors had a global infrastructure in mind. As the system evolved and spread, we hoped that the Internet and the technologies that support it could contribute to greater democracy, freedom of speech, and human rights worldwide. These are certainly values that will have a decisive impact on the shared future of billions of people and which can already be regarded as a basis for global empathy and culture. There is no guarantee that a single, global society will emerge from the widespread use of the Internet and ICTs. Indeed, such an outcome can only result if there is widespread agreement and determination that such an outcome is desirable.

Reston, VA, USA   Vinton G. Cerf

Foreword by Franz Josef Radermacher

Franz Josef Radermacher is a German mathematician and economist. He is Professor of Computer Science at Ulm University and the Director of the Research Institute for Knowledge Processing/n (FAW/n). He is one of the Co-founders of the Global Marshall Plan Initiative that suggests a socio-ecological plan to eradicate poverty, increasing global wealth while protecting natural resources. He is a Member of the Club of Rome since 2002 and President of the Senate of Economy since 2010. He was a Member of the Information Society Forum of the EU Commission between 1995 and 2001, is a Member of the scientific advisory board of the Federal Ministry of Transport and Digital Infrastructure since 2000, Member of the advisory board of the Vodafone Institute for Society and Communication since 2012, holds Best Human Award 2012, and was awarded with an honorary doctorate from the International Hellenic University in Thessaloniki in 2013.

This book by Halit Ünver is an important contribution to a holistic understanding of globalization processes in the context of digitalization and the global information society. It derives from his doctoral thesis in Computer Science at Ulm University/Germany. It integrates years of analysis on global cooperation and conflict with a particular focus on cultural aspects of globalization against a background of global digital transformation. Halit Ünver is well equipped for this issue, given (1) his general interest in topics of global cultural diversity and cultural exchange, (2) the broad scope of his studies in computer science and economics, (3) his cultural background, with deep roots and contacts in Germany and Turkey, and (4) his cultural connection via family ties to Kazakhstan and Russia.

The driving topic of this book is the quantitative and qualitative description of the global information society. With view to recent years, the role of the (mobile) Internet as a basis for global communication and international development is a main issue. The leading question is whether people's access to modern communication tools is influenced by cultural background. The result of the book is clear-cut: All people more or less react similar to modern information and communications technology no matter their cultural background. There is also a clear indication of (fast) convergence concerning ICT usage worldwide since the beginning of the new economy in 2000. This means that differences concerning, e.g., Internet usage are eliminated much faster than differences in global material situations such as income.

Two forces seem to be driving the issue. On the one hand, as a consequence of Moore's Law, powerful (digital) technologies are available and inexpensive. On the other hand, modern ICT tools mean massive advantages for their users therefore almost everyone is keen to use them. ICT devices are fashionable and modern. But we should also take into account that even the forces which would like to feed cultural conflicts make use of ICT, i.e., this technology is used by people who do not want to see any type of (cultural) convergence on a global scale. However, in general, the technology provides so much that it is very often misused. Therefore, it may well be that society has to strongly manage and regulate ICT access and usage in order to avoid massive disadvantages for society.

The book of Halit Ünver focuses on convergence and the current divides in technology usage in the information age, particularly the most recent and most powerful one: the (mobile) Internet. Still, differences remain and in international politics we speak of divides. Halit Ünver addresses these divides systematically such as the *digital divide* or *digital inequality*, what they mean and how they are to be positioned. The question is: Where do these divides come from? Do they come from different cultural backgrounds? It might be that certain religions are negative toward these technologies. Of course, there could be divides between men and women or divides between the young and the old.

Most natural are, obviously, two key figures that characterize *digital divide*: one is income. This is due to the high cost of technology as a result of the latest adaptations and updates. If we think about the total cost of ownership to process and operate information there is always a cost. Therefore income is a principal limiting factor of ICT use, which is analyzed and measured on a global scale in detail in this book. Another important key figure is education. To make use of (high quality) ICT, certain intellectual abilities are needed. For instance, if a technology forces people to know different languages or to be literate, many people are excluded. When looking at more analytical matters including mathematical graphs and operators, one must possess the knowledge in making proper use of technology.

Certainly, in the early days of mobile communications, e.g., when SMS was the mode of mobile electronic exchange of texts, literacy was crucial. Therefore, education was a major factor in allowing access to ICT. Now, with greater availability of modern intelligent technology the need for certain formal competences of use is being reduced. Nowadays, Internet offers, to a limited extent, translation of texts into different languages. Very often, pictures can be used for communication. They are quite universal and culturally neutral in usage if we do not consider religious images or pictures. There is a certain tendency to at least make considerable parts of content available without many formal requirements in education. In parallel, global politics such as the *Millennium Development Goals* (MDGs) and the *Sustainable Development Goals* (SDGs) work in the direction of higher education levels and more balanced income situations everywhere, though there is a long way to go. These are reported all over the world, but how much of this information is reliable is another topic.

Besides culture, particular income and education are two key figures for considerable differences in usage of modern ICT. Here, an additional problem in understanding arises from the fact that (high) income and (high) education are also correlated, which means that richer people tend to be better educated and better educated people tend to have a higher income.

This book does an enormous work and effort as well as detailed analyses with profound statistical and mathematical methods to clarify all kinds of dependencies between the different issues. All in all, the author of the book can prove that culture and religion are not an essential factor in distinguishing the adaptation of modern ICT between different countries and regions in the world. The dominating factors are income and education. As a result of a multivariate statistical analysis presented in this book, it seems that income is more important as a separating dimension than education.

This book gives deep insights into global networking processes and a signal that can stimulate hope. At least for one of the most powerful technologies available in the world, we have similar attitudes worldwide and between all cultures, concerning usage. We see clear convergence in access and usage which is due to the relative low cost of technology, its huge enabling potential to mobilize people and the increasing emphasis on education as well as higher income worldwide. There is a hope that there is a chance for a good future for a majority of humankind through adapting the power of these technologies. So this can be a positively accelerating factor to achieve a kind of *global culture*, particularly by allowing greater and widespread interaction between people.

There are a lot of activities globally toward this convergence on the political, economic, and civil society level. However, there are increasing risks of a different nature, namely the use of this powerful technology for purposes that do not deal with global cooperation and the implementation of sustainability and human rights. The usage of modern ICT can possibly be used to support the interests of small groups of privileged people. It is therefore not clear how the future will be affected by technologies. To have a look at critical points is certainly of importance;

therefore, this book gives us two stimuli: an indication for hope and a critical view at potential new risks.

As the main supervisor of his doctoral work, I congratulate Dr. Halit Ünver on behalf of the Ulm University, the Research Institute of Applied Knowledge Processing (FAW/n) in Ulm as well as the Senate of Economy, of which I am the president. Because we were all highly interested in this work, we strongly support it and we are happy that the book is now available in English. We congratulate the author and wish him success going forward.

Ulm, Germany   Franz Josef Radermacher

Foreword by Rainer Wieland

Rainer Wieland is a German Lawyer, Politician, and Member of the European Parliament. Since 2009, he has been one of the Vice Presidents of the Parliament. He is a Member of the conservative Christian Democratic Union, part of the European People's Party (EPP). Rainer Wieland was chairman of Europa-Union Deutschland Baden-Württemberg (regional section of Union of European Federalists) between 2001 and 2013 and since 2011 has been President of Europa-Union Deutschland. As Vice President of EU Parliament, he has responsibility for buildings, budget, transport, environment-conscious Parliament, Ad Hoc Working Group on the implementation of security policies and strategies, relations with the French, Belgian, and Luxembourg authorities on the seat and places of work of the Parliament, Representing the President of the Parliament for Africa / ACP, Member of the Committee on Constitutional Affairs, Member of the Committee on Development. In 2009, Rainer Wieland received the Order of Merit of the Federal Republic of Germany which is the only federal decoration of Germany.

It is a pleasure for me, to write this Foreword to the book of Dr. Halit Ünver on the subject *Global Networking, Communication and Culture—Conflict or Convergence*. For the summary, in only a few sentences, of the core message advanced in his research, led to no hesitation on my part. Mr. Ünver's topic is of paramount importance, not only for me, as Vice President of the European Parliament, but for everyone. In today's global society, it is hard not to think of information and communications technology. On the contrary, it is part of our lives and its development process is still far from complete. I speak here deliberately of a *global society*, because the versatile technical possibilities available today make fixed spatial, cultural, or religious boundaries unthinkable.

The Internet and Internet-capable smartphones allow each one of us, in every conceivable place in the world, to build global networks and be in continuous contact with each other and exchange information over wide spatial distances, as well as exchange our cultural differences. Through ICT, distances are no longer definite barriers to communication, including differences in culture and language or government. Mr. Ünver has his finger on the pulse of our times and demonstrates clearly and comprehensively, with his elaborations, that communication media offers, more than anything else, the chance for establishing *global balance*. He states clearly that it is not cultural or even religious differences that hinder the use or access to these technologies. Rather, the divergent use of communication technologies results from differing conditions concerning prosperity and education.

Through his book, he supports the global demand of taking a path toward more balanced income and education levels, in line with the positions of the United Nations, OECD, and the European Union. Mr. Ünver illustrates that a cross-cultural convergence is not only possible, but already in full progress. Our task now is to bridge the serious existing global disparities in education and economic capacity. As Mr. Ünver outlines, which I believe in a very convincing way, there is absolutely no disparity with respect to the use of telecommunications media among the major world views, such as Christianity, Islam, Judaism, Buddhism, and Hinduism. In terms of information and communications technology, they are all well advanced. The acceptance of using this communication option is therefore a cross-religion as well as cross-cultural process and is important to promote. Furthermore, Mr. Ünver addresses the fact that with growing technology and highly developed methods of communication a change must be made in the context of Internet governance. He argues correctly that globalization, in connection with the way toward becoming a global information and knowledge society, could bring more distinctive participatory governance. In addition, this could encourage a more intense and well-functioning cooperation between countries, both in Europe and globally.

In this context, Mr. Ünver does not ignore the potential risks associated with globalization and related to ICT. He speaks here of a possible development in which communication technology and networks do not result in convergence, but lead to potential conflicts; a development in which these technologies could be used for purposes of cultural struggle or for state control. In this context, Mr. Ünver illustrates the importance of democratic structures for global governance as well as its processes and anchorages to prevent such threats.

I would like to thank Mr. Ünver for this outstanding work, which shows me once again how important it is to deal in-depth with information and communications technology, particularly in our current times. I can only agree with the conclusion that the advancement of globalization in connection with communication media offers a chance for us to provide a path toward a *global culture* of peace and mutual respect—a direction that we should be following. Similarly to Mr. Ünver, I also advocate the view that we should take advantage of the potential that Internet and mobile telephony offer in order to create more balance, not only in the spheres of politics, culture, and religion, but also economic, social, and environmental.

Brussels, Belgium   Rainer Wieland

Acknowledgements

I am extremely lucky and grateful to have met the right people at the right time in the right place, which in retrospect, belong or belonged to my own *puzzle* of personal and professional development. I would like to take this opportunity to thank all those who have actively supported and inspired me in the research and writing process of this book, which essentially is the English version of my doctoral thesis at Ulm University in Germany (2015) with some updates.

Special thanks go to Prof. Dr. Dr. Dr. h.c. Franz Josef Radermacher, who motivated me through his guidance and continuous inspiration over the long phase of developing this book and broadly supported me in framing this issue. In recent years, I have enormously benefited from his interdisciplinary understanding and motivating attitude. He has methodically shaped and greatly influenced the content of this work. Professor Radermacher, as my superior and doctoral supervisor, has always offered me flexibility and space to successfully finish the work, participate in higher-level teaching, complete a second degree in economics, and also to become active in the economic, political, and the public sphere, where I particularly can bring in my Turkish–German background. I would like to thank him for recommending me via the Vodafone Institute for Society and Communications to the Young Leaders Group 2012/2013 of the European Institute of Innovation and Technology Foundation, steered by the EU Commission and Imperial College London Business School. In this environment, I have gained initial experience in bridging the cultural gap between the technical community and policy-makers.

I am eminently grateful to Prof. Dr. Michael Weber, President of the Ulm University, who took over the second report of my doctoral thesis. Already as a student of information technology, Professor Weber's expertise on the subjects of media and computer science fascinated me. It always has enriched me to have him on board as an expert with excellent experience in research and teaching as well as university–industry cooperation.

I would especially like to thank Prof. Dr. Wolfgang Coy who has made, as an external expert from the Humboldt University of Berlin, important contributions to the nature of this book. I thank to both, Professor Coy and the Alcatel Foundation, for providing me with the opportunity to work on a scientific basis in an interdisciplinary environment such as the Ph.D. group *Computer Science, Culture and Technology*, on global issues of communication and to learn *to think outside the box*.

A very big thank goes to Dr. Thomas Kämpke (1957–2015), who, as a senior scientist at the Research Institute of Applied Knowledge (FAW/n) in Ulm, helped me in forming this work up until his last days. I will probably never forget his moral during his days in the hospital regarding my success. This is especially true for the phrase "We need to switch to Plan B" when he knew that he only has a few weeks left to live. I am very grateful for his consistency and attention to detail in supporting my work. I was supervised by Dr. Kämpke in almost all relevant tasks of this research work. It was a great pleasure to know Dr. Kämpke as a scientific *sparring partner* at my side, and I will always remember him in lasting memory.

I would also like to thank Prof. Dr. Franz Schweiggert and Prof. Dr. Frank Kargl, both from Ulm University, for their cooperation in the doctoral committee. It was always motivating and valuable for me to find the professional and human support of Prof. Schweiggert at Ulm University, as a student and scientist. He also provided assistance to me during my time at Daimler Corp. as a trainee. Many thanks also to Prof. Kargl who participated in the doctoral committee following the passing of Dr. Kämpke and supported me with technical questions about the Internet ecosystem.

A big thank you goes to Vinton G. Cerf, Vice President of Google, co-designer of the TCP/IP protocols and widely known as one of the "Fathers of the Internet" as well as Turing-Award Winner, for the conversations with him and writing a preface. I met Vint Cerf for the first time at the Internet Governance Forum 2014 in Istanbul. The initial talk with him immediately guided my thoughts on my subject to a distant horizon. On the way to this horizon not only the invention and historical development of the Internet can be found but also possible and uncertain development paths accompanied by the knowledge that the future could be very different from what we think about it today. My book is complete now but the topic is still in my mind and on my horizon.

The same thank you goes also to the Vice President of the European Parliament Rainer Wieland. About five years ago, we literally got to know each other high up in the sky and talked intensively about technology, politics, and society. I gladly accepted his invitation to Brussels and since then I thank Rainer Wieland for his continuous support in politics and society.

I would especially like to thank my colleagues at the Institute for Databases/Artificial Intelligence and the Research Institute for Applied Knowledge Processing (FAW/n) in Ulm for their support and enriching cooperation over many years. These include in particular Sabine Grau, Regina Simon, Elke Moc, Michael Gerth, and Jürgen Dollinger. At this point, a great thank you also goes out to former colleagues and students such as Tobias Rehfeld, Martin Borowiec, Sultana

Chatzivasiliadou, Julian Schweiger, and Anil Aslaner, who contributed significantly to the dissertation, especially in the area of database management and detail work. For a final proof reading, I would like to thank Dr. Anne Wenger, Alev Kayagil, and Christoph Baldauf.

Eminently I would like to thank my mother and my whole family who supported me through permanent motivation and patience. I am proud of my mother who followed my father from Sarıyahşi, a small village in the middle of Anatolia, to Ankara and then to Germany and managed to pave my way with love and affection. This also applies to my deceased father whose basic values have shaped me decisively in my early childhood and youth.

My last big thank to Inna and our little daughter Aliya, who was born in the final phase of the dissertation work. They have both made this crucial phase pleasant for me with their motivation and support. Furthermore, they showed me a totally different perspective on life and cultural exchange. They gave an additional meaning to my ground of being. May we all together live in a world characterized by sustainable development and convergence to a *global culture*!

Contents

About the Author

**Halit Ünver** is a scientist and policy adviser at the Research Institute for Applied Knowledge Processing (FAW/n). After studying economics and information technology, he earned a Ph.D. with a thesis on *Global Networking, Communication and Culture Conflict or Convergence* at the Ulm University, where he was a scientist at the Institute of Databases/Artificial Intelligence between 2009 and 2016. He is currently Visiting Researcher at the Leadership Excellence Institute (LEIZ) at Zeppelin University exploring the topic of *Leadership in the Digital Age*. He was nominated by the Vodafone Institute for Society and Communication to the Young Leaders Group 2012/2013 of the European Institute of Innovation and Technology, steered by the European Commission and Imperial College London Business School. Within this framework, he was awarded at the Annual Innovation Forum in Brussels 2013. He was a Member of the Alcatel-Lucent Foundation for Communications Research Ph.D. working group *Computer Science, Culture, Technology* between 2011 and 2014. He is a board member of the Ph.D. Network for Sustainable Economics. Since 2015, he has participated in the T20 process within the G20 presidency and was part of the task force *The Digital Economy* during the German G20 presidency in 2017. He is part of the Stiftung Mercator network of young German, Turkish, and European leaders from all sectors. Between 2005 and 2008, he was a Daimler AG student trainee in the fields of truck production, global IT management, and R&D.

Acronyms

ALR	Adult Literacy Rate
BRICS	Brazil, Russia, India, China, South Africa
EI	Education Index
FTR	Fixed Telephony Rate
GDP	Gross Domestic Product
GDPpC	Gross Domestic Product Per Capita
GSM	Global System for Mobile Communications
HDI	Human Development Index
HTTP	Hypertext Transfer Protocol
IAB	Internet Architecture Board
IANA	Internet Assigned Numbers Authority
ICANN	Internet Corporation for Assigned Names and Numbers
IEEE	Institute of Electrical and Electronics Engineers
IETF	Internet Engineering Task Force
IGF	Internet Governance Forum
ICT	Information and Communications Technology
IPR	Internet Penetration Rate
IRTF	Internet Research Task Force
ISDN	Integrated Services Digital Network
ISOC	Internet Society
ITU	International Telecommunication Union
IP	Internet Protocol
MBR	Mobile Internet Penetration Rate
MPR	Mobile Phone Rate
OECD	Organization for Economic Cooperation and Development
Pew	Pew Research Center
PSTN	Public Switched Telephone Network
RDI	Religion Diversity Index
SQL	Structured Query Language
TCP	Transmission Control Protocol

UN	United Nations
UNESCO	United Nations Educational, Scientific and Cultural Organization
VoIP	Voice over IP
WB	World Bank
WCF	World Culture Forum
WLL	Wireless Local Loop
WSIS	World Summit on the Information Society
W3C	World Wide Web Consortium
WWW	World Wide Web

Part I
Preparation of the Topic

Chapter 1
Introduction

Before anything else, preparation is the key to success.
(Alexander Graham Bell, 1847–1922)

This book is centered on the study of globalization and the diverse cultures representing the transition to a digital future in the context of sustainable development. The processes of globalization are apparent in several periods in human history, however remain modest when compared to the most recent global trends. If one looks at the evolution of humanity as a system, the emergence of the spoken language can be seen as a medium of communication for people as an all–important new basis for future social and technical innovations. The invention of script also meant an immense step forward with the letterpress laying the base for the global spread of knowledge.[1] Nowadays, modern information technology allows for more wide-ranging and bigger effects. Innovation in the field of information technology is currently considered as innovation behind almost all other innovations. It is the fastest process of innovation which has ever taken place in history.

In the late 19th and during the 20th centuries, the fixed telephony, as a local oriented communication form, contributed decisively to the present globalization process. Afterwards, the mobile phone, as a person oriented communication form, has further accelerated this process. Today's communications via the Internet and the World Wide Web are unique in comparison to former processes of this kind. In this respect, they operate at huge speed while bridging spatial and temporal distances, spreading data across the globe. Therefore, the worldwide information and knowledge society distinguishes itself by further increasing interlinking processes, enabled

---

[1] By the way, these inventions had also caused social and economical risks and threatened the people. For example, many writers had lost their jobs cause of the letterprin technology.

© Springer International Publishing AG 2018
H. Ünver, *Global Networking, Communication and Culture: Conflict or Convergence?*, Studies in Systems, Decision and Control 151,
https://doi.org/10.1007/978-3-319-76448-1_1

and promoted by modern ICT. In particular, Internet and mobile telephony and their *technological convergence* to the mobile Internet shape globalization processes at different levels in economy, science, politics, law, religion, media and social life. Facing the possibility of a sustainable development in the context of the transition to a digital future is a key issue. It remains to be seen whether humanity with its diversity of cultures, art and various (national and international) power structures is able to establish a global superorganism with a kind of global culture by means of world ethos.

1.1 Situation

The systemic dimension of humanity as a superorganism is especially expressed in communication processes. We are more than a collection, a horde of people or individuals [369]. We are a system of higher order. Today, the nation is the most highly organized structure or system consisting of a large number of people living in a restricted land area. There are also various cultural systems as a form of existence of many people from different states, sharing norms and values with clear nameable differences with other cultural systems. That people belonging to different cultural systems meet each other (e.g. at the airport/on vacation) or live together in a certain state.

The humankind as a whole is on the way towards becoming a linked up *hybrid human-technology system*, a so called global *superorganism* [162, 371]. The process of digital interlinking between human beings and technical systems through the use of ICT opens up new economic markets, new equilibriums and options in the international division of labor and cooperation.[2] New learning processes and environments in the field of science and education are changing our present thinking patterns, processes of knowledge generation and know-how transfer. Debates and campaigns at the political level are strongly influenced by the Internet and changing patterns of social interactions. Large sport events and concerts are watched simultaneously online by millions of people worldwide. The way we make music and listen to it as well as our travel behavior have changed fundamentally. Ultimately, the Internet and mobile telephony together with global mobility infrastructures offer enormous mobilization potential both for small and large groups of people. All this forms the basic structures of our culture including our careers, family and spare time.

At the same time, the Internet and mobile telephony bring corresponding, however, also some known and unknown risks with itself. Our privacy is partially no more a private matter where cyber crime and (mass) surveillance activities have increased to a disconcerting level. The speed of a majority of social processes has become faster, thereby increasing competition and potential conflict situations. Nevertheless, communication technology helps to improve many living conditions (well-priced level). It essentially contributes not only to modern life in developed countries, but also to development aid in impoverished countries. Finally, the Internet and mobile

---

[2]In future *cyber-physical systems* and the *Internet of Things* (IoT) will play a more important role in the organization of society.

telephony are elementary technical achievements which decisively characterize and form our globe culture.

1.2 Motivation

The motivation behind this book is to develop an understanding about the current social and economic transformation processes towards becoming a global information and knowledge based society. In particular, the empirical functional relationships between digital interlinking, economic performance and education level of countries contribute to an essential understanding of society and the globalization process. This is attributed to the fact that the level of income and education appear to have significant impact on the level of digitalization. An exact consideration of the empiric relation between the different worldviews or religions and the respective rates of usage or penetration of single technologies on communication is worthwhile. Religion as a determining aspect of culture has (always) influenced the human behavior or action and will certainly influence in future. Here, for example, the two largest worldviews, Christianity and Islam, but also Buddhism, Hinduism, Judaism and Atheism, are present in varying degrees in different countries around the world. In this case the motivation is to illustrate that technology and especially ICT is accepted and used cross–culturally.

Somewhat more difficult is the acceptance of a cross-cultural understanding of power and its distribution or hierarchy structure. Therefore, it is of interest how power and controlling issues are regulated in the area of worldwide technology like the Internet by looking to see the bargaining position of some influential countries when it comes to the *Internet governance ecosystem*. Finally, a determining motivation is whether we use modern communicative devices in moving towards a *global culture* or *world ethos* as a new basis for our (common and shared) future on earth, living together in harmony.[3] A *cultural convergence* which allows a (harmonious) cooperation and coexistence is desirable for 10 billion people on our future planet. That means, not an arbitrarily extensive adjustment and certain specification of different cultures. The diversity of cultures must be protected as a value. But not as a divisive and delimiting value, but rather as an element in a diverse mosaic, in which everybody will find a place and commonality within diversity. This (*Unity in Diversity*) will allow the necessary cooperation and coherence in resolving future conflicts in a fair and peaceful manner.

No one can predict the future for certain. However, the exploration of relationships in the field of digitization at national and cultural level is a useful work to better understand possible lines of development in shaping the future. With this work the

---

[3]Please note that the term harmony is considered in different ways in various cultures.

author is motivated to make a contribution to essential problems regarding the development of ICT and its social acceptance in the context of the role of different cultures worldwide.

1.3 Question

The first question looks at which overall social factors effect the interlinkage of humans and information technology in countries worldwide. What is the empirical functional relationship between the economic performance of a country, the level of education and its access rate to fixed telephony, mobile phone, Internet and mobile Internet? How have these relations developed worldwide since the beginning of the *new economy* in 2000 to over more than one decade? What is the empirical relationship that exists between the different ICT penetration rates? What is the functional relationship between these technologies and the various big worldviews or religions? What is the state of affairs in Christian countries compared to the situation in the Islamic world with respect to the use of digital technologies? How are the conditions in multicultural states?

Another central issue at the global level is the governance of power structures and control mechanisms on the Internet, World Wide Web, and telecommunications infrastructure. In this regard it must be investigated how some relevant states and big cultural systems do think about controlling issues of interlinking process through Internet and mobile telephony. How will different cultural systems most likely influence the Internet ecosystem (technically and politically) in future? Which bargaining positions do big states and cultures take up on the current *Internet governance* debate?

It is important to consider the specific cultural challenges that await the world in becoming a *global superorganism* and *hybrid human-technology system*. After all, the inclusion or exclusion of communication processes is regarded as a driving force in the global superorganism as an *autopoietic system*, which is not always an unproblematic perception.

Conclusively, in the context of the digitization, one must question whether this worldwide interlinking leads to *conflicts* between influential cultural systems or whether we experience a *convergence* to a sort of global culture on the basis of a world ethos. For this purpose, three basic scenarios for the future are discussed in the context of the *Club of Rome*: *collapse*, *social inequality*, and *balance*.

1.4 Contribution

Guided by this challenge this book determines (empirical) correlations between global (digital) networking, in particular fixed telephony, mobile telephony, Internet and mobile Internet, and cultural conditions. The interdisciplinary approach of this book is typical to complement the research field with insights into empirical functional relationships between overall social key figures in the fields of economy, education, religion and the status as well as the development of different country's digitization levels. The analysis' are based primarily on methods of empirical statistics, whereby *linear* and *non-linear*, *univariate* and *multivariate regression models* serve to make a best possible fitting of the data.

The studies on the relationship between economic performance and access rates to fixed telephony, mobile telephony, Internet and mobile Internet start in 2000 (the beginning of the *new economy*) and go over a decade until 2013. The analysis of the empirical relationship between the level of education of countries, as measured by literacy rate and Education Index, and access rates to different communication technologies also began in 2000 (*Millennium Development Goals*) and end in 2012.

A majority of the previous research (see Chap. 3) describe the relationship between the economic performance, the level of education and the use of ICT in general with linear (multivariate) regression functions and partly with logarithmic or exponential models. In this work additional non-linear regression models, such as the logistic function or functions according to *Planck's law* are used to describe the relationship of the various parameters more precisely. Furthermore, a multivariate regression analysis is used to characterize the relationship between the different ICT access rates as dependent variables, and the level of education and income situation as independent variables, which themselves were modeled by non-linear regression functions.

A system–theoretical approach was designed based on the so–called multistep procedure of numerical mathematics to examine, e.g. the relationships between the different communication technologies. With this approach, e.g. the physical coupling or decoupling of different communication infrastructures, with respect to access rates, are made plausible.

The studies on the global relationship between the processes of digitalization and major worldviews, such as Christianity, Islam, Judaism, Buddhism, Hinduism, or other religious affiliation including non-believers is a major contribution to the scientific field and general public. These investigations take place (only) for 2010, because homogeneous data over several years at worldwide and at country level are not available in this respect. The data available for 2010 is in this sense very significant, because the results seem to confirm the trends that can be derived from the other records. The convergence trends in the use of ICT can be observed clearly among the major worldviews.

Another contribution is the identification of the coherence between digital networking, power and control issues on the Internet and WWW. Gaining relevant information according to a multi–stakeholder model which is anchored in a global framework (*global governance*) is crucial. To this end, some bargaining positioning of key countries and groups of countries will be investigated.

In addition, the description of the digitally linked up humanity as a global superorganism occurs in the form of an autopoietic system as an intelligent human-technology-system. Here, cultural challenges are discussed such as the *balance* between the *central concentration of power* and the extent of *decentralized power distribution*, the *inclusion* or *exclusion* of people and states in global communication processes, and the capacity for *global empathy*.

The discussion of possible future scenarios in principle that our world could face on the way towards the digital future—especially in the context of sustainable development—is a crucial contribution. In placing special emphasis on the positive effects of the Internet and WWW, the access to ICT helps people around the world in accessing financial systems, improving levels of educational, promoting cultural exchange, and developing a better understanding of *global empathy*. These trends favor a worldwide cultural convergence and a movement to a kind of *global culture*. These discussions will also be made for the study of future scenarios not compatible with sustainability such as a worldwide two-class society (high social inequality) or an ecological collapse.

Chapter 2
Fundamentals

> *Basic research is what I am doing when I don't know what I am doing.*
>
> (Wernher von Braun, 1912–1977)

In this chapter the conceptual basis for this book are discussed. The terms *global networking, communication, culture, conflict* and *convergence* are often used very differently, which is why a description of these terms in the context of this book are necessary. This clarification of terms is the subject of Sect. 2.1. Subsequently, Sect. 2.2 introduces concepts that attempt to classify global networking processes quantitatively (*digital divide/digital inequality*) and qualitatively (*Internet governance, superorganism, and global culture*). The former is relevant for empirical analysis in Part II of this book, the latter is of importance for future considerations in Part III. Section 2.3 describes the essential key figures that are used in later Chaps. 4 and 5 for the empirical investigations. Finally, Sect. 2.4 describes data management and used methods from empirical statistics.

2.1 Terms

2.1.1 Global Networking

The term *global* has its origins in Latin (lat. *globus*: sphere), and refers to the world or globe [325]. Nowadays, it is also commonly used as a worldwide phenomenon that reaches the whole of a system. In a narrow sense this term is, for example, used in computer science or mathematics. Here, it is also for the visibility of variables (see global variable) in a particular programming language [215]. When speaking of

© Springer International Publishing AG 2018

H. Ünver, *Global Networking, Communication and Culture: Conflict or Convergence?*, Studies in Systems, Decision and Control 151, https://doi.org/10.1007/978-3-319-76448-1_2

global, one is referring to the opposite of *local* which means not world–wide and is usually referring to an area close to someone or something [325]. In this book the term local is of relatively minor importance. For example, the empirical analysis of the relationship between the degree of digitization in countries and their economic performance is examined in a global dimension. Also the proportion of people of a country with a particular religious orientation is studied in a worldwide perspective (e.g. share of Muslims in Germany compared to share of Muslims in Turkey) instead of looking at local differences (e.g. share of Muslims in Germany compared to share of Christians in Germany).[1]

The word *network* is a derivative of the concept of *net* and comes from systems theory [511]. It is used to describe the relationship or link between elements of a system. The changing connections describe dynamic aspects of a network where individual elements can be connected, included or excluded. In computer science, networks are often described with models of graph theory [409]. Corresponding indicators help to quantify the network (*degree of crosslinking*) or the networking form (*complexity*). The brain is a network that consists of billions of neurons with multiple links between one another. By using the word *net*, one is also referring to the Internet generally. The Internet is now sometimes referred to as a *nerve network* of humanity [367, 374]. Therefore the Internet is for humanity, figuratively, what the brain is for a human and can be called the *brain of humanity*, when you look to this system as so-called *superorganism humanity* (see Chap. 7). The latter will certainly evolve drastically. On the other hand, when speaking of a (socially) well–connected individual the term refers to their diverse network of established relationships with other people of a society.

The term *global network* thus signifies the increasing worldwide connections (in particular shaped by information and communications technology) of states and people in all social areas (eg. politics, business, science, communication, art and culture). This occurs between individuals as well as organizations, companies and countries. The global networking recently was driven mainly by the rapid progress in ICT, but also through the development of transport technologies and infrastructure. It effects the contact intensity between members and organizations both within and between different cultures. At this point there is a close link to the questions that are of particular interest in this book [369].

2.1.2   Communication

Generally speaking communication (lat. *communicare*: to share) is the process of exchanging information between people, animals, living organisms and/or machines [369, 371]. The spoken and encrypted (or coded) language is of central importance in mediation between people. Recognizable love, fear, sympathy gestures and facial

---

[1]One reason for doing so: there is no data available on a global scale which gives an insight about ICT usage rates in various ethnical groups in a country.

expressions are also relevant forms between people, though these are analytically difficult to detect and do not (yet) exist between machines. Language and writing are forms of communication which machines can better handle. In a standardized form, they are increasingly used both for communication between machines (for example protocols) as well as for communication between man and machine (for example user-interface) or between people. All forms of communications have in common at least one transmitter and receiver in (successful) communication processes between whom the exchange of data or information happens. In a special case, the transmitter can also be the receiver at the same time [260].

Today, the concept of communication is mainly influenced and further differenti-ated by the technical progress in the field of ICT. In this book, communication stands primarily for the exchange of information between humans and machines via the fixed telephone, mobile telephone, Internet and the mobile Internet. Access to these technologies enables communication and is a necessary condition for global com-munication in the context of this book. This is the subject of subsequent empirical investigations. Furthermore, it is important for the considerations in this book that without communication there is no exchange and therefore no *cultural emergence* or even *convergence*.[2] In this sense, communication is constitutive for the development of culture. Communication between different cultural systems in the world is long established and constitutively for the development of humanity [210]. Today we are witnessing the process of closer exchanges and increasing communication between people from different cultures all over the world.

2.1.3 Culture and Cultural System

The latin word *cultura* is a derivation of the verb *colere* and means *processing, main-tenance, development, cultivation* in English. It originally referred to agriculture (lat. *agricultura*). The term culture or culture system in this book, in the broadest sense, stands for an aggregated overall structure of a society with its economy, science, pol-itics, religion, media and art [494]. The British evolutionary anthropologist *Taylor* used the following definition for culture in his work "*Primitive Culture – Researches into the Development of Mythology, Philosophy, Religion, Language, Art, Custom*" in 1871 [449]:

"*Culture or civilization, taken in its wide ethnographic sense, is that complex whole which includes knowledge, belief, art, morals, law, custom, and any other capabilities and habits acquired by man as a member of society.*"

This definition of Taylor speaks of a society in a very broad sense, similar to the formulation by the "*World Culture Forum*" (WCF) in Dresden 2007 and 2009 [494]:

"*Culture is everything, that is not nature*".

---

[2]In a certain sense, these considerations are inspired by Luhmann's theory of social systems [265].

The definition in the present work, is based on the general concept of culture which also includes a mental formatting or programming of people in their social environment, especially children, but also of all people through concrete daily activities. In relation to groups, individual or family behavior and lifestyles – such as frequent visits to the theater or sporting activities – can also be referred to as culture. At this level, the individual use of modern information technologies is decided. For example, people can only use the Internet for watching videos or for intellectual research work. The mobile phone, for example, can be used as a means to meet with other people or to operate commercially.

The *United Nations Educational, Scientific and Cultural Organization* (UNESCO) is interesting here, because most of the data and information used in this book are from international organizations. UNESCO defines culture as follows [467]:

> *"Culture, in its broadest sense, can be viewed as the set of distinctive spiritual, material, intellectual and emotional aspects, that distinguish a society or a social group. This does not only include art and literature, but also life forms, the fundamental human rights, value systems, traditions and religious orientation."*

An essential difference between the definition of *Taylor*, *WCF* and *UNESCO* is that the UN refers to *basic rights*. In the context of this book one can ask the question whether the access to the Internet[3] but also to certain algorithms[4] can or should be declared as a human right, especially when access for all would not lead to a *resource conflict*.

2.1.4 Conflict

The term *conflict* originates from the latin *confligere*, which means *clash, fight*. The Norwegian mathematician, sociologist and political scientist *Johan Galtung*, who also is an important peace researcher with an international reputation, defines the term conflict from the point of view of system theory as follows [121]:

> *"A characteristic of a system in which there are incompatible objectives, so that the achievement of a goal would preclude the achievement of another"*

---

[3] A similar question arises in the context of the (global) financial system, where it is about whether an access to a bank account can or should be declared as a fundamental right of a person [91].

[4] Among the most important algorithms of the information society are the algorithms of search engines. On one hand they have to be protected as intellectual property (competitive advantage for the company), on the other hand they can be used to manipulate crowds. For this example, Google founders *Sergey Brin* and *Larry Page* made the following statement in their article *"The Anatomy of a Large-Scale Hyper textual Web Search Engine* [41]: *It is clear that a search engine which was taking money for showing cellular phone ads would have difficulty justifying the page that our system returned to its paying advertisers. For this type of reason and historical experience [...], we expect that advertising funded search engines will be inherently biased towards the advertisers and away from the needs of the consumers."*

However, even with a common target (for example, a place as a destination) one can come to a conflict (for example, regarding the selection of the route), if the route is very important, because it perhaps affects other goals or not explicitly articulated objectives. A typical conflict with a common goal would be access to resources when, for example, several people want certain goods, but do not have a sufficient number for all interested persons. Dealing with corresponding shortages requires prudence when damage is to be avoided. The resource communications infrastructure was limited available in the past (e.g. stationary telephony via landline). Conflicts arose when several people wanted to call someone at the same time from a telephone or waited for a call. This situation has now eased in many parts of the world due to mobile telephony and many relatively inexpensive devices.

Computer science captures the term *conflict*, for example in the modelling of petry nets. These depict a graph theoretical formalism for modelling processes with concurrent character and causal relationships [385] (e.g., if two or more than two concurrent actions are planning activities or need resources that exclude each other). A conflict in a petri net does *not* represent an *error*, but displays a necessary *decision between several alternatives*. Such models in computer science and also respective theories in decision and game theory [214] show that (social) conflicts are not unacceptable or refutable per se. Rather, they are a central part of life and represent a challenge that should be dealt with wisely.

In this sense, conflict situations always require decisions regarding the settlement of the conflict. For this, *Galtung* introduced the term *conflict transformation*, which he prefers over the term *conflict resolution* [123]. This term aims at explaining that overcoming a conflict often requires changes. This book follows the term of conflict defined by Galtung and investigates potential for conflict of global networking through ICT in the areas of economy, education, politics, media and religion. Identifying corresponding options for conflict transformation is a main task. This is not about a final fixation of solutions, but showing the transformation capacity meaning the ability to handle global transformations through global networking in a way that it is sustainable and acceptable for the majority of humans.

2.1.5 Convergence

The term *convergence* is used in various scientific disciplines such as mathematics, biology, media theory, politics or sociology. The term has its origins in the latin word *convergere*, which means *merge, meet, join*. In mathematics, for example, this term is used to describe the existence of a limit and the approach to such a limit in number sequences or functions [482]. In sociology, there is the concept of convergence theory, which deals with the development of a society towards an *ideal image* [286]. On a political level, the European Union uses the EU-convergence criteria according to the Maastricht Treaty to review the existence of appropriate economic and financial conditions of countries looking to join the eurozone, which is a monetary union of today 19 of the 28 EU member states [103].

The distinction between the terms of *technological convergence* [323] and *cultural convergence* [146, 290] is very important for this work. The German institute "Das Deutsche Institut für Normung e.V. (DIN)"[5] describes technological convergence as *increasing gearing of different technologies in complex systems* [89]. Generally, technological convergence describes the result of a process in which two or more different products or services from the technology sector merge together to become one product or service. This new product or service combines the advantage of (the old) products or services. The merging of mobile telephony and the Internet produced the mobile Internet. The smartphone is currently a very good example of such a technological convergence. Furthermore, the Internet has expanded in uniting classical services such as radio and television in one medium (*media convergence*). Other examples concern the usage of modern IT and chip technology in the areas of gene analyses and corresponding chemical synthesis processes. The European Commission defines the idea behind technological convergence in its "*Green Paper on Technological Convergence*" from the year 1997 as follows:

> "*The ability of different network platforms to carry essentially similar kinds of services, or the coming together of consumer devices such as the telephone, television and personal computer.*"

The term *cultural convergence* is primarily defined as an adaption of national cultures to a common (worldwide) culture, which can be more than the sum of all cultures through its emergence property (also see Sect. 2.1.3). In such a process national cultures change in segments. The similarities with the corresponding cultures increase. Such adjustments, e.g. in the context of supranational alliances, are possible on the basis of shared values, norms and worldviews and can lead to a harmonized coexistence. The usage of technology in various different forms (car, plane, letterpress, mobile phone, Internet, etc.) by different cultural systems represent a form of cultural convergence.

2.2 Concepts

2.2.1 Digital Divide and Digital Inequality

The definition by the OECD for *digital divide* in 2000 [317], referring to global circumstances, as addressed by Norris [313], states:

> "*The gaps that separate segments of society as well as whole nations into those who are able to take advantage of the new ICT opportunities and those who are not.*"

---

[5]Together with the newspaper "Die WELT" the DIN conducted the conference "Wirtschaft-Digital" (Economy-Digital) in November 2014. Essentially, convergence potentials in the area of industry 4.0 were discussed. Technical harmonization and standardization of product networks are examples of technological convergence.

In the past the different conditions for Internet access stood in the foreground of debate. These were also discussed by the *Information Society Forum* of the EU [357] and the *Forum Informationsgesellschaft/Forum Info 2000* of the German government [112], in the context of European or national conditions. By looking at the international conditions at the time, many initiatives for this subject such as political conferences, like the *World Summit on Information Society* [514] or the Internet Governance Forum organized by the UN, were carried out at different levels. Recently, the term digital inequality is used as a more precise description of the situation [87].

2.2.2 Internet Governance

Under the umbrella of the United Nations, a definition of *Internet governance* in 2005 stated:

> "*Internet governance is the development and application by governments, the private sector and civil society, in their respective roles, of shared principles, norms, rules, decision-making procedures, and programmes that shape the evolution and use of the Internet*" [509].

A special feature of Internet governance in its present form is the composition of the multi-stakeholder community including governments, technical and business community and civil society [31, 254]. This special form of organization and decision–making structure found concrete usage at the *World Summit on the Information Society* in 2003 and 2005, taken place in 2006 in the form of the annual *Internet Governance Forum* (IGF) [186]. This Summit was strongly supported by the US, EU and Japan. Other governments such as China, Russia, Iran and Saudi-Arabia still have an interest in the negotiations surrounding Internet governance ecosystem. This is especially within the traditional structures like those of the *International Telecommunication Union* (ITU). In this framework, non–governmental actors have little or no opportunities for participation.

2.2.3 Superorganism

Superorganisms are higher order systems which emerge from the coordination and cooperation of many individuals, which are viable on their own, and technology components [42]. A coordinated interaction of billions of cells occur, for example, with mammals, ants and bees [231]. Humans, organizations, businesses, and humanity as a whole are also included in the systemic paradigm [162, 285, 358]. The concept *superorganism humanity* includes a certain view on the basic functioning of these systems. In this context, the evolution of humanity is more of an *evolution of cooperation* [12–14], than a *evolution of competition* [71].[6] The development of humanity as a whole system is essentially comprehensible as an evolution of cooperation between people complemented by a mechanism of *cultural conformity* [162].

In general systems theory, living (social and cultural) systems are thereby differentiated from non-living (physical or technical) systems through the *autopoiesis* property [265, 267]. This property initially occurred in the context of studying bio systems [278, 279]. In this sense, autopoiesis is defined as the processes in a system that will bring a system evolution through *self-organization* and *self-recursive production*. Self-organization increases the order and reduces the *entropy* of the system [513]. Information and communication structures/processes play an increasingly bigger role here.

2.2.4 Global Culture

The concept of *global culture* refers to a culture system of humanity [289, 487]. The nobel prize winner *Amartya Sen* states that we belong to different contexts, not only one culture or one religion [416]. According to the position of the *world ethos* movement, [234, 239] there will not be a constitution of a worldwide culture without *religious peace* [232, 237]. The concepts of world ethos and global culture complement each other in a way that the global culture is partly based on the concept of world ethos.

Furthermore, there are approaches for the conception of a world culture such as the neo–institutional *world society theory* by *Meyer* [286, 287, 289, 290], who developed and used the *world system concept* [489–491] by *Wallerstein* as a basis. Both concepts are of the opinion that a global truth evolves across *cultural divides* and nation states which increasingly effects the cultural circumstances and development in each country.

The concept of a world society by *Luhmann* [264] comprises society as a whole including all possible subsystems. It emphasizes the special role of communication which constitutes and maintains the *functional differentiated subsystems* of a society [265]. This concept assumes that (worldwide) communication is the central *operation* of the world society. Hereby there is no communication outside of the society and no other options than communication in society.

According to *Burton*, the concept of a world culture is feasible by *converting* the *world society* into a *world community* with the help of communication. In order to do so, the relationship structures and cooperation networks from the areas of science, religion, language, and economy have to be integrated into the process of the expansion of functional systems [265] in addition to formal institutions [47]. This leads back to the world ethos concept and requires a *consensus of values* between

---

[6]Please note: this does not imply that competition in general is out of importance for human evolution.

different world cultures. This consensus can be achieved with a configuration of a common economic [236, 238], education [235] and religious culture [232, 237] in order to possess a global culture.

2.3 Key Figures

The empirical analysis in Chaps. 4 and 5 regarding the relation between economic performance, education, worldview and access to information and communications technology require global comparable data on the mentioned areas. The presented key figures and indicators below constitute internationally recognized reference data, which mostly come from sources of the ITU [192], World Bank [508], UNESCO [473, 475], UNDP [466], UNCTAD [463] as well as Pew–Research Center [338].

2.3.1 Information and Communications Technology

Fixed Telephone Rate (FTR)

Fixed lines are installed telephone lines that connect the terminal of the telephone subscriber to the so-called *Public Switched Telephone Network* (PSTN), and are registered with a telecommunications office [190]. The fixed telephone line has existed for 150 years and developed greatly. But the spread of the phone was not always as obvious as it is today. The time of the *lady from the bureau* is long gone. She stands for a lady who worked in a telephone switch at a drop-type switchboard in the beginning of the telecommunication age (but partly also until 1987).

The total number of fixed lines is composed of all active connections, such as the fixed telephone *Wireless Local Loop* (WLL),[7] *Integrates Services Digital Network* (*ISDN*) and public call boxes. The fixed telephone rate (FTR) of country i at time t is defined as the sum of the fixed telephone lines (FT), divided by the population size (POP).

$$FTR_t^i = \frac{FT_t^i}{POP_t^i} \tag{2.1}$$

Even though the FTR decreases globally (the reasons why will be further explained in Chap. 4, Sect. 4.1.1), the landline is still an important communication infrastructure. Despite the rapid growth in mobile phone devices complementing fixed lines, the landline connection remains important for voice traffic and as a basis for fixed broadband infrastructure.

---

[7](WLL) is a system that connects the user with the aid of wireless technology to local telephone stations. Used predominantly in rural areas, this technique has significant cost advantages [504].

The data on the number of fixed lines in countries consists of statements from the administrative authorities of telecommunications operators, national supervisory authorities or ministries responsible for information and communications technology. The data is reliable and comparable, especially when controlling authorities and ministries use the same rules as those of the International Telecommunication Union (ITU). The ITU is generally the reference or integration site. However, it provides no information about regional differences (within countries) in regards to fixed lines or the distinction between private and corporate connections.

The ITU verifies the data of the states to ensure consistency among the data of the previous years and among different countries. Inconsistencies of definition and comparison of years are noted accordingly. Missing data on fixed lines can be estimated based on the annual growth rate of recent years and corrected based on regional trends.

Data on fixed telephone connections is administrative data and refer to the telecommunications infrastructure. Till this day, they cannot be assigned to a gender. Some household surveys make statements on access to and the use of ICT. Such information can be partly used to analyze the differences between gender in terms of the use of certain technologies. The data on fixed lines is up to date and complete. There are very few cases where this data is incomplete at the country level (for example due to war or social unrest).

Changes in technology distort the traditional definition of fixed line. This usually refers to the connection of a subscriber to the network node of a service provider consisting of a copper wire. Also, voice services are provided increasingly through Internet protocols, such as *Voice-over-IP* (VoIP). This is a relatively new and affordable communication channel which is separate from the *Public Switched Telephone Network* (PSTN). Some countries also have started to collect data about VoIP.

As explained above, the data is collected and maintained by the International Telecommunications Union. The ITU records the data of fixed telephone lines by regularly conducting questionnaires sent to relevant public authorities for telecommunications, usually the regulatory agencies or ministries for ICT. If a country does not answer these questionnaires the ITU performs an investigation and completes the missing data, both by suggesting to update the webpage of a government and by searching for operator report.

Mobile Phone Penetration Rate (MPR)

The mobile phone penetration rate (MPR) of a country i at time t is defined as the sum of the mobile phone connections (MP), divided by the population size (POP).

$$MPR_t^i = \frac{MP_t^i}{POP_t^i} \qquad (2.2)$$

This indicator includes both the number of mobile phone subscriptions and the number of active prepaid accounts that have been used within the last three months. It does not include the connections via data cards or USB modems, connections of public mobile data services, private trunked radio, telepoint, radio paging and telemetry services [190].

The data for the MPR is available from constituted reports by the relevant authorities or telecommunications providers, which are levied at regular intervals, at least once a year [194]. In these surveys, it is important to distinguish between active and inactive connections. Inactive connections (accounts) will be deleted from the connection list after a certain period in which they are continuously inactive (usually after three months). This distinction is of major importance in countries where most connections are prepaid, which is usually the case in developing countries. The data for the MPR is not divided into rural and urban regions. Mobile communication is available almost everywhere worldwide and especially important in developing countries where landline infrastructure is not extended or marginally existing. Mobile phones increasingly replace fixed telephony in many countries. In many places, the mobile telephone connection rate is above 100%. The reason for this is due to the use of more than one SIM-card per person. The data of the mobile phone subscription rate counts as reliable, up to date and complete.

The data on the MPR is gathered via questionnaires which are conducted annually. The ITU sends these to the authorities responsible for telecommunications, usually regulatory authorities or ministries for information and communications technology. If a country does not answer these questionnaires, the ITU performs the respective investigations and fills in the missing data. At the household level, surveys allow insights on access to and use of ICT. Surveys are also used for the analysis of user behavior, depending on gender and other socio-economic variables.

Internet Penetration Rate (IPR)

The Internet penetration rate (IPR) of a country i at time t is defined as the ratio of the number of Internet users (IU) and population size (POP).

$$IPR_t^i = \frac{IU_t^i}{POP_t^i} \tag{2.3}$$

Internet users are defined as persons who use the Internet in the last 12 months regardless of location, connection or device [190]. The data collected is either based on surveys of national statistical offices or on estimates. In countries where many people use the Internet at work, at school, in Internet cafes or other public places, the increase in public access leads to a significant increase in users. In particular, this is true if there is only a small number of registrations for Internet connections and households with an existing Internet connection. Especially in developing countries, where the IPR is currently still at a rather low level, there are often more than one user per Internet connection.

An increasing number of countries capture the share of Internet users via household surveys. Surveys usually concern the corresponding proportion of the population of a certain age span (for example 15–74 year olds).The number of Internet users in the general population is then derived using statistical methods. A breakdown of this indicator by age and gender is possible in countries where data from household surveys were gathered. This is often the case in developed countries and currently also in a growing number of developing countries.

The records are reviewed by the ITU in order to avoid inconsistencies in data from previous years. In most developed countries and an increasing number of developing countries the percentages of Internet users are gathered via methodologically sophisticated household surveys by the national statistical institutes. In countries, where no surveys are carried out and no own assessments are delivered, the (ITU) estimates the percentage of Internet users. Here, several indicators (such as the number of fixed broadband lines, fixed lines, active mobile broadband connections and income) are relevant. Inconsistencies in definitions and between years and countries are noted accordingly. Inconsistencies of data can occur, for example in cases where countries conduct surveys for different age groups and apply the results to the entire population.

Mobile Broadband Rate (MBR)

The mobile broadband rate or mobile Internet users rate (MBR) is derived from the number of all mobile phone connections which have an active, wireless broadband internet connection via *Hypertext Transfer Protocol* (HTTP), divided by the population size [190]. Mobile broadband via prepaid phones are thereby only considered if an active data connection occurred over IP within the last three months. SMS or MMS, which are transmitted via (IP), do not count as active data connection. The mobile broadband rate (MBR) does not contain mobile satellite links or firmly established wireless connections. The ITU defines broadband as an average transfer rate of at least 256 kbit/s.

2.3.2   *Economic Performance and Social Balance*

Gross Domestic Product per Capita (GDPpC)

The economic strength of a country can be quantified by the (financial or monetary) value of all goods and services produced within a given time period in a country [274]. GDP per capita (GDPpC) is one of the main indicators of economic analysis, both in spatial and in temporal international comparisons. For international comparisons, the *purchasing power parity* (PPP) is used. In a common currency area, such as the eurozone, PPP is present when the goods and services in a given basket in different countries can be acquired for the same amount of money. Between different currency areas, exchange rates and progression over time must be taken into consideration.

The GDP is the aggregate value of all money-based economic activities and transactions. It is based on a clear procedure that allows comparisons over time between different countries and areas. The GDP is generally calculated by three methods (expenditure, income and production approach) [432] that all (must) lead to the same result. The GDP is an important indicator, however, it is increasingly accepted that the GDP concept must be modified, or other data and indicators are needed to complement the GDP if one wants to understand the overall economic situation of a country. Questions of this type were and are the focus of a number of international initiatives which reflect the changing social and political priorities, especially after the consequences of the world financial crisis in 2007/2008. In November 2007, the European Commission (together with the European Parliament, the Club of Rome, the WWF and the OECD) held a conference entitled *Beyond GDP*. An up to date view on the issues of GDP, which for instance were acquired by commissions, in which the two Nobel Prize winners for economic sciences *Joseph Stiglitz* and *Amartya Sen* were involved as chairmen, can be seen here [444]. The average GDP per capita, by definition, says nothing about the distribution of economic performance in a country among the people. Therefore an additional parameter of *social balance* will be considered further in Sect. 2.3.2.

The GDP has taken over the role of a proxy indicator for overall societal development and progress of a country in general, although many doubts remain as of recent [102]. Due to its nature and its purpose, GDP does clearly not deliver a reliable basis for policy debate for each question in the context of economic and social development. More importantly, GDP does not measure the sustainability, nor the performance of environmental protection, nor the degree of social integration in a country. These limitations must be considered when GDP is used as a measurement basis in policy analysis and debates.

Measuring Social Balance and Equity-Parameter

The *social balance* or the distribution of income is, in addition to GDP per capita, an important parameter for understanding the social and economic situation of a country. It is empirically and theoretically known that neither too much nor too little *inequality* are good for a country [155, 205, 510]. Rather, the inequality should range in a so-called *productive inequality range* [205], similar to today's most OECD countries. Measures for inequality include the *Gini-Index* [519], the *80/20-Relation* or the so called *Equity-Factor* [201, 202, 204], which will be discussed subsequently. Metrics of this kind are important for the so called *wealth quintet* [80] or for the *Better Life Index* of the OECD [322].

The equity parameter has, compared with the other metrics, the advantage that it not only provides a measurement, but also a corresponding standardized (income–) distribution function (of Pareto type). As part of the mathematical theory of equity, which uses a comparison of the lowest with the average income as the basis of the balance measurement, the degree of social balance can be determined to a certain extent over the so-called equity parameter (ε). The approach is inspired by the former EU definition of poverty and was developed over several years at the *Research Institute for Applied Knowledge Processing (FAW/n)* in Ulm and its environment [201]. Extensive mathematical derivations and results in this regard can be found in the works of Kämpke [155, 206]. Here, the following should be noted: Until circa 2004, one was deemed poor under the EU's definition of poverty if they had less than half of the average income of a specific country. Meanwhile this changed through administrative ways to 60% of the median income - a grave step [203]. This mathematical (social inequality) theory is based on a (uniform) factor ε like described above, that has a largely inherent *self-similarity* of the structure of a society. That is not only the case for the whole society, but also for every segment (quantile) of the x percent highest income. This means, that in this x percent of the highest income, the lowest income has a volume, which is ε times the average income of the x percent highest incomes [359].

The defining equation for the equity-parameter is therefore:

$$Lowest\_Income = \varepsilon \cdot Average\_Income. \tag{2.4}$$

This is true for all segments of the x percent highest incomes in a country $x \in [0, 1]$. Such an equation is empirically (of course) only approximately true. Such an *optimal* adjustment function to existing data points of the actual distribution of income according to the minimization of the sum of the quadratic deviations is then derived as an equity distribution. From a mathematical viewpoint, the defining equation is a differential equation for the distribution function F of the income. Hereby the function $F(x), x \in [0, 1]$ yields a value for every x that describes, about how much commutative income $F(x)$ the x percent lowest incomes of the population have on hand. The corresponding curve is also the Lorenz curve of the corresponding income distribution [205]. The corresponding differential equation is:

$$F'(x) = \varepsilon \cdot \frac{(1 - F(x))}{(1 - x)}, 0 < \varepsilon \le 1 \tag{2.5}$$

with the exact solution (*Lorenz-Curve*):

$$F_\varepsilon = 1 - (1 - x)^\varepsilon \tag{2.6}$$

Specifically, this is a Pareto distribution [206]. As described, a best fit of the empirical data (according to the method of the least squares) to the solution of the differential equation provides the (further-used) equity parameter of a country. For example, with an equity parameter of $\varepsilon = 1{:}1{,}2$ (83%), the income of 67% of the population is below the average income. 75% of the population have a below average income with an equity factor of 1:2 (50%). When the equity moves forward to 0 (more inequality), almost every citizen has an income that is below average.

2.3.3 Education Level

Adult Literacy Rate (ALR)

The skill of reading and writing is a core competence for humans. It is usually learned in early childhood education from entry to the first year of school and is generally completed by the end of primary education. The *literacy rate* is a measure that compares the population of a country, who have writing and reading skills, in relation to the total population [471, 475]. Theoretically it yields a value between 0 and 1, or 0 and 100%. In general, the literacy includes numeracy like simple arithmetic calculations (for example adding). In practice, a value of 100% can not be achieved because of the biological limits of the literacy rate, for example, infants, babies and toddlers. The *adult literacy rate* (ALR) by definition excludes people under 15 years from data collection.[8]

Empirically, for example, almost the entire adult population of the Scandinavian countries, as well as the affluent smaller states and city-states, are alphabetized. The lowest literacy rates are found in economically poorer states. Geographically, many states on the African continent belong to this group. For political reasons, states where dictatorships and (civil) war-like conditions prevail often have lower literacy rates.[9] However, this does not apply for the former Soviet Union or the former COMECON countries. Table 2.1 roughly lists countries with relatively very high, average and low ALR.

People that are not able to read and write, are called *analphabet*. There is a difference between so called primary, secondary and functional analphabetism [171]. Analphabetism is not used for the analysis concerning the relationship between the use of information and communications technology and the education level. It should be mentioned that the development of technology has the potential to make it more available to individuals who are analphabets. For example, the usage of a mobile phone hardly requires any alphabetization, and is therefore also suitable for analphabets. However up to now, the intensive use of the Internet requires at least basic literacy skills.

Education Index (EI)

The *Education Index (EI)* is one of three equilibrated indices (health, education, economy) of *Human Development Index (HDI)* [464]. The HDI is a parameter for measuring the general performance or potential of a country, which is developed by the Pakistani economist *Mahbub ul Haq* working alongside Indian economist *Amartya Sen*. In contrast to GDP it also considers health care and education. For obvious reasons, GDP and HDI are positively correlated [273].

The EI is a combined measure of *expected* and *average school time* of individuals in a country [466]. The expected school time results from the expected number of years a child will spend in school or university. Statistically possible repetitions of classes are included. From a mathematical viewpoint, the expected years of schooling is of a *predictive* character. Accordingly, the average duration of school is given by

---

[8]Because of empirical reasons, there sometimes may be small differences between countries in data collection, because their authorities partially include humans from 10 years (for example Argentina) in their statistics, not 15 years like the usual or only include people over 15 years with at least 5 years of school education, like in Great Britain [475].

[9]The different education systems that exist among most countries will not be considered for the corresponding level of education in this work. Rather, the decisive factor is the output (e.g. literacy rate) which a country delivers with its education system.

Table 2.1 Adult literacy rates in 2010

Ranking	Land	Literacy rate (%)
1	Finland	≈100
2	Norway	≈100
3	Luxembourg	≈100
4	Liechtenstein	≈100
⋮	⋮	⋮
95	Turkey	95.3
96	China	95.3
97	Mexico	93.4
98	Brazil	90.4
⋮	⋮	⋮
200	Senegal	39.3
201	Ethiopia	39
202	Chad	34.5
203	Afghanistan	28.1
204	South Sudan	27

the average number of years of school education for individuals above 25 years. In the calculation of the EI, the average and expected duration of education are weighted equally. Equation 2.7 is the formula for the determination of a dimension index for the HDI [465].

$$Dimension\_Index_{country_i} = \frac{Current\_Value_i - Minimum\_Value_{all}}{Maximum\_Value_{all} - Minimum\_Value_{all}}$$

(2.7)

The following example shows the calculation of the EI for Ghana for the year 2010:

- Average school years in Ghana are 7 years, expected are 11 school years
- Maximum value for average school years = 13.3 years (across all countries)
- Minimum value for average school years = 0 years (fixed value of 0 years across all countries, i.e. no school attendance at all)
- Maximum value for expected school years = 18 years (across all countries)
- Minimum value for expected school years = 0 years (fixed value of 0 years across all countries, i.e. no school attendance at all)
- Maximum value for the (combined) EI = 0.971 (across all countries)

Calculation of the EI for Ghana:

- Index average school years $= \frac{7-0}{13.3-0} = 0.526$
- Index expected school years $= \frac{11-0}{18-0} = 0.611$
- Education Index $= \frac{\sqrt{0.526}-0}{0.971-0} = 0.584$

The EI is currently an internationally-recognized measure of comparing different levels of training at country level. Since 2010, the determination base for the EI has changed such that it is calculated as described above. Previously, the EI was determined as a weighted combination of alphabetism of the adult and adolescent population. The work [273] investigates the effects of the change of determination base for the HD and EI. These are not of importance for this work due to minimal changes.

Government expenses on education and research contribute both to absolute and relative GDP and the level of education. These public expenditures include school expenses, colleges and universities and other public and private institutions that offer educational services. Here there are very large differences worldwide, in particular when it comes to expenses for ancillary services such as social services (for example Bafög in Germany) [319]. The size of the private education sector determines further substantial distinctions.

2.3.4 Worldview

Religion Diversity Index (RDI)

The *Religion Diversity Index (RDI)* calculates the level of diversity of the eight large worldviews (Buddhism, Christianity, Hinduism, folk religion and superstition, Islam, Judaism, and other religions as one group including people that do not identify with any faith) through an index for countries, regions and worldwide [336]. The RDI is a version of the *Herfindal–Hirschman-Index*, which serves as a common and frequently used index for measuring the concentration, for example, of human or biological populations. The RDI is calculated in a three step process according to the method of the *Pew Research Centers* from Washington [336, 338].

Step 1: The proportion each of the above-mentioned eight religious groups compared to a country's total population is squared and summed up. If, for example, almost the entire population of a country belongs only to one religious group (for example Afghanistan), it yields a value of 10.000 ($=100^2$). If the religious groups are proportionally distributed uniformly in a country, it yields a value of 1.250 ($= \sum_{i=1}^{8} (\frac{100}{8})^2$). This value represents the maximum possible diversity when eight groups are taken into consideration.

Step 2: The value of a country from Step 1 is inverted so that a low value represents a small religious diversity and a high value a high religious diversity. For this purpose, the value of step 1 is subtracted from the maximum possible value that represents no religious diversity. If the population of a country belongs to one religious group exclusively, step 2 yields a value of 0 ($10000 - 10000$). If the eight religious

Table 2.2 Step 1 of the calculation of the RDI for the US, year 2010

Religion	Share of total population (%)	Squared share
Christians	78.3	6.132, 3
Muslims	0.9	0, 8
Non-religious	16.4	269, 8
Hindus	0.6	0, 3
Buddhists	1.2	1, 3
Folk religions	0.2	0, 0
Other	0.6	0, 4
Jews	1.8	3, 4
Total	**100**	**6.408, 3**

groups are equally distributed in a country, we obtain the value of 8,750 ($=10,000 - 1,250$), which is the maximum possible value. As indicated in Table 2.2, the value for step 2 is:

$$10.000 - 6.408, 3 = 3.591, 7 \tag{2.8}$$

Step 3: The determined value in step 2 is now divided by 875 to obtain a final value for the RDI in a scale from 0 to 10. Ultimately, a RDI of 0 means no religious diversity and a value of 10 represents a high religious diversity. This means all religious groups are represented with an equal proportion in the total population.

Step 3 for the calculation of the RDI of the US for example yields:

$$\frac{3.591, 7}{875} = 4, 1 \tag{2.9}$$

The data for determining the fraction of religious groups in a given country is taken from the study *The Global Religious Landscape: A Report on the Size and Distribution of the World's Major Religious Groups as of 2010* [334]. Throughout the course of this report, the Pew research team compiled and analyzed information on the composition of religions in each country from over 2,500 sources (including population censuses, demographic assessments, investigations of the general population and other demographical studies). Censuses are the most important sources for the report and cover 45% of the world's population in 90 countries. Large–scale demographic reports are common sources for covering the additional 43 countries which account for 12% of the world population. For the further 42 countries, with a coverage of 37% of the world population, investigations in the general population are crucial sources. Data for 175 countries and 95% of the world population could be evaluated with help of all sources together.

In order to ensure the comparability of statistics between countries, the inclusion of an individual to a religious group is not captured theologically but sociologically in the report of the *Pew research teams*. The study assigned these individuals to the religion to which they identify themselves with. All persons are assigned to a religion based on their personal testimony, regardless of whether they perform in their common religious practices [336, 337].

2.4   Data Management and Statistical Methods

In this book, data management is essentially based on the reference architecture of data warehouse systems according to work of *Bauer* [22]. This authoritative architecture allows an abstract view of data management systems. Figure 2.1 demonstrates the adjusted within the meaning of this work reference architecture which is explained in the following segments concerning data procurement, storage and processing as well as with regard to the data analysis with the help of statistical methods and tools.

2.4.1   Data Procurement, Storage, Processing and Analysis

In the case of a data management system, *monitoring* is a task in which the relevant source data are observed with regard to the changes that are carried out to them. This function is performed by so-called monitors which react to certain *triggers* (time or event driven) for changes in source data. The task of monitoring is not used in this book, because the relevant data from the fields of information and communications technology (ITU source data) [192], economy (source data of the World Bank) [508], education (UNESCO and UNDP source data) [466, 473, 475] and religion (Pew–Research Center source data) [338] are collected on annual base.

Step 1 in Fig. 2.1 is the *extraction component* from the so-called ETL–process (*extraction, transformation* and *load*). The extraction component has the function to procure data from the sources. Among other things, there are questions regarding the exchange of data, for example, formats to be considered. Furthermore, normally certain subsets need to be extracted from the source data. In this case *separation* is of importance, which is the selection of the data required for analysis purposes from the total amount of available data in the respective database. The decisive factor here is also the question of time of extraction. In this work, this aspect is done as needed, whereas a periodic, event-based or immediate extraction is generally possible. After extraction, the needed data for analysis is stored in the workspace.

Step (2a) in Fig. 2.1 is the so-called *transformation component*. In doing so, the data obtained from different sources must be standardized in a certain way in order to make valid processing possible. The uniting and processing of the data is carried out in this component. This process can also be described as *migration*. At first in this step (2a), the customizations of the data types, the conversion of codings (UTF, 8, ISO 8859 etc.), the standardization of dates (*DD-MM-YYYY* instead of *YYYY-MM-DD*), the conversion of measurement units and the combination or separation is realized.

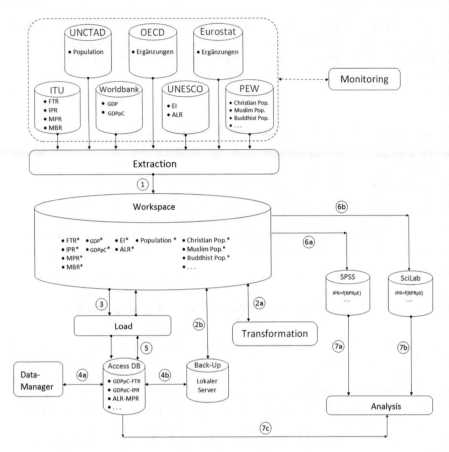

Fig. 2.1 Reference architecture of the Data–Management–Systems

Also the attribute values, plausability of data and the correction of faulty or duplicated data is processed. The tasks which are performed in the transformation-component can be divided into three classes.

(i) *syntactic* transformation: The changes refer to formal aspects like renaming of column name, sign conversions or customization of units. The syntactic transformation generally deals with the change of the data format with regard to data structure.

(ii) *semantic* transformation: The data become restructured so that it can be interpreted in other data models. The corresponding vocabulary of the data model is transformed in a common semantics. For example, some countries are identified due to corresponding international country codes, whereas different data sources use partial different country codes. The conversion of units is also a part of the semantic transformation.

(iii) *content-related* transformation: Irrelevant data sets are corrected or deleted, duplicates and faulty data sets removed.

In principle, the plausibility of the data is checked in the transformation component. Step (2b) in Fig. 2.1 serves the safeguarding of data as a backup on a server. After the tasks of the transformation component have been executed, data records that are suitable for analysis are located in the workspace. The actual data analysis is carried out in a later step (7) since no data sets are brought together or combined from different sources yet.

The *loading up* component in step (3) passes the prepared data from the workspace to the corresponding database (Access DB). Furthermore, this component is responsible for the historicization of the data in a certain regard. This means old data is not simply deleted or overwritten but rather provided with a timestamp and stored. A change at a data set therefore has the consequence that two data sets exist with a different timestamp.

In step (4a) the corresponding data are brought together for analysis purposes from different sources. For this the database language *Structured Query Language (SQL)* is used. The following SQL code (see Listing 2.1) provides an example of the uniting of adjusted data on the GDP per capita and the Internet penetration rate (IPR) (see Listing 2.1). The combined data is then saved on a server as a backup in step (4b).

Listing 2.1 SQL example code: Combination of data regarding GDP per capita (GDP)and Internet user rate (IPR), year 2000

```
SELECT J.[Country], J.[CountryCode], GDP.[2000] AS GDP_2000, J.[IPR_2000], J.[Pop_2000]
FROM(
SELECT DISTINCT IPR.[Country Code] AS CountryCode, IPR.[Country Name] As Country,
IPR.[2000] AS IPR_2000, Pop.[2000] AS Pop_2000
FROM IPR, Pop WHERE Pop.[Country] like '*' \& IPR.[Country Name] \& '*' )
 AS J, GDP WHERE GDP.[Country Name] = J.[Country];
```

Step (5) is used to assign the records necessary for the regression analysis to the workspace. The corresponding data is exported from the database program *Access* as a *Excel*-file. Afterwards the data is loaded in step (6a) into the statistics program *SPSS* or in step (6b) into the numeric program *SciLab*.

In step (7a) or (7b), regression analysis is performed by using the numeric program SciLab or SPSS, in which some queries (e.g. the categorization of the countries after economy performance) take place in step (7c) directly at the Access DB.

Listing 2.2 SciLab example code: Non-linear regression with 3-parameters logarithm

```
// Initialization
Z = [X; Y]; // Arrange data by line
a_Start = [1; 0; 0]; // Column vector as start parameter
XX = linspace(min(X),max(X),301);

// Calculations
function [e] = G(a,z)
e = z(2) - a(1) * log( z(1) - a(2) ) - a(3); // Ideal equation w/o(!) least squares
endfunction

// Calculation of solution parameter and corresponding passing value
[a_End,error] = datafit(G,Z,a_Start);

function w = Result(x)
w = a_End(1) * log(x - a_End(2)) + a_End(3);
endfunction
```

```
// Output — numerical
a_Start
a_End
error

// Output — graphical
plot2d (X,Y, style = −5, rect = [min(X)−5000 min(Y)−1 max(X)+5000 max(Y)+10]);
fplot2d (XX, Result , style =  5);
xtitle ( 'Regression_with_3−parameters_logarithm_' , 'GDPpC_[ Intl ._$]' , 'IPR '),
```

2.4.2 Descriptive Statistics

It is the decisive feature of descriptive statistics that only statements concerning the
data set are made themselves. Descriptive statistical parameters only describe what
applies to the participants of a survey or observation. Generally one uses *simple
parameters* for the description and evaluation of data like *mean*

$$\bar{x} = \frac{1}{n} \sum_{i=1}^{n} x_i \tag{2.10}$$

or *median*

$$\tilde{x} = \begin{cases} x_{\frac{n+1}{2}} & n \text{ odd} \\ \frac{1}{2} \left(x_{\frac{n}{2}} + x_{\frac{n}{2}+1} \right) & n \text{ even} \end{cases} \tag{2.11}$$

but also indicators such as *standard deviation* σ or *variance* σ^2

$$\sigma^2 = \frac{1}{n-1} \sum_{i=1}^{n} (x_i - \bar{x})^2 \tag{2.12}$$

or *correlation coefficient* (see Sect. 2.4.4).

One can deduce the smallest or greatest value of a data set or a distribution by *minimum* or *maximum* (extremum). Dispersion summarizes various metrics in descriptive
statistics and stochastics, which describe the spread of values in a frequency distribution or the probability distribution of a suitable location parameter. The dispersion
of the frequency distribution is called *standard-error*.

The *span* calculates the difference between the greatest and the smallest value
of a data set. The span is charged to only two extrema, but nevertheless a measure
frequently used for dispersion since this size is very easy to calculate.

Standard deviation is one of the most important and the most used measures of
dispersion. The *variance* is the square of the standard deviation.

2.4.3 Method of Least Squares

The *method of least squares* is a procedure to fit a function f or curve of a candidate class efficiently to existing data. One would describe the relation between two or more variables approximated with the help of this function. Essential questions concern the choice of the respective candidate class of functions. Normally the classes summarize functions which are defined by the variations of some parameters, such as normal distributions, dependent on mean value and variance. As a rule, the aim is a good compromise concerning the quality of the *adaption* (generally many parameters are helpful) and the avoidance of *over-fitting* (few parameters are helpful).

Be a data set $(x_1, y_1), \ldots, (x_n, y_n)$ given, the *least square error* then can be described for the connection $y = f(x)$ with the following:

$$\text{LSE} = \sum_{i=1}^{n} (y_i - f(x_i))^2 \tag{2.13}$$

This square error has to be minimized. This is an attempt to describe the relation between two variables by a linear function and the usual case of the determination of correlations. In reality, one would hardly come across a perfect linear context, because measurement errors or measurement deviations are enough to resolve an actual linear connection. Many phenomena in natural sciences, economics or social sciences are of complex nature and require non–linear functions for the description of relations. Furthermore, many processes have so called *feedback effects* (negative/positive) that generally trigger a non–linear relationship. In this book usually a fit with up to four parameters from typical candidate classes, such as logistic functions, is used.

The *Levenberg–Marquardt algorithm* is named after *Kenneth Levenberg* and *Donald Marquardt* [299]. This is a numerical optimization algorithm, which is used to make a solution of the generally non–linear fitting problem using the method of least squares. The algorithm combines the *Gauß–Newton method* with a technique which forces descending function values. The Levenberg–Marquardt algorithm is considerably more robust than the Gauß–Newton method because it converges with a high probability also at *bad* starting parameter values. However, convergence is not guaranteed here. In addition, the algorithm is often slower than the Gauß–Newton method at initial values which are near the minimum.

2.4.4 Correlation Analysis

A *correlation analysis* is a relatively easy method to derive a value of the statistical connection of two interval-scaled variables. By calculating the correlation (*correlation coefficient*) the (non-directional) linear relation of these variables can be examined. In this sense, non-directional means that no level of cause-effect can be taken.

Therefore, it is also referenced as a bivariate connection. In the present book the following *Pearson–Bravais*-correlation coefficient

$$r_{XY} = \frac{Cov(X, Y)}{\sigma(X)\sigma(Y)} = \frac{\sum_{i=1}^{n}(x_i - \bar{x})(y_i - \bar{y})}{\sqrt{\sum_{i=1}^{n}(x_i - \bar{x})^2 \cdot \sum_{i=1}^{n}(y_i - \bar{y})^2}} \tag{2.14}$$

is calculated to determine the strength of the linear relation [330, 331]. A correlation analysis should be preferred to a simple linear regression analysis if, for factual reasons, no statement is made or can be made about the presumed direction of the relation between two variables.

2.4.5 Regression Analysis

In the subject of *big data* and *analytics*, which are currently discussed within the research community and the public, there is a trend to move away from finding reasons for relations to rather recognizing connections statistical nature. This process is sometimes described with the picture *from causation to correlation* [69]. The correlation and regression analysis provides an important approach for this procedure. Regression analysis is a collection of statistical analysis methods based on different models (linear and nonlinear), methods (for example, least squares) and applications (find contexts) [106]. The aim of this analysis method is to determine a relationship between an *independent* and a *variable* (univariate or multivariate situation). It is particularly used if relations are to be described quantitatively. The earliest form of regression was the method of least squares (french: *méthode des moindres carrés*), published in 1805 by *Legendre* and 1809 by *Gauss* [127]. Both used this method to determine the orbits of the planets around the sun based on astronomical observations. The main issue was dealing with measurement errors, which posed a major problem. Gauss published a further development of the theory of least squares in 1821 [125] and 1828 [126].

Many synonyms are used conceptually in the regression analysis. The dependent size is also called a regressand, endogenous or to be explained variable. The influencing variables are called regressors, explanatory variables, exogenous variables or predictors.

2.4.6 Simple Regression

The dependence of a variable on an independent variable can be examined with the support of the simple regression analysis. The relation can be described and explained quantitatively and brings up three primary questions: [15]

1. How strong is the influence of the independent variable on the dependent?
2. How does the dependent variable change as there is a change in the independent?
3. How does the dependent variable change in the time sequence and thus ceteris paribus also in the future?

In contrast to the correlation analysis (see Sect. 2.4.4), a *clear direction* of the relationship between the dependent and independent variable is assumed in the regression analysis. For many questions, the direction of the relationship is largely unique. If one examines, for example, the volume of sunscreen sales, this will be closely related to the hours of sunshine and the temperature. For other questions the direction is not as clear. *Regression cannot show a causality.* Rather, it is important to establish an assumed direction of the relationship theoretically well.

In this work it is supposed that the economic performance of a country has influence on the respective access rate to technology in a country, such as the Internet user rate. The explanation for this lies in the reality of life (access to and devices for telephony and Internet cost money) as well as in the findings within the empiric data that came up while analyzing them (see Chap. 3).

The aim of the regression analysis is the calculation or estimation of a regression equation. Generally this equation can be written as follows:

$$y_i = f(x_i) + \varepsilon_i \tag{2.15}$$

Here, x is the independent and y the dependent variable. To examine how well the found regression function describes the data with the help of the method of the least squares, two goodness of fit measures are calculated:

1. The *coefficient of determination* R^2

$$R^2 = 1 - \frac{\sum_{i=1}^{n} (f(x_i) - y_i)^2}{\sum_{i=1}^{n} (\bar{y} - y_i)^2} \tag{2.16}$$

indicates how well the regression line fits the empirical data. This is determined with the help of the *residuals* - mentioned above. In a regression analysis the coefficient of determination R^2 is used, which is the ratio of explained variance to the total variance. R^2 has values between 0 and 1. No variance of the data is explained by the model with a value of 0, while a value of 1 explains the whole variance. Both values are highly improbable. In most cases, the *corrected* R^2 value is additionally used, which has the characteristic of being larger, the more independent variables we have, regardless of whether these independent variables deliver a significant contribution to the explanation.

2. The *F–statistics* checks whether the coefficient of determination R^2 has originated (only) by chance or from the connection of the data. The F-statistic is displayed when carrying out a regression analysis by the corresponding statistics software. To check whether the coefficient of determination has originated (only) by chance, the calculated F-statistics must be compared to the theoretical value from the likelihood distribution. In general one can read the comparison directly from the

given significance (*p–value*). Here it is valid that this value should be under a value of 0.05 to be able to assume that a statistical (significant) relation is given between both variables.

2.4.7 Multiple Regression

Only bivariate correlations between two variables can be examined with a correlation analysis and simple linear regression. If one would like to instead, investigate the relationship between several variables, there is the so-called multivariate and multiple regression analysis. In the *multiple analysis* a dependent and several independent variables are examined. The dependent variable is a vector in the multivariate regression analysis. Multivariate regression modeling is always appropriate when two predictors present a whole range of measurements that are to be regarded as dependent and can be expected to be correlated. An important application are repeated measures. In this work, the multiple regression method is used, simply because the data is not obtained from such measurement repetitions.

The following are the basic principles of multiple regression with a concrete example, represented in [292]. There is a dependent variable Z (for example, Internet penetration rate, IPR) and two independent variables (for example, per capita GDP and level of education). Furthermore, it is assumed that a simple regression analysis for the relationship between the Internet penetration rate and the GDP per capita provides a non-linear function $f(x)$. The function $g(y)$ corresponds to the non-linear relationship between the Internet penetration rate and the level of education. The corresponding multiple regression model with the corresponding weights A and B is:

$$\boxed{Z = A \cdot f(x) + B \cdot g(y)} \tag{2.17}$$

The underlying functions $f(x)$ and $g(y)$ is a logarithmic 3 – parametric type provided, so:

$$f(x) = a + b \cdot \log(x - c)$$
$$g(y) = d + e \cdot \log(y - h)$$
$$\Rightarrow Z = A \cdot (a + b \cdot \log(x \cdot c)) + B \cdot (d + e \cdot \log(y - h))$$
$$\text{with a, b, c, d, e, h const.}$$

Now it is based on the method of least squares, to minimize the function $\varphi(A, B)$:

$$\sum_{i=1}^{n}(z_i - A \cdot f(x_i) - B \cdot g(y_i))^2 = \varphi(A, B) \rightarrow min \tag{2.18}$$

For this we calculate the partial derivatives of $\varphi(A, B)$ in accordance with Eqs. (2.19) and (2.20). These equations are set to zero to find the appropriate minimum.

$$\frac{\partial \varphi}{\partial A} = 2 \cdot \sum_{i=1}^{n} (z_i - A \cdot f(x_i) - B \cdot g(y_i)) \cdot f(x_i) \overset{!}{=} 0$$

$$\Rightarrow \sum_{i=1}^{n} z_i \cdot f(x_i) = \left(\sum_{i=1}^{n} f^2(x_i), \sum_{i=1}^{n} g(y_i) \cdot f(x_i) \right) \cdot \begin{pmatrix} A \\ B \end{pmatrix} \tag{2.19}$$

$$\frac{\partial \varphi}{\partial B} \overset{!}{=} 0$$

$$\Rightarrow \sum_{i=1}^{n} z_i \cdot g(y_i) = \left(\sum_{i=1}^{n} g(y_i) \cdot f(x_i), \sum_{i=1}^{n} g^2(y_i) \right) \cdot \begin{pmatrix} A \\ B \end{pmatrix} \tag{2.20}$$

The result is a system of equations, which delivers the regression weights A and B, after its resolution according to the following procedure.

$$\begin{pmatrix} \sum_{i=1}^{n} x_i^{*2} & \sum_{i=1}^{n} x_i^* y_i^* \\ \sum_{i=1}^{n} x_i^* y_i^* & \sum_{i=1}^{n} y_i^{*2} \end{pmatrix} \cdot \begin{pmatrix} A \\ B \end{pmatrix} = \underbrace{\begin{pmatrix} \sum f^2(x_i) & \sum f(x_i)g(y_i) \\ \sum f(x_i)g(y_i) & \sum g^2(y_i) \end{pmatrix}}_{=N} \cdot \begin{pmatrix} A \\ B \end{pmatrix}$$

$$= \underbrace{\begin{pmatrix} \sum z_i f(x_i) \\ \sum z_i g(y_i) \end{pmatrix}}_{=\beta} = \begin{pmatrix} \sum z_i x_i^* \\ \sum z_i y_i^* \end{pmatrix}$$

$$\boxed{\Rightarrow \begin{pmatrix} A \\ B \end{pmatrix} = \beta N^{-1}} \tag{2.21}$$

2.4.8 Test Functions

The considered class of (test–) functions is extended as necessary during the analysis of the empiric relations between, for example the economic performance, level of education as well as the populations share with a specific worldview and the level of digitalization. The selection of such (test–) functions is primarily due to subject-specific or interdisciplinary skills or insights. In addition to the linear description of relations, typical forms of growth or saturation behaviour will be modelled on such functions. Furthermore, it may happen that certain models for growing– and inhibiting behaviour are required. The justification for a broad class of functions is the accuracy with which later the influence of the independent variables can be simulated on the dependent variable in the multiple regression analysis. As mentioned above, however, the phenomenon of over-fitting is to be avoided.

For clarity reasons, all the used (test–) functions are defined here. In Sect. 4.1.1 the relation between the GDP per capita and the fixed-telephony rate is analyzed. There it will be exemplary described, how, starting with the linear function, the class of functions will be extended gradually.

The *linear* regression function is the simplest example and one of the most applied when it comes to the modelling of the relationship between dependent and independent variables. Firstly, it is in the nature of things that people bring many phenomena in a linear relationship, such as trip time and travel distance on a path. The simple linear regression model has the following form:

$$f(x_i) = a + b \cdot x_i \tag{2.22}$$

The modeling of relationships based on *logarithmic* functions in science is a widely used tool.[10] For example, the logarithmic modeling in economics is used when certain rules are based on the assumption that the benefit of a good decreases with increasing consumption. This law is called as 1st *Gossen'sches law*. Here the 3-parametric logarithm is given by:

$$f(x_i) = a + b \cdot \log(x_i - c) \tag{2.23}$$

The *Monod function* adjustments (*Monod I*) relates to an approach that was originally used for predicting the growth of microorganisms depending on substrate concentration. From an initial linear increase in the Monod function it then passes into a constant. The *Monod I* test function is:

$$f(x_i) = \frac{a \cdot x_i}{b + x_i} \tag{2.24}$$

An extension of the *Monod I*-function is the *Monod II*-function, which is defined as follows:

$$f(x_i) = \frac{a \cdot c \cdot x_i}{(b + x_i)(c + x_i)} \tag{2.25}$$

The *logistic* function originally stems from biology, where it was used to observe the growth of bacterial cultures. Here, the growth is considered over time, which is initially with very few bacteria and a linear growth (initially low). Later, with the number of increasing bacteria, an exponential growth follows. Finally, space constraints limit growth. This results in an S-shaped curve with a constant final value. The logistic function is defined as follows:

---

[10]Logarithms tend to *compress* high values and to *stretch* low values. Therefore, a logarithmic model gives more weight to developing countries while an exponential model gives more importance to developed countries.

$$f(x_i) = \frac{a}{1 + b \cdot e^{c \cdot x_i}} \tag{2.26}$$

The *MaxLog*-function has the following form:

$$f(x_i) = a \cdot e^{(b - c \cdot \log(x_i) - d \cdot \log^2(x_i))} \tag{2.27}$$

Planck's radiation law originally describes the intensity distribution of the electromagnetic energy and power or the density distribution of all the photons as a function of wavelength or frequency that are emitted by a black body – an ideal source of radiation at a given temperature. So again, there is the dependence of the radiated power of its frequency or wavelength. The associated function initially grows as x^b (original with $b = 3$ or $b = 5$), then has a maximum, which (only) depends on the temperature and then drops exponentially. In this present book, the parameter b in contrast to the original model is not on $b = 3$ or $b = 5$ fixed but also optimized, thus indicating the following 4-parametric test function.

$$f(x_i) = \frac{a \cdot x_i^b}{c \cdot e^{d \cdot x_i} - 1} \tag{2.28}$$

For an *exponential* function, the exponent of a power is considered a variable. Such functions have the property that changes their functional value in equal intervals by the same factor. They are particularly well suited if growth or decay processes are to be modelled, such as population growth or compound interest.

$$f(x_i) = a + b \cdot e^{c \cdot x_i - d} \tag{2.29}$$

Generally, one finds *power* functions in physics and engineering for various applications. For example, the distance traveled with an initial speed and a constant acceleration is a quadratic function of time. One calls this relationship also a path-time law of uniformly accelerated motion. Furthermore, *Stefan-Boltzmann-law* states that the radiating power of a black body depends on its temperature. The radiation power of a black body is proportional to the fourth power of its absolute temperature according to this law. The power that must be applied to overcome the resistance of a car is determined by a power function of 3rd degree. This means, eight times more power is needed to drive with double speed. Due to these physical examples and due to the comparatively simple mathematics of the class of power functions, it will be used for the empirical studies in this book. The case underlying 3-parametric exponential function is of the following type:

$$f(x_i) = a + b \cdot x_i^c \tag{2.30}$$

The *hyperbola* is defined as the set of all points of the plane for which the absolute difference between the distances to two given points, the so-called hot spots, is constant. We use the function

$$f(x_i) = \frac{1}{1 - a \cdot x_i}. \tag{2.31}$$

A generalization is the inverse function

$$f(x_i) = \frac{a}{b + x_i} \tag{2.32}$$

2.5   Expert Interviews

Discourse analysis and expert interviews relate to Part III *power issues and reflections on possible future developments* of this work. Disclosure analysis is the analysis of legal texts, strategy papers, official information, action plans and transcripts of the analyzed organizations (such as the *Internet Governance Forum*) or of states. The interviews were conducted qualitatively with individuals, e.g. including *Vinton G. Cerf* or *Wolfgang Kleinwächter*. Vint Cerf has given very valuable information on topics such as Internet Governance and superorganism humanity, discussed with him at the Internet Governance Forum 2014 in Istanbul and 2015 in Brazil. Furthermore an interview with Prof. Wolfgang Kleinwächter was conducted, who deals with the topic of Internet Governance in the cultural context and international environment.

Chapter 3
State of Affairs

> *We can only see a short distance ahead, but we can see plenty
> there that needs to be done.*
>
> (Alan M. Turing, 1912–1954)

The year of science 2006 in Germany was under the subject of computer science. It is
an initiative of the *FederalMinistry of Education and Research* which was developed
and realized together with the *Initiative Science in Dialogue* (WID), *Society for
Computer Science* (GI) and numerous partners from science, economy and culture
[48]. During this period, it was socially recognized and acknowledged that computer
science and services affect almost all areas of life, more than any other scientific
domain. Additionally, the science year of 2014 was entitled as *The Digital Society*
[49].

Projecting initiatives such as the *computer science year* or the *The Digital Society*
in Germany to a global level, it becomes clear that in the context of globalization the
way towards a digital future will be a major challenge. Generally, globalization processes have complex structures. Several scientific disciplines deal with this subject,
which (often) also has interdisciplinary characters. If we go back to the importance
of global networking processes from Chap. 2, the disciplines of computer science,
economics, communications and cultural studies are involved in all cases. The study
of demographics (national and worldwide) is of central importance when it comes to
a holistic understanding of current and future globalization processes. The number
of people is a primary determinant of many these processes and influence a variety of
basic human needs. Climate research is an example of an important component of a
holistic approach to model globalization effects, because the climatic conditions are
particularly important to the character of our future. However, climate research is not
the subject of the present book and remains largely unmentioned. Climate policy is
only discussed in Chap. 8 under three broad scenarios for the possible convergence
of a *global culture*. More specifically: Climate issues as a part of issues relating to
natural resources play a fundamental role in the scenarios of *global two-tier society*
and *ecological collapse*.

© Springer International Publishing AG 2018
H. Ünver, *Global Networking, Communication and Culture: Conflict
or Convergence?*, Studies in Systems, Decision and Control 151,
https://doi.org/10.1007/978-3-319-76448-1_3

Central importance for this book is the state of research in relation to global access to information and communications technology, in particular for fixed telephony, mobile telephony, Internet and mobile Internet (see Sect. 3.1). Since beginning the discussion on access to ICT in the 1990s, the term *digital divide* characterizes the research environment [313, 403]. Recently, the term *digital inequality* is used as a more precise description of the conditions, based on the discussions on the topics of social inequality and balance [87, 454, 525].

The subjects of *Internet governance* and its specific design become important as humanity develops towards becoming a global *superorganism*. This is in the context of power issues and reflections on possible future developments of the global networking processes. It is also about the question of how it is possible for a networked world society to establish this type of *global culture* as the basis for peaceful coexistence of various cultures (see Sect. 3.2).

3.1   Empirical Analysis of the Global Spread of ICT

In the 1990s, the term *digital divide* originated to describe computer technology based ownership or non-ownership, and perhaps more importantly, the opportunities for harnessing this powerful technology [313]. Past and current research on the analysis of the digital divide usually have a descriptive character [17, 60, 85]. These studies describe, in general, the digital divide as a function of the amount of access to ICT, on the individual level of ICT use or in response to the *digital skills* of people in respective countries. Usually demographic, economic or education specifications are used to model relationships.

The extensive work in *"The Deepening Divide"* [84] provides a central point in this research on the digital divide. It distinguishes between four types of digital divide (see Fig. 3.1) and explains why the digital divide widens and also deepens in technologically strong societies.

Motivational access refers to the incentives to use ICT in general and the Internet in particular. *Material access* refers to the physical and economic opportunities to access the Internet either at home, school or work. *Skills access* refers to the substantive capacity of Internet use, which parallels the concept of *media literacy* (also *digital literacy* or *information literacy*). *Usage access* describes the use of digital media for a particular purpose of information, communication, transaction or entertainment. Related differences in the expression of different usage levels result in inequalities in the information and knowledge based society.

A large number of empirical studies on the subject of the *digital divide*, particularly for Internet use, are carried out. A few of these stress that the *income level* [7, 9], *income distribution* [119, 522], *education level* [148, 217], *population size* [351] and *urbanization* [17] are essential determinants of the digital divide.

This book focuses on an empirically-functional study of the global digital divide for fixed telephony, mobile telephony, Internet and mobile Internet in the context of economic performance and the education level of countries with regard to *material*

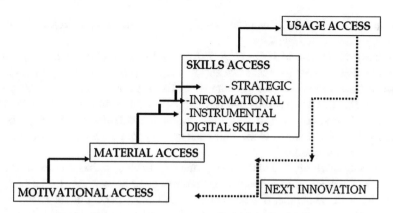

Fig. 3.1 Cumulative and recursive model on digital divide according to van Dijk

access and *skills access*. It is implicitly assumed and empirically relined that the higher the level of prosperity is in a country or a culture, the higher or greater the physical or material access to information technology. At the same time it is believed or made plausible that the digital literacy of a society is greater when the education level is higher. In addition, the access rates are examined for these four technologies from different worldviews.

3.1.1 Global Networking, Economic Performance and Education Level

In a sense, the (global) deployment and development of various ICT as fixed-line telephony, mobile telephony, Internet and mobile Internet have similar course. It can be the case that a particular type of access (e.g. fixed-line) has spread somewhat slower worldwide than another ICT (e.g. mobile phone). The empirical studies and functional relationships in part II illustrate these different patterns of communications technology distribution. These studies are starting in the year 2000 and ending in 2012/2013.

In the area of telephone networks, previously, the focus was primarily on the disparity between rural and urban areas in a country. This was because such a distinction was crucial for the economics of establishing telephone networks in rural areas [56, 147]. In 1963 *Jipp* published the article *"Wealth of Nations and Telephone Density"* [200]. His interest was to discover in which context the level of prosperity of a country is associated with the proliferation of fixed telephony. In particular, he was keen to find out how far or to what extent the spread of telecommunications could increase the level of prosperity. He shows a high correlation between the level of development of telecommunications infrastructure in a country and its level of prosperity. More precisely, a high positive correlation between the fixed telephone rate and per capita GDP. This result is very well known with the present use of the term *Jipps-Curve* or *Jipps Law*.

A study carried out in 1968 (*GAS–5 Handbook*) by the *Consultative Committee on International Telephone and Telegraph* (CCITT) [56] illustrated the correlation between GDP per capita and the number of fixed telephone lines per 100 people for about thirty industrialized- and developing countries in the years 1955, 1960 and 1965. The study showed very high correlation coefficients for all years. An investigation for the same relation for each country individually over several years came to a similar conclusion showing high correlations.

The study of Hardy in 1980 is one of the first known studies that examined the potential impact of telecommunications on economic growth [147]. The GDP per capita data and the number of telephone lines and radios per capita for 15 industrialized and 45 developing countries based on a regression analysis were examined (with a certain time delay). This work concluded that the landline rate is closely connected with the GDP per capita. At the same time the landline rate at time t has a positive impact on the GDP per capita at time $t + 1$. The distribution of radios incidentally has little connection with per capita GDP.

In the works of Gille (1986) [129], Saunders et al. (1994) [402] and Mbarika et al. (2003) [282] a (multiple) linear regression analysis is applied for nearly 120 countries to examine the relationship between the fixed line rate in a country and various other variables. This variables include the investment in telecommunications infrastructure, number of employees in the technical field, time of waiting for the acquisition of a fixed-line connection, and per capita GDP. The main finding in this work is the significantly higher correlation that per capita GDP has in comparison to other studied variables. Furthermore, it was shown that investments in telecommunications infrastructure alone are not effective to accelerate the GDP growth. Such types of investments and the resulting increase in landline rate therefore has no effect to GDP. Additionally investments in other socio-economic factors (e.g. education) have to be made simultaneously.

Dutta (2001) describes that improved telecommunications infrastructure leads to faster and wider dissemination of market information, e.g. used for administrative and commercial purposes [94]. On the other hand, increased economic activity can increase the demand for improved telecommunications infrastructure for coordinating and monitoring the corresponding governance and market information. In each case, 15 industrialized and developing countries have been studied by GDP, GDP per capita, total number of fixed lines and number of fixed telephone lines per inhabitant. The results suggest that there is substantial evidence of a positive relationship between the telecommunications infrastructure and economic performance. The Granger-causal relation [136] shows from the economic performance in the direction of telecommunications infrastructure. This result applies to both industrialized and developing countries.

Shiu and Lam (2008) [421] undertake a study where they use the *telephone density* (as the sum of the fixed telephone lines and mobile telephone lines per inhabitant) and examine its relation to the economic performance of 105 countries between 1980 and 2006. The countries have been divided into different economic categories according to the logic of the *International Telecommunication Union* (ITU). Empirical analysis shows that there is a reciprocal relationship between per capita GDP and telephone density in the category of so-called *high income* countries and European countries. For all remaining economic categories (*upper-middle-income, lower-middle income, low income*) as well as the continents of Africa and Latin America, the influence of economic performance on the landline rate is greater than the influence of the landline rate on economic performance. Empirical results also show that the distribution of telephone lines are largely disconnected from economic performance in developing countries.

Generally, it is expected that a low level of education counteracts or slows down the dissemination of information and communications technology [44, 131, 339, 472, 478, 479]. This is especially true for ICT such as the Internet, because an intensive use of the Internet (still) requires the ability to read and write [92, 479]. The use of the phone or mobile phone, on the other hand, requires minimal literacy skills and only the need to identify a sequence of digits [282, 398, 479]. The traditional fixed-line telephone in recent years has lost its importance, at least as a pure communication tool for the transfer of spoken language. This results from mobile communication whereby the introduction of the so-called prepaid card has become cheaper [191]. Meanwhile, conventional mobile phones are used worldwide and smartphones are beneficial devices in organizing life in more developed countries [194]. In addition, mobile telephony has significant advantages over the fixed telephone in regards to location independence and individual assignment. Fixed-telephony is localized and usually allocated at home or in a specific space. Households today make do without landline access.

The mobile telephone has spread steadily since the early 1990s with the development of mobile phone standards *Global System for Mobile Communications (GSM)* and enshrined itself culturally in today's society worldwide [55]. Since the 1990s, the improvement of the network coverage and the reduction of costs has contributed to the decisive spread of mobile telephony globally [398]. In 2005, Kauffmann [212] and van Dijk [84] reported that the mobile telephone access rate still varied greatly in different regions and countries of the world. In his book *"Cell Phone Culture: Mobile Technology in Everyday Life"* [130] Goggin writes that in 2006 the mobile phone has become a central cultural technique where society now developed a type of *cell culture*. Mobile telephony has been considered a hopeful technique due to its ability in reducing the digital divide, particularly between industrialized and developing countries [17, 84, 212]. Unlike in the case of the digital divide that is linked with an individual possessing or not possessing a computer or Internet connection, the use of the mobile phone does not require a great deal of (technical) expertise. Furthermore, there is neither a high education level nor a landline infrastructure, which is on a low stage in developing countries, required to use a mobile phone [17].

Generally the research on the dissemination of ICT observe that the so-called *early adopters* tend to be younger, wealthier and highly educated in comparison to the average population [60, 82, 124, 313, 394, 525]. Individual studies, like that of Wareham and Levy from 2002 [492] show that although there is generally a positive correlation between income and mobile telephony use, there is no significant relationship between the age or education and mobile telephony usage.

Kiiski and Pohjola (2002) [217] study show that in developing and OECD countries, tertiary education has a positive (statistically significant) impact on ICT dissemination. At the same time, this study shows, as well as the work of Hargittai (1999) [148], that education has no significant impact on the spread of ICT in these groups of countries.[1] In addition, Norris (2001) [313] illustrates that education has no significant effect on the uptake of ICT, both in industrialized and in developing countries. The research of Baliamoune [19] examined the empirical relationship between *literacy rate* and ICT use in 1995 and the relationship between the *Education Index* and the ICT usage in 1999. The study concluded that there was a positive, but hardly significant, correlation between the two indices. Possible reasons for this may be due to the presence of non-linear relationships or the bad quality of data. The present book makes a significant contribution to this point in which investigations are made (also) based on non-linear regression models.

For electronic communications, the Internet can be compared in significance to the (mobile) telephone network for voice communication. By definition, an Internet user is a person who has access to the Internet without considering the type of Internet access or quality of connection [190] (see Sect. 2.3.1). The study of Porter and Donthu [343] from 2006 shows that the highest growth rates of global Internet use are a result of older and less educated individuals who belong to a minority or have a low income. The absolute level of Internet use of these demographic groups is not surprisingly lower than the corresponding levels of the general population. The study *"The Internet and Knowledge Gaps: A Theoretical and Empirical Investigation"* from Bonfadelli (2002) [38] recalls that educated individuals are using the Internet more actively. Internet use is more information-oriented, whereas the interests of less-educated groups is particularly directed towards entertainment features of the Internet.

The work of Pratama et al. (2012) [347] examined the relationship between the Internet user rate and the so-called *Human Development Index* (HDI) from 2000 to 2010. The HDI reflects the level of prosperity, education and health level of countries in an aggregate measure. The results support a general popular thesis that the *digital divide* between developed countries and the developing world still exists and continues to grow. The main results of this study show a positive relationship between the HDI and the Internet user rate where this relation has become slightly stronger since 2000. This study further concludes that the growth rates of Internet use in countries still under development are higher than in developed countries. This corresponds in a certain respect to the so-called *leapfrogging effect* that describes the (technological) catch-up of developing countries relative to the (technical) progress in industrialized countries. Another finding is that the Internet user rate is growing

---

[1]Note: In this context it is expressed that the education level and the diffusion of technology in developed and OECD countries is generally high.

slower in countries that had a large change of the HDI between 2000 and 2010 compared to countries with a small HDI change. If one looks at the digital divide with regard to the Internet user rate in 2010, significant inequalities between developed and developing countries are existent [347]. Within this book, these conclusions of Pratama are relativized slightly in favor of the developing countries.

Andres et al. describe in 2010 [9] that low-income countries have a steeper Internet diffusion curve than high-income countries. Although this result is obvious, since low-income countries can skip technological developments (*leapfrogging*), it must be mentioned that the distinction of only two income levels limits the statement potential. In 2013 Zhang mentioned [522] the presence of a positive relationship between per capita GDP and Internet user rate, however there is also a negative relationship between income distribution (measured by the *Gini-Index*) and Internet usage. Richer countries (in terms of higher per capita GDP) thus have a higher rate of Internet users in this trend. Here it should be noted that higher average income corresponds with a more *balanced* income distribution [206]. This is not discussed in the results of Zhang [522]. However, no explanation is given to the relationship between per capita GDP and Internet usage, taking into account detailed mathematical functions. Hargittai [148] demonstrated that the level of education and the English language is corresponding to an increase in the number of Internet hosts in countries. Kiiski and Pohjola [217] show that average education level is a positive indicator for Internet hosts per capita. Another study of Chinn in 2007 [60] demonstrates that the extent to which the Internet user rate depends on education level is relatively low. In this book these results are partially detected as not applicable depending on the examined year between 2000 and 2012.

The social barriers and disparities that exist with access to Internet leads to the development of the *digital inequality* concept [119, 525]. This is better suited conceptually to the multi-faceted problem area than the mere physical access question [84, 87, 525]. As a term, digital inequality gives a more sophisticated approach to describe the *digital divide*. The original use of the term *digital divide* focused on physical access to the Internet, representing a binary approach. It however largely ignored the complexities of the subject as described above. Tsatsou speaks up for the term *digital divides*, due to the divisions between nations and split between different groups of a nation or society [454]. His work examines the research on the digital divide in terms of the interaction between technology, society and politics coming to the conclusion that binary approaches which insinuate a dichotomy of *information poor* and *information rich* are not sufficiently satisfying the digital divide. Instead, it is an approach that recommends the involvement of different socio-demographic characteristics and the resulting differences in Internet use based on gender, income, race and place of origin. This approach produces different grades of digital inequality.

3.1.2   Global Networking and Worldview

As described in Chap. 1, the focus of this book is primarily the analysis of the global access to ICT in the context of countries and cultures. If one understood religion as an essential part of the differentiation of cultures (see Chap. 2), then the question arises: at what level the respective major worldviews (*Christianity, Islam, Judaism, Hinduism, Buddhism and other worldviews including atheism*) have access to ICT? It must be noted that the general literature analyzing the distribution of (modern) ICT and their relationship with different worldviews are often of different character than the chosen approach to the subject and the questions studied in this book. Most relevant research work deals with the issue of how far religions and ethical values are affected through the use of ICT and media and how the respective world religions are represented in the Internet. There is very little research on a global scale (also due to the limited availability of data). But there are isolated empirical studies regarding the situation on the subject in major industrialized countries.

A study by the *Pew Research Center* from 2011 notes [335] that about 79% of the people in the US who are active in a religious community use the Internet. The proportion of people who are not active in a religious community and use the Internet is about 76%. Thus, there is no significant difference between religious and non-religious groups regarding Internet use in the US. Despite many advantages of the Internet, as in the field of economy and education, those surveyed responded that the Internet tends to have a more negative than a positive impact on the moral values of a society, such as religious values [339].

In 2014, Downey measured the effect of education and Internet use on religious affiliation using data from the *General Social Survey* [92]. His work found out that Internet use is associated with decreased probability of religious affiliation in general. In particular, this work explored the effect of increases in college education and Internet use on religious affiliation. Downey believes that this is the first study to show that Internet use is associated with disaffiliation, and to estimate the magnitude of the effect. He has identified *three factors* that are statistically associated with religious affiliation: *religious upbringing, education* and *Internet use*. He found that the effect of religious upbringing on college graduation is small, positive, and borderline statistically significant. So even if religious upbringing affects college graduation, it does not explain the negative relationship between college education and religious affiliation or the decrease in religious affiliation over time. Finally, it is easy to imagine mechanisms by which the experience of attending college might decrease religious affiliation. One can imagine two ways Internet use could contribute to disaffiliation. On the one hand, the Internet provides good opportunities for people living in homogeneous communities to find data about people of other religions, and to exchange information with them. On the other hand, for people with religious doubt, the Internet provides access to people in similar circumstances all over the world. Conversely, it is harder (but not impossible) to imagine plausible reasons why disaffiliation might cause increased Internet use.

In contrast, Wilson's book on *"The Internet Church"* [502] in 2000 states that the Internet offers Christianity the possibility to reach every human being on earth. Although the Internet can potentially provide such an opportunity for the future, Wilson has not considered the necessary condition for a global spread of Internet connection sufficiently. That is, at the time of the statement, this was not true in the formulation chosen.

In 2007 Horsfield compares the impact of the Internet on Christianity with the impact of printing on Christianity [168]. He claims that the printing era during the Reformation had formed a *new construction* of Christianity. This construction did not promote an establishment of the Christianity as this religion was postulated in the Bible or books. The lives of people led to a kind of interactive mix of experienced reality, written and the narrative. In the same way, but in an enhanced form, Internet today has an impact on the design of different worldviews. From a theological point of view, research in the future will be well less oriented towards examining the question of how far world religions are represented in the Internet world. Instead, the question will ask how the Internet affects the character and message of world religions. From a technical point of view, the global challenge is firstly, the ability to offer all world religions full engagement in the Internet world. Ultimately, from my point of view, it seems like that the possibility of *cross-cultural convergence* in the future will be determined by the *technical capabilities* and the *theological knowledge* of our societies around the world.

The article of Deitrick [78] describes ways of how modern online technologies can be used to give students a better understanding of Chinese religions and religious texts. It presents (adaptive) tasks, which give beginners an insight into the requirements of the Chinese religion, while at the same time offer the opportunity to learn, read and write Chinese characters.

There are few studies in regards to the relationship between *social networks* and *religious practice*. These few activities focus primarily on two aspects of the topic. Firstly, the aspect of *Privacy*. Obviously, adults who have access to the Internet and are members of a social network give data regarding their religious preference or affiliation award [36, 37]. The second aspect deals with the relationship between religion or *religious practice* and the use of social networks [335, 425]. The main result is that religious people are less likely to be a member of a social network in comparison to others. If they are a member of such a network the frequency of use is low. This religiosity is connected to a lower participation in online communities. It is not entirely clear why an active religious life reduces acts to take part in online communities [294]. It seems that active religious affiliation leads, in principle, to a person's stronger social orientation in local environments and participation in civil society organizations compared to people without such a religious orientation [137, 417, 505]. This could mean that these people are more stretched in social terms than others. These individuals are looking less for something new or unknown, and prefer personal contacts with peers than virtual contacts.

In modern society, people increasingly tend not to belong to a religious institution or organization even though they may continue to adhere to and practise a particular faith [54]. If society continues to be influenced by a specific religion, we can assume that this influence will also affect the *Internet culture*. On the other hand, religion in the Internet world reflects human trends in connecting their online with their offline activities. [169].

The paper *"The Unknown God of the Internet"* of Leary and Brasher from 1996 [249] provides an overview of how religion is manifested and affected by public stakeholders in the online world. Technology can thus be used as a form of empowerment for people of different worldviews, if one understands how religious communities imagine and prefer Internet use [54]. This is an important basis for a global discourse concerning the shaping of global media.

According to Barzilai, the Internet is a *cultural technique* and includes a number of different *sophisticated technologies* for a variety of cultural contexts [25]. However, the Internet is a cultural technique and consequently forms the culture of a society. Conversely, the society itself (also) shapes the culture of the Internet and its community. Understanding the respective religious cultures of Internet users remains an important basis when it comes to understanding the values and usage behaviour of Internet users [54]. To reduce the digital divide in countries, particularly in developing and emerging countries, actors such as religious communities positioned outside governmental processes play an active role in mediation [318, 429].

3.2 Power Issues and Reflections on Possible Future Developments

3.2.1 Internet Governance

In 1994, the Internet pioneer Robert E. Kahn described the leading role of governments in the process of the Internet evolution in his article *"The Role of Government in the Evolution of the Internet"* [208]. He pleaded that the involvement of governments in many areas of the Internet will be crucial in a long-term perspective, although the Internet is basically decoupled from any government in the world. The Internet was considered, from an infrastructure perspective, a type of *information superhighway* in the 1990s. The role of politics was primarily in the expansion and provision of Internet access at the time [219]. Given the enormous economic and political significance, which the Internet has nowadays, this role has changed. Therefore, Internet governance and the role of policies are becoming increasingly significant.

Internet governance is a research area with increasing importance for national and international relations. The complexity of the Internet architecture and its international structure requires the exploration and development of Internet governance issues across national boundaries technically, politically, economically, legally and culturally. Until now Internet governance had *nongovernmental governance* mechanisms [16]. This relates to a hybrid (global) development and regulatory complex with stakeholders in state, industry, academia and civil society according to a *multistakeholder-approach* [222, 254]. Internet governance therefore refers today to the (international) control of the Internet in a dialogue or in an interaction between the state and society as a whole [226, 281, 515, 517].

The *"Information Infrastructure Project"* at Harvard University [149] in the early stages of Internet research addressed the coordination of Internet management, in particular the management of the *Domain Name System*. The project staff offered advisory capacity to leaders of the US government [207].

The term Internet governance was originally used to identify the institutional and political problems arising from global coordination of Internet domain names and addresses [96]. In this context, ICANN was established with the strong participation of civil society actors in 1998, whereby the term *co-regulation* evolved parallel to emphasize the interaction that takes place between states and civil societies [223, 270]. ICANN represents a unilateral establishment of a global system, which was founded by the request of the US government and is still dominated by the US with the technical communities involvement in the foundation process. However, many governments are unhappy with the special role and participation of the US government in the Internet governance ecosystem.

The civil society side represents the view that the Internet is more apolitical. It is a *technical substructure* with implications for the *political superstructure*. This special view of reality cannot be justified if one considers the latest impacts of the Internet in all areas of life. Today, the consensus is that the ICANN functions are multi-partitioned into coordinative, technical and regulatory or policy issues [250]. Because ICANN is a private organization linked with the US Department of Commerce, a new debate has emerged in the field of communication concerning the role of private actors in global politics. Some governments make an analogy of the Internet to telecommunications, because it represent names and numbers of a resource that should be managed by state [165].

The Internet regulation has particular aspects related to various applications and protocols. This field is distinguished in relation to the regulation of conventional ICT. On the other hand it is important to understand that the evolution of the Internet has occurred in the context of conventional ICT [448]. This aspect is now reinforced by the impact of *technological convergence* [323]. As a consequence the Internet transforms itself into a system which has the capability of combining telecommunications, radio, television and mail [100]. The Internet also changes the self-contained or autonomous role of other media. In a certain respect it is regarded as a global public good, which is why state regulation tasks gain importance. This also holds good for the growing significance of the Internet for business and politics.

A second stage in the evolution of Internet governance [196] are the outcomes of the *World Summits on Information Society* (WSIS) organized by the United Nations in 2003 (Geneva) [516] and 2005 (Tunis) [517]. The WSIS was a platform on which the world's governments had the opportunity to deal with the Internet governance ecosystem. Here, a wide range of civil society organizations, stakeholders and academics were actively involved [301]. A certain *rivalry* between the ITU and ICANN was already apparent in these years [220, 221]. The rivalry between ICANN and ITU culminated in the establishment of the *Internet Governance Forum* (IGF) in 2006. The interdisciplinary *multi-stakeholder-approach* of the IGF, in which state actors, economy, academia, technical community, and civil society were represented. Historically, the source of this approach is in the telecommunications policy [324].

Considering recent discussions on Internet governance, on the one hand, the design, management and development of protocols are stressed at a technical level [79]. On the other hand, the role of civil society, Internet users and their personal responsibility as well as self-control is emphasized to establish a functioning and viable Internet regulation for a majority of society [197]. It is currently obvious that the Internet world both in terms of infrastructure and hardware as well as in the area of applications and services is dominated by the US government and US corporations. The unilateral control of the DNS system by the United States was a problem from the perspective of many states. One of the main reasons for this is that the US government had the exclusive means and capabilities to generate and delete *online objects*. It was observed from time to time that entities of the US government or courts have forced global registries to remove domain names from the address system. For example, many registrars around the world were forced to delete the domain in the case of Wikileaks and .iq (before the 2nd Gulf War) [350].

In principle, two worlds or cultures meet each other when it comes to Internet governance, namely the *technical world* and the *political culture*. The governments of the world all have a (national) idea about how they want to embellish the Internet and how they wish to participate in the global Internet governance debate. In a legal sense it is also their responsibility to safeguard the interests of their own population. Alliances across cultural boundaries and countries result from similar interests or views at the global level [255]. On the one hand, authoritarian or totalitarian states have a common understanding of the configuration of Internet governance. These states want to monitor the content of the Internet. In other words, they want to perform more and better controls than today [35]. In contrast, democratically oriented states want the status quo of the multi-stakeholder-model and are guided by the historical success of the Internet as an open system. It should be noted that many work on exploring Internet regulation or control focus on authoritarian or totalitarian states [35, 74–76, 199]. But there are also some works that focus on regulatory measures in the Western world, particularly in North-America and Europe [199, 271].

In March 2014 the US government announced through the *National Telecommunications and Information Administration* (NTIA) that in response to the given worldwide criticism since about a decade, they are willing to give up the inspection and monitoring of critical Internet resources such as addresses and protocol parameters by ICANN or IANA [316]. An adequate oversight of ICANN functions had central importance in the transition of these functions to another (international) organization or multi-stakeholder-community. During and after the transition things have to continue and function as usual [248]. The ICANN has three essential functions:

1. Management DNS Root Zone
2. IP address allocation
3. Administration protocol parameter registries

However, in addition the NTIA has indicated in its announcement in March 2014 that any proposal for a possible solution must find the broad support of the international community and noted four key points [316]:

1. Support and expansion of the multi-stakeholder-model
2. Need in ensuring the security, stability and resiliency of the DNS
3. Needs and expectations of global customers and partners of IANA services must be met
4. The *open* Internet needs to be assured.

The NTIA also announced that they will not accept a proposal which includes a control by a (single) government or an international organization. Several models for the transfer of ICANN functions were under discussion such as the widely discussed 3x3 model. The separation of the three IANA functions was proposed by various actors [225]. Another proposal focused mainly on names and includes a comprehensive restructuring of the eponymous feature, where two additional or two other organizations help to determine regulation [302]. On October 1, 2016, the contract between ICANN and NTIA to perform the IANA functions officially expired. The IANA stewardship transition is completed now.

Looking to the future, a central challenge in terms of Internet governance is clear. What do various large states or cultural systems in the world understand under a national and international perspective on Internet Governance? Against this background, the technical requirements of the global system Internet as well as the cultural needs and sensitivities of different countries and its people need to be understood and considered. It would be better to transform the current *multi–stakeholder–model* to a *multi-cultural–stakeholder–model*. In the views of the author, this includes a cultural dialogue of the actors in government, business, academia and civil society on the further evolution of the Internet governance ecosystems. This discourse needs to focus on respective state, business education and civil society cultures. It seems to be a difficult, but sustainable process. One of the fathers of the Internet, Vint Cerf, states here: *"It remains to be seen whether such a vision can be realized but it is fair to say the multistakeholder, cooperative, and collaborative nature of the Internet's development has been a major source of its resilience and its ability to absorb new applications and players since its conception 40 years ago and should form the basis for its future evolution"* [58].

3.2.2   Superorganism Humanity

Superorganisms arise from the coordinated composition of many independent and viable organisms in order to create something bigger [42]. Of key importance are so-called *autopoietic systems*, which were initially described in research by Maturana and Varela for the study of biological systems [278, 279]. Around the 1990s, among others, Luhmann with *Social Systems* [265] and with *"The Society of Society"* [267] as well as Robb with *"Cybernetics and Suprahuman Autopoietic Systems"* [390] and

Hufford with *"All Autopoietic Systems Must Be Social Systems"* have expanded the theory of autopoietic systems to the description of social systems. One is encouraged to describe networks of social processes, which space their own components and competencies in the area of social systems. The world society can be seen as an autopoietic network of *self-generating* components, and thus as a living system or *superorganism humanity*.

About 100 years ago in the late 19th century Spencer described his *"Principles of Sociology"* [430] based on the idea that society is an organism finding many analogies between structures and functions of biological organisms and social systems. This worked stressed, in particular, the internal processes of *integration* and *differentiation* as forms of division of labor. Spencer believed that *consciousness* or the *nervous system* of society are reflected through the (democratic) institutions and the government; he took global information networks and any brain-like structure out of consideration.

In his work *"Living Systems"* [291] the systems theorist Miller developed in 1978 a construction scheme that can be used for the analysis of any (biological or social) living system, for example a cell (system 1st order), a body (system 2nd order) or a society as a whole (system 3rd order). This theory describes a list of must have functional components for a functioning *metabolic process* and for a functional *nervous system* of society. These abstract functional components primarily include the processing of resources and information, but also protective functions, and capabilities of learning and decision-making.

The futurologist and systems theorist Rosnay used the term *macroscope* in 1979 as a perspective allowing us to describe the big picture, on the condition that *makroscope* and *microscope* complement and not replace each other [396]. The *macroscope* was used to examine the abstract functional components, or according to Miller, the material, energy and information flows that are effective at the level of *superorganism humanity*.

Another level of global application of the *organic perspective* is the *Gaia hypothesis*, stating that the earth is itself a living organism [263]. The earth is thus a complex dynamic system which generates (autopoietic) responsive mechanisms caused by human activities. These responses are based on the principle of self-organization and stabilize the biosphere. To model this hypothesis, Lovelock and Watson have designed a computer simulation (*Daisy World*) of a hypothetical simple planet with only two living beings, white and black daisies. The aim is to produce complex behaviours of organisms based on the interaction of very simple mechanisms [262]. The simulation resulted in the planet as a self regulatory system through adapting, balancing and stabilizing temperature fluctuations produced by different solar energy incidents through the multiplication or reduction of corresponding types of creatures. Later, this computer model was extended to include other species. The outcome of the extended model is as follows: The greater the self-regulating forces of the planet, the greater the *biodiversity*. These were significant arguments in protecting the biodiversity of the earth even if the model had been exposed to criticism. One criticism is that the similarity of the model to Earth is vague. For example, a specific mortality rate has to be set, which then has an impact on further growth. In reality, demographic indicators such as birth, death and migration rate are only partly controllable under certain conditions where these numbers can change only in the long-term. This can be done as long as no massive environmental or social disasters occur. These figures result from a complex interaction of prosperity level and education standards, cultural circumstances and especially, respective real-world politics. For some critics of Gaia hypothesis as Hern and Russel, humanity appears more to be like a *parasite* or *tumor*. The world population explosion and its significant impact on the consumption of resources hinders the achievement of sustainable development [156, 399].

A more concrete description of the *superorganism humanity* can be found in the book *"Metaman: The Merging of Humans and Machines into a Global Superorganism"* of Stock from 1993 [446]. Here, Stock refers primarily to the part of humanity and activities which result from the interaction between communication, trade and travel. At the time, Stock was referring to industrialized world and urban areas in developing countries. The remaining regions and countries in the world were regarded as peripheral, with the indication that the superorganism expands and, in connection with the proliferation of communication technologies, offers new opportunities for all people and regions. Here, the term Metaman is used for what is (currently) described with term global civilization and world culture (see Sect. 3.2.3).

This kind of theory, which advances the dialogue on *superorganism humanity*, opens up the possibility of modelling societies as *complex adaptive systems* (CAS) [293]. However, this approach is not well researched or developed even though it has a number of useful concepts and methods. There is no single integrated theory for these issues, neither for biological organisms nor for social societies. From the cybernetic systems theory there are attempts to model societies with a holistic approach [162]. Cybernetic models generally describe centralized, hierarchical (biological or social) systems. However there are approach principles as *self-organization, autonomy, decentralization* and *interaction* between multiple agents. These include the *"Living Systems"* theory of Miller (1978) [291], the theory of *"Autopoiesis"* by Maturana and Varela (1974, 1980) [278, 279], the *"Perceptual Control Theory"* (PCT) of Powers (1973, 1989) [345, 346] and also the theory of *"Social Systems"* or *"The Society of Society"* of Luhmann (1995, 1997) [265, 267].

The interest in such theories is growing steadily, in particular since the rapid development of information and communications networks which can be considered a nervous system for the social organism [133]. The *global brain* can be understood as the *neural network* of the *superorganism humanity* which is a self-organizing network of all (digitally) connected people and machines in the world. It works together with the information and communications technology as a brain [162].

The descriptions above demonstrate the complex network of communication links made up of individuals, equipment and software systems [446]. Challenges and problems, desires and expectations that can not be completely solved or served by a single agent will be propagated to other agents along the communication links in the network. These agents contribute their own experiences and knowledge in solving problems and overcoming challenges. If the problem persists, it is necessary that it is circulated until it is somehow completely dissolved or all possible options have been exhausted and the problem is recognized as (currently) unsolvable. The spread of knowledge regarding the challenges, needs and expectations in the global social communication network is a complex process which satisfies the principle of self-organization [159, 267, 390]. Radermacher describes this analogous to signaling in human brains by activating neurons, which characterize thinking patterns [356, 358]. It also raises questions about the *consciousness* of *superorganism humanity*. Furthermore, the question is how far this can be something else or could be more than the sum of the consciousness of all human beings together (digital consciousness) [353, 356]. The exchange processes in the global network also usually result in a change of the existing network by strengthening useful links and weakening less useful connections, which is changing the network to another state [20]. Thus, the *superorganism humanity* has a better *adaptive capacity* and the ability to *learn* or to tackle new challenges [162, 358].

Current research on the description of *superorganism humanity* include a reference to the role of information and communications technology, especially the *Internet* or the *Internet of Things* (IoT). Such kind of systems comprise distributed self-organizing planetary additive intelligence of all humans and machines [164]. This is aimed especially at the *Internet of Things* or *Internet of Entities* and thus at a hybrid: *human-technology-system*. It is also about the concept of *technological singularity*, which is under attention in particular with reference to the work of Kurzweil [241]. Technological singularity was originally inspired from the natural sciences idea refer-ring to pioneering work in electrical engineering, computer science and cybernetics as by Shannon [420], Turing [457] and Wiener [500, 501]. The paradigm posits that human individuals will have reached a type of super-intelligence by technological changes or *enhancements* and/or humans will be replaced by super-intelligent tech-nologies [407, 483]. This paradigm sees the future of human culture determined by the perspective of the development of artificial intelligence and robotics [32, 241]. It will come at the same time as the emergence of a new form of general machine intelligence or artificial general intelligence [332].

In a way, the development of *superorganism humanity* is very closely coupled to the paradigm of technological singularity. In addition, this coupling is also closely related to Moore's Law [297, 298]. It describes the exponential growth of the per-formance of chips in a certain period of time. Specifically, the law states that the number of transistors on a given chip area is to double within 18–24 months and the costs for the new generation of chips remain relatively constant in comparison to the older generation. In other words, the current generations of chips are available in 18–24 months at half the price today. This has come true in the past 50 years where this law has become the basis of the incredible innovative nature of the ICT field in the 21st century. Such innovations have never been seen anywhere before [39, 240, 242]. Note, however, that Moore's law has natural barriers placed at the atomic level, which in turn means that an increase in the performance of chips can not be continued indefinitely, at least not according to the current state of affairs.[2] On the

---

[2]This fact could be drastically changed, e.g. by capabilities of quantum computers, which are different from binary digital electronic computers based on transistors by using quantum-mechanical phenomena.

contrary, powerful chips will be very cheap in future. That's why chips will not be a limiting factor for the expansion of the Internet of Things.

3.2.3 Global Culture

A *global culture* includes essential elements of all cultures of the world. Therefore, a single culture does not dominate, rather the world culture and its underlying values are characterized by the principle of *cultural diversity* with increasing interaction and interdependence of different culture systems [146]. The development of a world culture should be understood in parallel to the establishment of a global ethic. Both concepts are complimentary. The *"Global Ethic"* principle of Küng [233–239] could be a fundamental building block for a peaceful coexistence of world cultures and for the sustainable development of world society. From the authors point of view a global culture is implausible without a (networked) world society, since a world society will not automatically produce a global culture.

For a long time, the linkage of society to national structures has prevented the understanding and analysis of all societies as a unity. The link with national structures arises initially from the perspective of designing national states as independent entities and then in conjunction with the *dependency theory* of Frank [114–116] and Galtung [120, 122], *world-systems theory* of Wallerstein [489, 490]. These theories take into account the interdependence between countries. However, this perspective takes up no independent mechanisms of *global total correlations* and hence does not consider implicit mechanisms of action due to the emergence of global potential. Such emergence processes and macro properties that occur are irreducible; that can not necessarily be attributed to the relations between the parts at the micro level [153, 154].

Wallerstein assumed that the global society is shaped by *divergences* (in this book also called *conflicts*) between different regions of the world [489]. The world is composed from a *center* and different *peripheries* caused by forces of the global market. Between the *centre* and the *peripheries* are differentiated forms of inequality characterized by different economic actors and roles, resulting in social inequalities and conflicts between the centre and peripheries.

Originally, the developed neo-institutional *world society theory* of [286, 287, 289, 290] had the *world system concept* of Wallerstein [489–491] as a base. Both schools of thought share beliefs that a global reality forms across cultural borders and beyond nation states increasingly effecting the cultural realities and developments in each country. In opposition to the notion of Wallerstein, Meyer developed a way of looking at the world society represented by a *convergence thesis*. The convergence hypothesis was taken up by the work *"Convergence and Divergence in Development"* of Meyer in 1975, where he provided substantial evidence of convergence [286]. After Meyers work almost all societies grow and modernize in the economic sphere leading to the modernization of other areas of society. In particular, the field of education and its global development provide evidence for the theory of convergence

[379]. Education development aims to principles such as justice and development, in particular from the view of neo-institutionalists. Education is seen as a means to convey both the national welfare as well as individual development. Not surprisingly a trend towards expansion is reflected in the education systems of most countries [386]. Meyer explains why different societies in a nationally organized world are structurally similar to each other [288]. He refers to this as a kind of *structural isomorphism* despite significant differences in access to resources, religions and traditions. Thus, the nation-state is a product of a worldwide network of power, trade and competitive relationships. Meyer uses the term *world polity*, which has cultural components as well as the term *world society* [288].

A global cultural order is not only a world culture to be understood as a set of cultural values, but also as a constitution of objectified culture. The *global culture* is more than the simple composition of ideals and cultural values spread within societies by individuals and their communication exchange. The content and structure of a global culture is influenced in a reciprocal process by states, organizations and individuals. The approach explains a high degree of isomorphism in structural and political programs (i.e. constitutions) of different nations. In the field of education one can find increasingly standardized curricula and, with regard to digitization, there are increasing rationalized data acquisition systems. The *Universal Declaration of Human Rights, gender equality, environmental protection, standardized health systems* etc. are other good examples of increasing isomorphism according to Meyer. These are also examples that point to the existence of a global culture that is increasingly codified by international treaties and diffuses accordingly in respective countries. At the same time, however, this indeed results in a massive decoupling of formal structures and real actions, whether through lack of resources or lack of adequate global models in the local context. Pressure on nation states to be guided by such models is exerted on the one hand, by transnational organizations (UN, World Bank, EU, as well as NGOs) and on the other hand, by local actors (citizens and their associations).

A basic allocative problem of the concept of global culture is the *statelessness of world society*. This association problem with possibly internal contradictions, inconsistencies in the national cultural models at appropriate locations and at the global level can be solved by organizations such as the United Nations and related international bodies. In this context it should be noted that the current format of the UN does not necessarily have a democratic order at the global level. Joint focus of all nation states should be based on the central role of institutions designated to (inter-) cultural rules. This would give national actors a global collective meaning and value and integrate them in a common frame. This frame corresponds to the values of all cultures and does not violate any cultural sensitivities. The same pressure is used at the level of nation states by transnational models (for example, UN, World Bank, EU, as well as NGOs) and local actors (citizens, associations).

Luhmann sees the *world society* [264] including all possible subsystems in a society, which are constituted and maintained as *functionally differentiated social systems* [265] based on communication. He promotes a concept of society as a social system, which includes all actions through communication and has no doubt that the socio-cultural revolution has already created a global society. For Luhmann there is only a single globe-spanning society system; the world society concept is understood as a type of *operatively closed autopoietic social system* [172, 390], which includes all held communication processes. In modern society, global communication is specifically recorded by ICT. It assumes that (global) communication is the central operation of the (world) society where there is no communication outside society. There are no other operations than communications within society. Luhmann is setting the basis of the world society on autopoietic systems [278, 279] and because this world society has its limits in communication, today's global society is, according to Stichweh [439–443], significant with its *singularity*, because no other society or world societies exist. Thus, Stichweh indicates simultaneously that today all societies are world societies.

Beck has partly represented a parallel but partly controversial opinion over Stichweh, with an ambiguous concept of society in his sociology of the second modernity [26]. This parallel states that we already live in a global society, but at the same time a controversy is encouraged, because this world society is specifically marked by its social *non-integrity* [27]. This *non-integrity* results from the fact that on the one hand, the world society forms the sum of all nation states and on the other hand, transnational organizations and actors have emerged which is why Beck uses the term *world societies*. From 1986 Beck used the term *risk society* referencing the distribution of scientific-technically produced risks in society. He pointed out that risks are always a result of social reconstruction process as long as one understands the society as a living and evolving system. Given the ongoing globalization processes Beck has expanded his theories of *risk society* to the *global risk society* [28].

Considering the opinion of Baylis and Smith to Beck's *world societies* in the context of current globalization processes, the interdependence and interaction of regional societies increases. These interactions do not form a world society as a whole as long as there is no supranational or worldwide constitution. Only when a global normative framework or a system of global governance has been established with a cosmopolitan law, then the transformation of the nation-state structures is complete and a world society can be obtained. Every person can invoke these cosmopolitan rights regardless of citizenship.

Even after Habermas a global society is emerging. However, the world community does not have a sufficient degree of *social integrity* [141]. In this development process Habermas is concerned about the cosmopolitan solidarity being *undermined* by the dynamics and global expansion of a specific functional system [142]. In 2011, Habermas understood the EU as a *higher-level political value-based community* [143]. He viewed this as a decisive step in the direction towards a politically consistuted world society. Here, Radermacher has a similar view by considering the EU as *model in miniature* for globalization [359, 367, 371]. The design of globalization should proceed similarly to the EU's design if a peaceful coexistence of the international community is the goal [365]. After Burton this must succeed in the *conversion of the world society into a world community* [47] by integrating global relationship structures and important cooperation networks in the fields of science, religion, language and economy additionally to the formal institutions into the processes of the expansion of functional systems [265].

Küng [233, 234, 239] discuss a *global ethic* that involves the path from a world society to a world community and establishment of a *global culture* which refers to the culture system of all human beings. An important *prerequisite* for the constitution of a *global culture is digital networking in the context of global sustainable development*. This is especially true under conditions of modernity. At the core of a *global culture* and global ethic there exists a consensus of values between cultures with the definition of a common economic [236, 238], education [235] as well as religious culture [232, 237]. Complicated interactions arise when questioning whether a person is rich or poor, educated or uneducated, religious or not religious. These are distinctions that have a significant impact on how people behave in different societies and in accordance to the investigations of the Nobel laureate Sen, these differences can be found in all cultural contexts [416]. Sen states that we belong to many contexts, not only to a specific culture or religion.

Dahrendorf spoke of *ligatures* in which we are incorporated individually [70]. Any direct close global exchange in context to global culture must respect the values of different cultures, whereas overcoming poverty can be regarded as a necessary condition on the way towards a global culture. The common goal is to ensure that no human being has to worry about suffering from hunger [342]. In this context, the Nobel Peace Prize winner Yunus believes that poverty already belongs in a museum [520].

Huntington has doubts on the idea of a global culture. He describes these in detail in his book *"Clash of Civilizations"* [173–175]. At the same time, it must be mentioned that Huntington is very often misunderstood in this context. The idea of a universal global culture based on the Western model is very unlikely. His idea was that a global culture will continue having conflicts in this world, which would, however, not be ideological but cultural. Huntington considered culture (only) from the perspective of eight major religions and derives from this juxtaposition possible conflicts in the future. However, the versatility or transformative ability of cultures are central factors against the theses of Huntington.

Part II
Empirical Analyses of the Global Spread of ICT

Chapter 4
Global Networking: Relation to Economic Performance and Education Level

The empirical analysis in this book on the spread of global information and communications technology (ICT) mainly start in the year 2000. This include the dependence of ICT on the country economic performance and education levels. The reason for analysis began in 2000 is attributed to the data situation regarding the key figures. Furthermore, this year has a historical significance in the computer world as the *Millennium bug (Y2k)* appeared and lead to the misinterpretation or misrepresentation of the year in computer systems which only used the last two digits [62]. This could have had dramatic impacts on the management of computerized business and social processes. Secondly, 2000 was the year when the big *Internet bubble* burst, also known as *dotcom bubble* [327]. The excessively high short-term expectations on the new medium Internet and the resulting excessive prices for businesses and start-ups in this area from speculative activities led to a massive slump in global markets. However, the resulting burst could not stop the rapid rise in the global spread of mobile phone usage and Internet access. The era of the global *new economy* had begun.

In 2000 the *Millennium Development Goals (MDGs)* adopted at the United Nations set out global objectives to ensure primary education for all children around the world. As part of the MDGs, among other development goals, the benefits of ICTs should also be guaranteed for developing countries within the framework of a global partnership for development until 2015. The *Sustainable Development Goals* (SDGs) were developed to keep up and extend the MDGs between 2015 and 2030 and are a collection of 17 interrelated goals. The SDGs not only have a higher number of objectives, but also developing and developed countries on the focus. The MDGs have only partially been achieved, but there has been great advances in the area of modern global communication options.

The subsequent analyses extends over a decade in which the development of the respective technologies could be well understood and the data basis is reasonable.

© Springer International Publishing AG 2018
H. Ünver, *Global Networking, Communication and Culture: Conflict or Convergence?*, Studies in Systems, Decision and Control 151,
https://doi.org/10.1007/978-3-319-76448-1_4

4.1   Global Networking and Economic Performance 2000–2013

The first good news in 2000 was that the feared problems of the Millennium bug did not appear [62]. Aside from a few problems with some of the applications, the mobile networks and telecommunications infrastructure were hardly affected. The feared Y2K-problem went unnoticed by a majority of the population certainly due to the intense preparations that had been made.

The hypothesis presented in this book is that the scope and quality of digital networking is primarily related with economic potential or performance in comparison among states. More specifically it is related to the *gross domestic product* (GDP) and even more so with *GDP per capita* (GDPpC) of the related countries. The GDP and in particular GDPpC, as shown above in Sect. 2.3.2, belong to the major indicators for economic analysis, both in spatial and in temporal international comparisons.[1,2]

The following sections are devoted to the empirical relationship between economic performance, in terms of GDP and per capita GDP (GDPpC), and the fixed telephone access rate (FTR), the mobile phone access rate (MPR), the Internet penetration rate (IPR) and the mobile Internet penetration rate (MBR) of countries. The definition of each access rate is only briefly described in the following sections as they had been previously discussed in detail in previous Sect. 2.3.1. The factual context between the investigated variables, e.g., GDPpC and IPR, is obvious. An inclusion of persons or countries into worldwide communication processes is expensive and the associated infrastructure requirements are high [9, 17, 60, 84]. This is especially true before the appearance of mobile solutions, because terrestrial networks were required [402, 421]. Thus a degree of prosperity is a prerequisite for participation in communication. Conversely, a good technical basis in communications infrastructure promotes the growth of economic wealth in a country [363]. This is well analyzed in the field of development aid [147], such as the positive effects of *Grameen Banking* in Bangladesh [452].

---

[1] In November 2007, the European Commission, together with the European Parliament, the Club of Rome, the World Wildlife Foundation (WWF) and the Organization for Economic Cooperation and Development (OECD), organized a conference entitled *Beyond GDP* [102]. The aim of this conference was to develop a better measure of a country's economic situation beyond GDP measurement. In 2009, the idea was extended to *GDP and Beyond*.

[2] "Whether it comes to the distribution of voting rights in the International Monetary Fund (IMF), or to the allocation of funding from the structural and cohesion funds of the European Union: GDP sizes and purchasing power parities (PPP) play an important role and are guidelines in many decisions for the comparison of macroeconomic variables. Instead of exchange rates they also serve as conversion factors for converting the relevant amounts into a single currency and simultaneously eliminating price differences between countries. In addition to this function as a spatial deflator (a measure of inflation, which is the ratio of the nominal gross domestic product and the real gross domestic product per year) macro-economic indicators can be derived from the purchasing power parity comparisons of price levels between countries" [113].

Table 4.1 Descriptive statistics for fixed telephone rate (FTR) and GDP per capita (GDPpC) in 2000 and 2013

	2000 (N = 172)						2013 (N = 175)					
	SP	Min	Max	MV	MD	SD	SP	Min	Max	MV	MD	SD
GDPpC	90741	270	91011	10884	5711	14364	141961	603	142564	17752	11842	21233
FTR	73.05	0.02	73.07	19.21	10.32	20.19	62.98	0.02	63.00	16.98	13.86	16.32

4.1.1 Fixed Telephony and Economic Performance

Fixed telephony is the oldest technology investigated in this book. The FTR is defined as the ratio of the number of fixed telephone lines to the population in a country (see Sect. 2.3.1). Table 4.1 shows the minimum (Min), maximum (Max), span (SP), mean (MV), median (MD) and standard deviation (SD) for GDP per capita (GDPpC) and fixed telephone access rate (FTR) in a worldwide perspective in the years 2000 and 2013.

The number N of the analyzed countries increased from 172 in 2000 to 175 countries in 2013. If one describes the worldwide digital inequality in the context of FTR and the average GDPpC with the help of the SP and SD as easy dispersion measures, it turns out that the inequality in the use of the landline connection has decreased between 2000 and 2013; inequality regarding the average GDPpC has increased.[3]

The maximum for the FTR in 2000 was 73% (*Switzerland*). This value decreased to 63% until 2013 (*Korea*). This was a result of the triumph of mobile devices (see Sect. 4.1.2), which suppressed landline communication while representing an alternative mode. In this regard, mobile communication limits the landline expansion due to the absence of infrastructure investments. The minimal value 0.02% stayed the same in 2000 (*Dem. Rep. Congo*) and 2013 (*Central African Rep.*). Such a low value means that in these countries only a few wealthy individuals, businesses and government agencies continued to use landlines. Not surprisingly, the minimum value comes both from African countries.

It is also interesting that the global average of the FTR of approximately 19.2% in 2000 decreased to just 17% in 2013, pointing to the increasing influence of mobile phones. In contrast, the median value increased from about 10.3% in 2000 to almost 13.9% in 2013 and is a result of catch up process of countries. This median increase can also be attributed to the FTR that had dropped significantly in a few wealthy countries. Further reasons and explanations are discussed below.

The minimum value of the GDPpC in 2000 was 270 Intl$. (*Dem. Rep. Congo*), which was almost doubled to 603 Intl$. until 2013 (*Central African Rep.*). Although

---

[3]Please Note: The measurement of social inequality is more advanced compared to the measurement of digital inequality. For example, there is the well-known *Gini–Index* or the so called Equity–factor (see Sect. 2.3.2). However, these measures with one parameter only give a rough indication. The data basis developed by *Picketty* is much more accurate, especially in significance for the top income level [341].

it is random in which the poorest country in a given year has the lowest FTR, it is nevertheless an indication of the relationship between the GDPpC and FTR. The maximum value is 91.000 Intl\$. in the year 2000 (*Qatar*) and about 142.000 Intl\$. in 2013 (*Macao*). The richest states are small countries. In 2000, the global mean was almost 11.000 Intl\$. and increased to almost 18.000 Intl\$. in 13 years. The median increased from 5.711 Intl\$. to almost 12.000 Intl\$., and therefore became relatively stronger than the mean, similar to the trend of the FTR.

To determine the empirical relationship between economic performance and the fixed telephone rate (FTR) of a country, a regression analysis is performed. In this analysis the FTR is the dependent and the GDPpC the independent variable, which is used for explanatory reasons. Considering the point cloud for the year 2000 in Fig. 4.1 bottom, one can recognize that states with a low GDPpC have a low FTR. The larger the GDPpC values, the greater the FTR. However, the FTR reaches its maximum value if one is going for higher GDPpC, but then decreases the greater the GDPpC. The latter seems to correlate with the fact that high GDPpC bear certain effects in small countries like *Qatar* or *Kuwait*, which obtain wealth through selling valuable resources (oil/gas) without having the characteristics of a highly technical society.

Figure 4.1 top shows the point cloud for the year 2013. A first observation is that the data of 2013 does not reach the values from 2000 (also see Table 4.1). This is caused by the general decrease of the FTR. Furthermore, it is noticeable that the data in 2013 are more strewn along the GDPpC axis than in 2000. This means that the income is more widely spread out (also see Table 4.1). The basic statement of the year 2000, that states with low GDPpC have a somewhat lower FTR, also applies to 2013. This is also valid for the following statement: The bigger the GDPpC, the higher is the FTR. Furthermore, it again comes to a slight lower FTR for the highest GDPpC in 2013.

The optical impression about the data provides a guide for possible classes of (test–) functions (see Sect. 2.4.8), with which the relation between GDPpC and FTR can be modeled. In Fig. 4.1, one can study the trend of selected curves (*Linear, Logarithm, Monod1, Monod2, Logist, MaxLog, Planck*), which are used in the modeling of the relation. The selection process of this class of (test–) functions will be described later. For a best fit function out of this set of functions the mean square error (MSE) is used as minimization aim. The MSE and the coefficient of determination (R^2) for the separately tested functions of the years 2000 and 2013 are listed in Table 4.2.

In Eq. (4.1) is the (test–) function with four parameters, which is a peculiarity of the so-called *Planck's* function. It provides the best fit for the regarded class of functions, both for the year 2000 and 2013.

$$FTR_t = \frac{a \cdot GDPpC^b}{c \cdot \exp^{d \, GDPpC} - 1} \qquad (4.1)$$

The parameters values a, b, c and d for the respective year, are found in Table 4.3. To obtain an understanding of the range of (test–) functions used in the regression analysis for the relation between the GDPpC and the FTR it is described in the

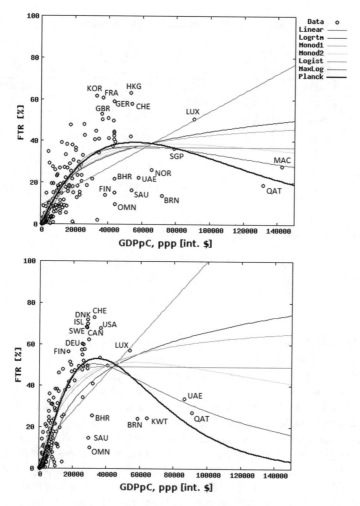

Fig. 4.1 Fixed telephone (FTR) as function of GDP per capita (GDPpC), year 2000 (bottom) and 2013 (top)

following, how the amount of functions is successively extended, to get the best fit function for the data. Those functions are to be found in Table 4.2, whereby the mathematical description of all the used functions can be found in Sect. 2.4.8.

First, a linear function (as easiest way) with a shift in the ordinate is considered, which clearly isn't adequate for the coherence between the GDPpC and the FTR. For the sake of completeness, it is nevertheless tested. The rather low linear correlation between GDP and the FTR is also confirmed numerically (see Table 4.2), because the MSE has the highest value in comparison to other functions.

Literature also shows (see Chap. 3) that the relation between economic performance and technology use can generally be described with *logarithmic* models in a

Table 4.2 Mean Square Error (MSE) and coefficient of determination (R^2) for regression $FTR = f(GDPpC)$

	2000 (N = 172)			2013 (N = 175)		
	MSE	MSE/N	R^2	MSE	MSE/N	R^2
Linear	40314	234.4	0.422	30231	174.7	0.348
Logartm	22748	132.3	0.674	19917	113.8	0.570
Monod1	22001	127.9	0.684	19493	111.4	0.579
Monod2	19605	114.0	0.719	18425	105.3	0.602
Logist	17750	103.2	0.745	17994	102.8	0.612
MaxLog	17302	100.6	0.752	18418	105.2	0.603
Planck	**15852**	**92.2**	**0.773**	**17707**	**101.2**	**0.618**

Table 4.3 Parameters for the function $FTR = f(GDPpC)$

t	a	b	c	d
2000	5.392	1.458	98409	$4.195 \cdot 10^{-5}$
2013	$5.402 \cdot 10^{-5}$	1.222	3.287	$1.949 \cdot 10^{-5}$

relatively satisfying way, which is why this function is also tested here. One explanation for a logarithmical model of these coherences is the law of *diminishing marginal benefit (Gossen's law)*.[4] In order to generalize the logarithmic model (*Logartm*), a shift on the abscissa and ordinate is carried out respectively. The logarithmic model is theoretically unlimited and doesn't fit to this situation, because the FTR has the practical upper bound of 100%. This is why the next obvious functions for modeling this coherence are a logarithmic model with a *saturation (Monod1)* and with a *restraint (Monod2)*.[5] Saturation applies if the function reaches a limit value and runs at this level consistently. In the case of inhibition, after reaching a saturation limit an additional lowering of the function values occurs. Over long periods of time adaption and diffusion of technology can behave in an *S-shaped* trend which is, mathematically speaking, also called *sigmoid curve*. In the following, this is labelled as a *logistical* model (*Logist*). Ultimately, two functions (*MaxLog, Planck*) are used and can lead to a decrease after a rise and satiation (e.g., in a S-shape).

The studies based on this approach for all years between 2000 and 2013 show the same results. The best fit function, according to the mean square error (MSE), for the relation between GDPpC and FTR for all these years is given with Plank's function. Figure 4.2 shows the development of the best fit function for the coherence between GDPpC and FTR from year 2000 (blue) to 2013 (black). The analyses for

---

[4]The law is named after the German national economist *Gossen* (1810–1858). According to the first law of Gossen, the benefit of a good decreases every single time it is consumed until satiation is reached. This additional benefit is also called marginal benefit.

[5]This is not arguing against the general literature, as the general logarithmic model can fit well in practice for a corresponding issue.

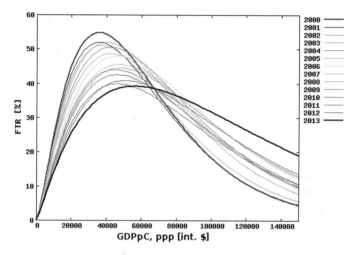

Fig. 4.2 Development of the fixed telephone rate (FTR) as a function of GDP per capita (GDPpC), from 2000 to 2013

the selected functions conclude that the FTR has decreased in the interval [0, 65000] Intl$. between the years 2000 and 2013. The FTR has increased in the interval [GDPpC > 65000$] Intl$. in the same time. This result stems from the optimization of the fitting function and is not necessarily true for the existing data. A more detailed look at the data is given below.

Overall, the global average FTR decreases between 2000 and 2013. However, the data shows, that the global FTR rises until 2006, but decreases continuously since then. This decrease is a result of the success in mobile phones (see Sect. 4.1.2). The sum of both accession rates to telecommunications has in fact grown continuously. The global FTR, as described earlier, was at 19.2% in 2000. In 2013, this value decreased to almost 17%. The absolute number increased from about 1 billion in 2000, to about 1.16 billion landline connections in 2013, which is a growth of 16%. The reason for the decrease of FTR on the global level is mainly due to population growth. The world population grows faster than the number of fixed lines in this time. While the FTR decreased by about 2% between 2000 and 2013, the global population increased by almost 20% in the same time.

Figure 4.3 shows the best fit function, the point cloud and the center of mass[6] (center c) of this point cloud for the years 2000 (blue) and 2013 (black). The arrow (red) shows the shift of the center between both these years. The result show that the average global GDPpC increased while FTR decreased simultaneously. This complies with a shift to the lower left of the complete data cloud. In a graphical depiction the shift is seen between C2000 and C2013 with their associated coordinates in the diagram.

---

[6]The focus area of a point cloud on a 2-dimensional place is defined as the point of intersection of the mean values from the x- and y-data.

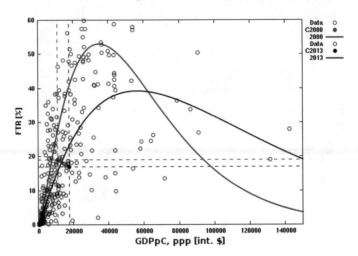

Fig. 4.3 Data, best fit function and center shift for FTR = f(GDPpC), year 2000 (blue) and year 2013 (black)

$$C2000 = (11114, 19.21) \rightarrow C2013 = (17752, 16.98)$$

Furthermore, one can see the point of intersection of the best fit functions for the years 2000 (blue) and 2013 (black), which is at about 65.000 Intl$.

On the national level, the change of the FTR is to be examined in more detail because there are various possibilities such as how the FTR can increase or decrease in a country. The FTR in country i at the point in time t is defined as

$$FTR_t^i = \frac{FT_t^i}{POP_t^i} \tag{4.2}$$

where by FT_t^i is the number of fixed lines and POP_t^i is the population size in a country i at the point in time t. Both, FT and POP can grow or decrease between the point in time t and $t+1$. If one defines the percentage change of the FTR between the time t and $t+1$ in a country i as $\Delta FTR_{t,t+1}^i$, one gets a positive value $\Delta FTR_{t,t+1}^i$ ($\Delta FTR \uparrow$) as a result, if the following inequation applies:

$$\frac{FT_t^i}{FT_{t+1}^i} > \frac{POP_t^i}{POP_{t+1}^i} \tag{4.3}$$

Corresponding, one gets a negative value for $\Delta FTR_{t,t+1}^i$ ($\Delta FTR \downarrow$), if the inequation (4.4) applies:

$$\frac{FT_t^i}{FT_{t+1}^i} < \frac{POP_t^i}{POP_{t+1}^i} \tag{4.4}$$

Fig. 4.4 Number n of countries with positive (blue) and negative (red) change in landline rate between 2000 and 2013 (above); Number of countries with a change of landline rate by maximal $\pm 1\%$ (below)

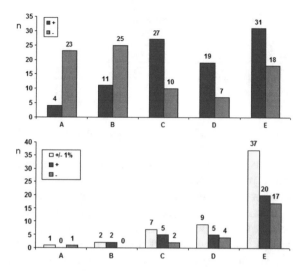

Figure 4.4 shows national categories according to the economical performance per capita (GDPpC), and the number of countries with a positive (blue) or negative (red) value for ΔFTR between 2000 and 2013. The bar graph (above) shows, that in the groups with highest GDPpC (category A) respectively high GDPpC (category B), the absolute number ($n_{A-} = 23$, $n_{B-} = 25$) of countries, whose FTR decreased, is the highest.[7] Simultaneously, group A and B have the lowest number of countries, whose FTR increased ($n_{A+} = 4, n_{B+} = 11$). In the poorest category E, the number n_{E+} of countries with an increased FTR, is absolutely the highest, while 18 countries in category E (poorest)($n_{E-} = 18$) have a decreased FTR. The middle GDPpC category C has 27 countries with increased, and 10 with decreased ($n_{C+} = 27, n_{C-} = 10$) FTR. In category D (poor), 19 states increased ($n_{D+} = 19$), and 7 decreased ($n_{D-} = 7$) its FTR.

Table 4.4 summarizes the results for the number of countries according to economic performance: categories A to E with positive and negative ΔFTR. *Iran* had the highest increase in the FTR from 14.4% in 2000 to 38.2% in 2013 despite its growing population from about 60 to 70 million people. The four Scandinavian countries *Denmark* ($\Delta FTR \downarrow = 34.4\%$), *Finland* ($\Delta FTR \downarrow = 41.2\%$), *Norway* ($\Delta FTR \downarrow = 27.3\%$) and *Sweden* ($\Delta FTR \downarrow = 27.6\%$) had the highest decrease in FTR. This massive decrease is by far not only explainable by the growing population, which has its maximal value at 5%. The main reason is loss of fixed lines.[8]

The fact that the growth in FTR can partly be compensated by the demographic growth (especially in developing countries), gives the FTR the ability to be on a constant level over the time. Concerning this topic, we provide the following

---

[7]The classification of countries by economic capacity in different categories can be found in the Appendix A.1.

[8]Reasons may be due to the cost of alternative technologies as well as the fact that several fixed lines can be bundled technically in order to reduce costs.

Table 4.4 ΔFTR according to country categories

t	A	B	C	D	E
$\Delta FTR \uparrow$	4	11	27	19	31
$\Delta FTR \downarrow$	23	25	10	7	18

considerations: $\Delta FTR = 1\%$ is regarded as a neutral trend, while Fig. 4.4 (bar graph to the right) shows the number n of countries with increased and decreased FTR by the maximal value $\pm 1\%$ in the economical categories A, B, C, D and E. Theoretically, the value for a neutral development is $\Delta FTR = 0\%$, which means, that the number of fixed lines increases/decreases (relatively) identically to the growth/drop of a countries population. In reality, there is no country worldwide with a ΔFTR of exactly 0% between 2000 and 2013, which is why the value of $\pm 1\%$ is chosen and accepted as a neutral trend because countries such as *India*, *Philippines*, *Nigeria*, *Egypt*, and *Bangladesh* fall in the range of this neutral trend which are enough big and under development. A neutral trend only took place in one of the countries of category A (*Oman*), and two of category B (*Equatorial Guinea*, *Portugal*). The more one looks at the poorer countries (categories C, D and E), the more likely they are to see a neutral trend. It is known that the population continues to grow in poorer countries, whereas it is stagnating or decreasing in richer states. One can see that in poorer countries the population growth strains the fixed telephone infrastructure and the (absolute) growth of fixed lines is mostly compensated by the population growth. The consequence is that the FTR is resting on an (almost) constant level.

In summary one can say that the FTR has decreased in richer countries (category A and B), while countries with an average GDPpC (category C) increased their FTR. In 75% of the poorer categories D and E, the change in FTR was at most 1%. This is the case, because poorer countries (partly) still invest in landline infrastructure. This growth in a sense is mostly eliminated by the population growth. The decrease of the FTR in richer states is primarily explained with the success of mobile communication over the last decade.

Now, the relation between the growth of GDPpC and growth of FTR between 2000 and 2013 can be examined. Figure 4.5 shows the different rates of growth for the corresponding GDPpC categories A, B, C, D and E between 2000 and 2013. This examination concludes that the FTR has decreased by 11.3% in the GDPpC category A and (average) 4.1% in GDPpC category B. The average and lower GDPpC categories C and D gained an extra average 3% respectively 3.5% to their FTR. The lowest GDPpC category E has an average growth of 0.7% in their FTR. Richer countries don't expand their landline infrastructure any more (to a certain extent), or rebuild it in a certain technical form. This is due to a greater amount of bundled fixed lines. The average and poorer states (considering the GDPpC) continue expanding their landline infrastructure further, growth of the FTR is being compensated or inhibited by the population growth.

Fig. 4.5 (Total) growth for GDPpC (blue) and FTR (red) between 2000 and 2013

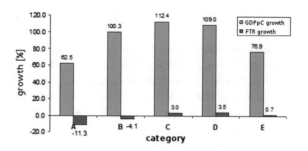

Table 4.5 Descriptive statistic of mobile phone rate (MPR) and GDP per capita 2000 and 2013

	2000 (N = 177)						2013 (N = 178)					
	Sp	Min	Max	MV	MD	SD	Sp	Min	Max	MV	MD	SD
GDPpc	90636	270	91011	10955	5702	14434	131154	603	131757	17898	11764	19760
MPR	79.68	0.01	79.69	15.89	4.46	22.63	233.07	5.61	238.68	101.48	106.09	38.77

One of the theses discussed in this book is that the *convergence in the ICT* sector and in other technical-related areas is taking place distinctly *faster than wealth itself in general*. Although the gap between the rich and the poor will only be compensated slowly, if at all, *there is a fast catch up in the usage of ICT regarding the whole world population. At least, this thesis applies to the given numbers regarding the fixed telephone rate.*

4.1.2 Mobile Telephony and Economic Performance

Mobile telephony is the second-oldest technology, after the fixed telephony, under the technologies examined in this book. The analysis of the relation between the GDP per capita (GDPpC) and the mobile phone penetration rate (MPR) occurs analogously to the investigation of the landline access rate in the previous section. Table 4.5 illustrates the values span (Sp), minimum (Min), maximum (Max), mean (MV), median (MD) and standard deviation (SD) of the descriptive statistics of the data for GDPpC and MPR.

In 2000, the average MPR worldwide was just under 16%, although there are countries in which the MPR is clearly above 50% (*Finland, United Kingdom, Iceland, Israel, Italy, Luxembourg, Norway, Austria, Sweden, and Singapore*). The ten countries with the lowest MPR in the same year are *Afghanistan, Bhutan, Eritrea, Guinea-Bissau, Iraq, Comoros, Micronesia, Niger, Principe, Sao Tome, and Tuvalu*. In 2003 the MPR worldwide is higher than the FTR for the first time. In *Germany* in 2000, the fixed telephone had to give up its role as the leading communication medium to the mobile phone. However, for the sake of fairness it must be said that often several telephone lines are bound to one fixed-line port, such as ISDN terminals (see Sect. 2.3.1).

One can describe the development of worldwide digital inequality with regard to the MPR between 2000 and 2013 with the help of the dispersion measures span and standard deviation. Thus, both values increase with this in mind (see Table 4.5). Indeed, it is in such a way that in 2013 much more people used mobile phones than in 2000 and that the average MPR in 2013 was just above 100%. This value is derived from the adjusted data (step 4a data management, see Sect. 2.4). According to the source data of ITU the worldwide MPR in 2013 was about 95%. In the same year, the countries *Argentina, Bahrain, Botswana, Estonia, Finland, Gabon, Hong-Kong, Italy, Kazakhstan, Kuwait, Libya, Lithuania, Montenegro, Oman, Austria, Poland, Qatar, Russia, Saudi Arabia, Singapore, Uruguay and United Arab Emirates* had a MPR of more than 150%. These include states such as the oil-rich Arab countries. The Scandinavian countries are also here, as in other areas of IT usage, global leaders.[9] It is also interesting that the median value in 2013 is higher than the average value of 100%, even though the average in 2000 is 16%, about three times higher than the median. It is rare that the data in this context has a median that is higher than the average value. This is an indication of the significant catching up of developing and emerging countries in mobile phone usage.

For the analysis of the functional relationship between mobile phone usage and the economic performance of countries, a regression analysis is used. The average mobile phone penetration rate (MPR), which determines the number of active mobile phones with a SIM card in relation to the size of the total population of a country (see Sect. 2.3.1), is the dependent variable. The independent (and thus also the explanatory) variable is again the average GDP per capita (GDPpC).

Looking at the data for the year 2000 in Fig. 4.6 (diagram below), one can notice that the point cloud has a similar scattering as the data for the relationship between the GDPpC and the FTR (see Sect. 4.1.1). In this diagram one can recognize that states with a low GDPpC have low MPR. The greater the GDPpC values become, the greater the MPR values. Moving in the direction of higher GDPpC, the MPR first reaches a maximum and afterwards drops again. It is not surprising in this initial situation that the *Planck* function again returns a very good adjustment of the relationship between the GDPpC and the MPR in 2000 (within the considered class of functions and according to the method of the squares).

Figure 4.6 top shows the data points for 2013. The first observation is that the data concerning 2013 reaches significantly higher values for MPR than the data in 2000 (see also Table 4.5). Another observation is that the MPR is over 100% in 2013 in many countries. This means that on average there is more than one mobile terminal per person in a country. Some people have, for example, a private mobile phone as well as a commercial one. Furthermore, people in developing countries own more than one mobile phone for financial reasons to call up as cheap as possible in the

---

[9]In this context, reference is made to the current debate in Finland which wants to abolish the cursive handwriting in primary education. The blockletters are retained and writing with the computer keyboard is to come into the curriculum.

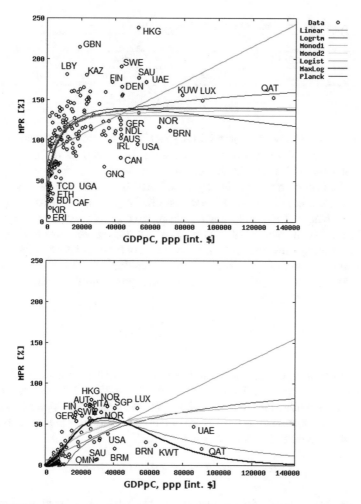

Fig. 4.6 Mobile phone rate (MPR) as a function of GDP per Capita (GDPpC), year 2000 (bottom) and 2013 (top)

respective net. Further on, the basis statement valid for 2000 claiming that countries with low GDPpC have a lower MPR than states with high GDPpC, has not changed much. Overall all states have increased their MPR, so that the catch-up in the mobile phone usage is very distinct worldwide.

The main difference between the data for the fixed telephone rate (FTR) and MPR in 2013 is the fact that for the MPR, the countries with higher GDPpC have only a slightly lower MPR than the countries in the medium GDPpC area.[10] This can be explained through the fact that the states with the highest GDPpC between

---

[10]Reminder: For the FTR, the countries with the higher GDPpC have (significantly) lower values than the countries in the medium GDPpC area.

Table 4.6 parameters for Planck function $MPR = f(GDPpC)$

t	a	b	c	d
2000	$2.001 \cdot 10^{-7}$	2.240	5.998	$6.271 \cdot 10^{-5}$
2013	0.708	−1.429	−1.176	0.052

2000 and 2013 have invested more into the mobile communications than into fixed network compared to the rather poorer counties. But this does not mean that the poorer countries did not invest into their mobile network.

Figure 4.6 also points to the history of selected functions such as *Linear* (red), *Logrtm* (blue), *Monod1* (pink), *Monod2* (gold), *Logist* (green), *MaxLog* (brown) and *Planck* (black). The best fit function from the class of (test) functions is to be determined using the method of least squares. For this purpose, one minimizes the sum of the mean square errors (MSE), wherein the coefficient of determination (R^2) is a quality measure for the adaption of individual functions. The 4-parametric function in Eq. (4.5) with the parameters a, b, c, and d (expression of the Planck function) deliver the best fit for 2000.

$$MPR_t^i = \frac{a \cdot (GDPpC_t^i)^b}{c \cdot \exp^{d \cdot GDPpC_t^i} - 1}$$

(4.5)

After verification of the particular functions for each year between 2000 and 2013, it turns out that the relationship between the MPR and the GDPpC only in 2013 the *MaxLog*–function (see Sect. 2.4.8) delivers a better adaptation than a Planck function.[11] The corresponding (best approximation) function is of the type of Eq. (4.6) for the year 2013 with the parameters a, b, c and d.

$$MPR_t^i = a \cdot \exp^{b - c \cdot log(GDPpC_t^i) - d \cdot log^2(GDPpC_t^i)}$$

(4.6)

Table 4.6 shows the values for parameters a, b, c and d for each function for the year 2000 and 2013.

Table 4.7 shows the different fitting functions and the associated values for the sum of mean square errors (MSE or MSE/N) and the coefficient of determination (R^2). A clear distinction can be seen here in comparison with the relationship between GDPpC and fixed telephone rate (FTR) (see Sect. 4.1.1, Table 4.2). The functions *Logrtm, Monod1, Monod2, Logist, MaxLog* and *Planck* in particular all allow a similarly good adjustment in accordance with the least square method for 2000 as well as 2013. Here one can summarize the three functions *Logrtm, Monod1* and *Monod2* to a class and the other three functions *Logist, Maxlog* and *Planck* to another class. Within these classes the adaption of the particular functions is similarly well, see the affected coefficient of determination (R^2).

---

[11] The class of functions is expanded as in the previous section.

Table 4.7 Sum of the square of errors (MSE) and coefficient of determination (R^2) for regression $MPR = f(GDPpC)$

	2000 (N = 177)			2013 (N = 178)		
	MSE	MSE/N	R^2	MSE	MSE/N	R^2
Linear	48565	274.4	0.439	209077	1174.6	0.281
Logartm	32312	182.6	0.627	147529	828.8	0.439
Monod1	32674	184.6	0.622	148515	111.4	0.489
Monod2	30556	172.6	0.647	148515	105.3	0.489
Logist	21324	120.5	0.754	150714	102.8	0.482
MaxLog	21471	121.3	0.752	**148380**	**105.2**	**0.491**
Planck	**20788**	**117.4**	**0.760**	151586	101.2	0.479

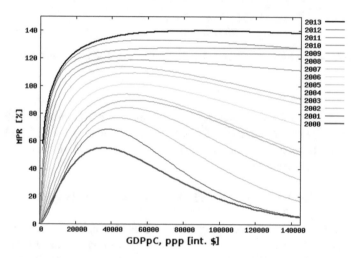

Fig. 4.7 Development of mobile phone rate (MPR) as function of GDP per capita (GDPpC), year 2000–2013

Mobile phone usage has become very widespread between 2000 and 2013 and has developed well over all GDPpC ranges. In Fig. 4.7 one can study the development, per year, of the best fitting functions for the relationship between GDPpC and MPR from 2000 (blue) till 2013 (black).

The result shows that the MPR has increased in all areas of GDPpC from year to year. It should be noted that this is not the case for the development of the relationship between GDPpC and FTR (see Sect. 4.1.1, Fig. 4.2). Furthermore this result is not only an artefact of the adaptation in the optimization process. Rather the investigation of the database laid for this optimization delivers the same result.

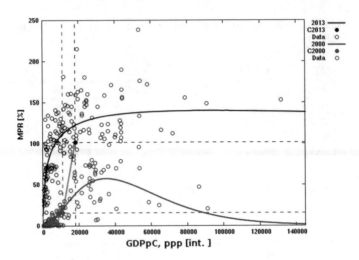

Fig. 4.8 Data, best fit function and focal shift for MPR = f(GDPpC), year 2000 (blue) and 2013 (black)

Figure 4.8 shows the data points, intersections of the averages of GDPpC and MPR (center C) and the best fitting function for 2000 (blue) and 2013 (black). Here the red arrow shows the shift of emphasis between these years ($C2000 \rightarrow C2013$) in suitable x- and y-coordinates.

$$C2000 = (10955, 15.89) \rightarrow C2013 = (17898, 101.48)$$

Based on Fig. 4.8 one can recognize that the MPR has increased over all GDPpC areas and accordingly in all countries. This can be seen on one hand in the higher course of the curve for 2013 (black) as the curve course for 2000 (blue). On the other hand the data points for the year 2000 (blue) are not or barely overlaid with data points for 2013. In relative terms the shift in emphasis is stronger upwards (in the direction of MPR) than to the right (in the direction of GDPpC). *This study confirms to some extent the hypothesis that the catching up in the use of ICT and in particular, mobile phone usage, is faster worldwide than the process in respect to general prosperity.*

In the previous Sect. 4.1.1 regarding the relationship between GDPpC and fixed telephone rate, one can find an analysis pertaining to the influence of demographic growth (worldwide and in countries) on the FTR. This is because trends were to be observed which pointed to a sinking of the FTR. Such an analysis is omitted in this section, because the MPR has increased in all countries and thus the size and/or the growth/shrinking of the population has no significant influence on the MPR.

If one performs a growth analysis in the period from 2000 to 2013 for the GDPpC and the MPR in the suitable GDPpC categories A, B, C, D and E (subdivision see Appendix A.1) where category A stands for the richest and category E for the poorest countries, one is able to examine to what extent the GDP growth and the MPR growth

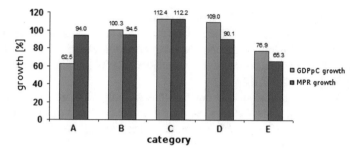

Fig. 4.9 (Total) growth from 2000 until 2013 for GDPpC (blue) and MPR (red)

are connected within and between these categories. The bar graph in Fig. 4.9 shows on the x-axis, the respective GDPpC category and on the y-axis the growth of GDPpC (blue) and MPR (burgundy).

Hence the highest growth of GDPpC and MPR took place in category C (medium GDPpC). Both growth rates lie in category C with about 112%. In the categories A (highest GDPpC) and B (high GDPpC) a similarly high MPR growth of approx. 94% and 94.5% is to be observed, whereby the GDPpC growth lies in category A with about 62% and in category B with approx. 100%. The GDPpC growth is in category D (low GDPpC) 109% with a MPR-growth of 90%. In category E (lowest GDP) was held the slightest MPR growth, namely by about 65% with a GDPpC growth of nearly 77%. The result indicates that the GDP and the MPR growth have a strong coherence (high correlation), especially if one does not consider the richest countries (category A). This is also understandable if one minds that the richest states can not show such a high growth rate performance as economically weaker states, because they start from a clearly higher level of GDPpC.

How can this success in overall GDPpC categories, also known as the *mobile miracle*, be explained in a worldwide perspective? The prepaid card is an important business model for the mobile market. The concept of a prepaid card has also asserted itself because the advantage for the rate supplier is that he does not have to grant a loan to the customer if he takes up the offered service. The customer in contrast is protected against non-transparent and cumulative costs, at least insofar as they exceed the prepaid amount.[12] In this business model, the customer grants the rate supplier an interest-free loan, which, as with vouchers for some tariff providers, expires if the customer does not use it. In July 2006, there were 1.5 billion prepaid contracts worldwide at about 2.3 billion mobile contracts in total. In African countries more than 90% as well as *South American* and *Eastern Europe* with about 80% are prepaid cards of all mobile contracts. In 2006 in *Germany*, 37.7% of all mobile phone owners had a prepaid card. In 2012 these were about 41.8%. The use of prepaid systems is tied together with the fact that many people still have no access to the financial system,

---

[12]Note: Previously, prepaid cards were well suited for criminal misuse, because the customer was not required to provide personal information. Since 2004, providers are only allowed to unlock prepaid cards after specifying the name, address and date of birth(§111 Abs. 1 TKG).

even to the simplest forms of a bank account. In 2007, Vodafone, one of the world's biggest mobile phone providers, recognized this early and overcame this bottleneck by successfully establishing the *mobile banking system M–Pesa* in Kenya.[13] This was followed by Apple's worldwide initiative in launching its mobile financial service, ApplePay in September 2014.

A hypothesis presented in this book is that we will experience a convergence in the usage of ICT. This thesis is confirmed so far for the mobile phone usage, as that in 2013 the worldwide MPR average had reached about 100% and that the MPR growth between 2000 and 2013 over all GDPpC areas took place at a similarly high level. This is also due to the fact that in many countries there is more than one mobile phone connection per person and the availability of prepaid systems has favored the mobile phone usage in poorer regions of the world. In other words: The world population makes calls using mobile connections and has moved from local-oriented communication (fixed telephony) to the person-oriented communication (mobile phone).

4.1.3 Internet and Economic Performance

Within the connection between the economic performance and the Internet penetration rate (IPR) it is initially of interest how the worldwide distribution of (absolute) economic potential GDP (purchasing power parities, PPP) is related to distribution of worldwide Internet usage. Beside the considerations in the previous Sects. 4.1.1 and 4.1.2, the question is answered on how big the percentage of a country is in the world economic capacity (world GDP) and which percentage a country has worldwide of Internet users (world IPR). The absolute population size is factored out in these considerations. Figure 4.10 on the x-axis represents the economic potential of a country as a portion of the world GDP. The y axis measures the extent of Internet usage of a country as a portion of the world IPR in the year 2000 (bottom diagram) and 2013 (top diagram). The findings illustrate a linear relation between the share of a country in the world GDP to the share of a country in the world IPR. There is a high correlation in the year 2000 as well as 2013.[14] The two-parametric linear model is from the Eq. (4.7) and is defined as follows:

$$Value\_World\_IPR_t^i = a + b \cdot Value\_World\_IPR_t^i \qquad (4.7)$$

Here the parameters a and b have the following value for the years 2000 and 2013 (Table 4.8).

---

[13]Note: M–Pesa was introduced as a system for the processing of basic functions of money transfers and private non-cash payment via mobile phones without the need for regular bank accounts. It was developed by Kenyan mobile company Safaricom in cooperation with Vodafone.

[14]Note: The concept correlation can be used here, because the relation in this example is linearly modelled.

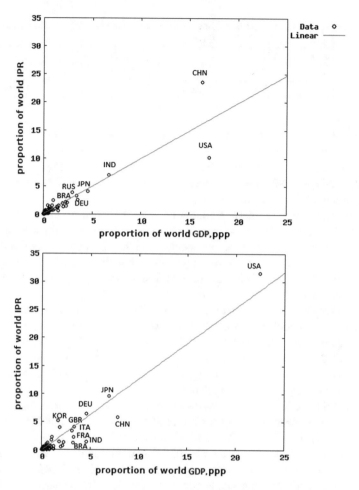

Fig. 4.10 Country's Proportion of world IPR as a function of the percentage of world GDP, 2000 (bottom) and 2013 (top)

Table 4.8 Parameters for the regression $Value\_World\_IPR = f(Value\_World\_GDP)$

t	a	b
2000	−0.156	1.275
2013	0.005	0.992

Thus the United States accounted for about 23% of world economic output and about 30% of Internet users worldwide in 2000. If one divides the states into one group above and below the regression line (red), the *USA*, *Japan*, *Great Britain* and *Korea* are located above and states like *China*, *India*, *Brazil*, *Russia* (so-called BRIC states) but also *France* and *Italy* are seen below the line.

Table 4.9 Mean squared error (MSE) and coefficient of determination (R^2) for the linear regression $Value\_World\_IPR = f(Value\_World\_GDP)$

	2000 (N = 177)			2013 (N = 181)		
	MSE	MSE/N	R^2	MSE	MSE/N	R^2
Linear	91.65	0.52	0.924	110.81	0.61	0.852

As Fig. 4.10 shows, there is a high correlation in the year 2000. This is in such a way that the relation after a 13-year development in 2013 still can be the linear model, i.e., that still a high correlation is given. The following Table 4.9 shows the mean squared error (MSE) and the coefficient of determination (R^2) for the (linear) relationship in 2000 and 2013.

This analysis points to a narrow relative between the economic power of a country and its Internet use. This is also valid if one ignores the population with regard to the economic power. In a per capita consideration the differences between rich and poor countries are of course still much greater than is the case when looking at the economic power of states. This is also due to countries with greater populations such as *China* and *India*.

It is interesting in this context that in 2013 all the *BRIC* countries are above the linear regression model and *USA*, *Japan* and *Germany* lie under it. This development is a consequence of large populations of BRIC countries leading to an increased IPR between 2000 and 2013 significantly. Since the alignment of communication is still faster than economic performance, the size of population of *BRIC* states leads to a correspondingly large Internet population.

China and the *US*, also known as the G2 or *Group of Two*, who compete for the supremacy in the global economy, are exemplified separately. In 2000 the US, with a portion of 22.5% of world GDP, and with a portion of 31. 5% in the global IPR, was still clearly before China. At this time China had a portion of 7.8% in global GDP and 5.8% of global IPR. The massive economic progress of the *People's Republic of China* during the recent years clearly affects the number of its Internet users. In 2013 the US had a portion in global GDP of 17.1% and a portion global IPR of 10.1%. In the same year China had almost an equal share of global GDP with 16.3%, but with 23.6% share of global IPR more than twice as high as US. In Fig. 4.10 (chart above) one can see China and the US on the x-axis almost at identical heights, however with China above and the US below the regression line. These statements relativize somewhat in favor of the United States, considering that China's populations is four times larger than that of the US.

At this point, one can examine the asymmetrical distribution of the Internet population in relation to the distribution of the world's population to countries. Figure 4.11 for the year 2000 shows the proportion of a country to the (global) Internet population (A) relative to its share of the world population (B). It is clear that the distribution of the world's population is asymmetric due to country size. By analyzing this data, the (different) "representation strength" of a country in Internet population is clear. The horizontal (green) bar corresponds to a ratio of $A/B = 1$, which means that a country is as strongly represented in the Internet world as it is by population in the

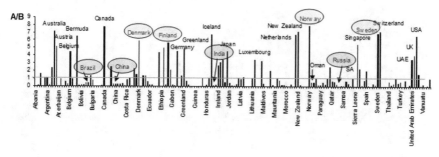

Fig. 4.11 Share of a country of the Internet population (A) in relation to its share of the world population (B), 2000

world. Thus, a "symmetric" distribution with regards to two variables are illustrated however there can be no talk of this yet. It is striking that there are only a few countries in 2000 represented over proportionately high in the Internet world relative to their proportion of the world's population. Unsurprisingly, these countries include the four Scandinavian countries of *Denmark, Finland, Norway* and *Sweden*. These states are about 7–8 times more represented in the Internet world than their population percentage (relative) to the world's population. The Anglo-Saxon countries of *Australia, Great Britain, Canada* and the *US* also have a very high ratio. European countries are particular including *Belgium, Germany, France, Iceland, Luxembourg,* the *Netherlands* and *Austria* which are represented over proportionately on the Internet compared to their proportion of global population. The remaining European countries are adequately represented. The African countries illustrate a relation not proportional to their population. This disproportion is also valid for the economic power of African states. In addition, the BRIC countries in 2000 are also far below average in representation of the Internet world.

Figure 4.12 illustrates the relationship for 2013. This is to ensure that the axis scaling here is different compared to the year 2000. This figure takes into account that the total population of Internet users is significantly higher than in 2000. The increasing balancing effect of A/B moving in direction of $A = B$ can be seen well. The proportions are much more balanced, although there are still many countries that are clearly underrepresented in proportion to their share of the world population in the Internet world. The states that are over proportionately represented in 2013, are essentially the same countries as in 2000, however, the difference is that only India is under represented within the BRIC countries.

Overall, it is clear that wealthier countries in 2013, especially the Western world, are still over proportionately represented in the Internet world in proportion to their share of the world population. Nevertheless, between 2000 and 2013 the rest of the world caught up considerably in this respect pointing to the cultural convergence of Internet use.

In the following section, investigations are continued by analogy with the previous Sects. 4.1.1 and 4.1.2. Table 4.10 illustrates the GDPpC and the IPR values of the descriptive statistics. The appropriate values of the GDPpC were already discussed in previous sections in detail.

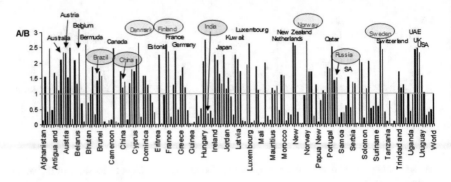

Fig. 4.12 Share of a country of the worldwide Internet population (A) in relation to its share of the world population (B), 2013

Table 4.10 Descriptive statistics Internet penetration rate (IPR) and GDP per capita (GDPpC), year 2000 and 2013

	2000 (N = 177)						2013 (N = 181)					
	Sp	Min	Max	MV	MD	Sta	Sp	Min	Max	MV	MD	Sta
GDPpC	90741	270	91011	10919	5732	14351	141961	603	142564	18484	11653	21728
IPR	51.99	0.01	52.00	7.97	1.98	12.91	95.65	0.90	96.54	41.57	39.20	29.02

The number N of the countries studied is 177 in 2000 and 181 in 2013. The IPR has in 2000 a maximum of 52% (*Norway*) and this value increases till 2013 to approximately 96.5% (*Iceland*). The IPR minimum value of 0.01% (*Dem. Rep. of Congo*) in 2000 increases slightly and is almost 1% (*Eritrea*) in 2013. These are again the African countries, as in the FTR and MPR, which have the lowest values for the IPR. The average increases over the same period from nearly 8% to about 41.6%. It is interesting that the median of 39.2% in 2013 is nearly at the same level as the average value, although the median of nearly 2% in 2000 is on a four times lower level as the average. This is a consequence of the significantly higher balance of the proportions till 2013, as further noted in the context above.

We consider further, in detail, the empirical relationship between GDP per Capita (GDPpC) and (IPR) per country. Significantly greater differences are to be expected here as above in considering the economic output and mobile phone usage of each state. This relationship is no longer linear. We use a regression analysis, wherein the IPR is the dependent variable and the GDPpC the independent, i.e., the explanatory, variable.

Examining the point cloud of the relationship between GDP and the IPR in 2000 in Fig. 4.13 (left), it is evident that states with lower GDPpC have a lower IPR. The greater the value of the GDPpC the greater the IPR, and for much greater values of the GDPpC it falls again. A similar observation applies to the relationship between the GDP and the FTR (see Sect. 4.1.1). This is not surprising in this respect, as nearly every Internet access in 2000 occurred via a fixed network connection

Table 4.11 Parameter for Planck's function, $IPR = f(GDPpC)$

t	a	b	c	d
2000	$4.675 \cdot 10^{-12}$	3.168	1.903	$8.465 \cdot 10^{-5}$
2013	0.357	0.795	14.253	$1.068 \cdot 10^{-5}$

by dial–up (modem). Since the mobile Internet access hardly existed worldwide, the fixed network was a necessary condition to obtain Internet connection.[15] As a result, the best adaptation (from the previously discussed functions) delivers the Planck's function (as well as within the connection between GDPpC and FTR) for the description of the relationship between GDPpC and IPR in 2000. The Scandinavian countries including *Denmark, Sweden, Norway* and *Finland* have one of the highest IPR rates worldwide in 2000. This is in addition to *Australia, Germany, Great Britain, Japan, Canada, Korea, Switzerland, Singapore* and the *US* with high IPR values. There are countries such as *Bahrain, Brunei, Kuwait, Oman, Saudi Arabia*, and the *United Arab Emirates*, but also surprisingly *Luxembourg*, which has a high GDPpC. Furthermore, in relation to this huge GDPpC height they have a relatively low IPR compared to the remaining wealthier states. These are primarily the oil states that enforce a form of Planck's function for the relationship between GDPpC and IPR. Reason for this is partly a prosperity "borrowed" from nature by selling oil and gas instead of prosperity based on an internationally competitive, diversified productive economy. Latter would require massive dissemination of technology and Internet usage and innovation in this area.

The diagram (above) in Fig. 4.13 shows the relationship for the GDPpC and the IPR in 2013. The values for the IPR in 2013 have increased significantly compared to 2000 (see Table 4.10). The statement from 2000 stating low GDPpC have a rather low IPR is also valid for 2013, despite of catching-up processes. Furthermore the obvious correlation is valid that higher GDPpC values cause higher IPR values. While the correlation between GDP and IPR in 2000 has a very similar curve to the relationship between GDPpC and FTR, this similarity has decoupled with respect to the curve shape. This is initially a sign that in 2013 many users worldwide utilized the Internet without a landline due to the accessibility of mobile phones. This is examined below in systems theory approach to determine the interactions of key figures in more detail (see Sect. 4.4.2).

In Fig. 4.13 one can see the course of the tested functions Linear (red), Logrtm (blue), Monod1 (pink), Monod2 (gold), Logist (green), MaxLog (brown) and Planck (black). The 4-parametric shaping of the Planck function in Eq. (4.8) returns the best fit for the year 2000 and 2013.

$$IPR_t = \frac{a \cdot GDPpC^b}{c \cdot \exp^{d \cdot GDPpC} - 1} \tag{4.8}$$

---

[15]The interaction between the access rates and the various technologies are investigated in the Sect. 4.4.2 below.

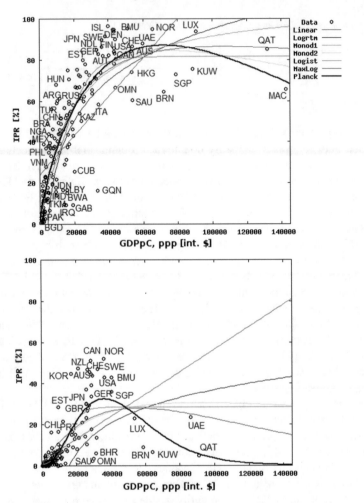

Fig. 4.13 Internet penetration rate (IPR) as function of GDP per capita (GDPpC), year 2000 (bottom) and 2013 (top)

Table 4.11 shows the values for the parameters a, b, c and d in Eq. (4.8) for 2000 and 2013. Table 4.12 lists the results of the regression analysis with the values for the mean squared error (MSE or MSE/N) and the coefficient of determination (R^2) for all functions tested for the years 2000 and 2013. The procedure for selecting the amount of functions to be tested is analogous to Sects. 4.1.1 and 4.1.2. As already mentioned, in 2000 the Planck function fits the best way followed by the functions *Maxlog* and *Logist* with a similarly good adjustment. In 2013, the Planck function again has the best fit, but also the rest of the tested functions, except for the linear, describe the relationship similarly well. This points to the fact that the functions with

Table 4.12 Mean squared error (MSE) and coefficient of determination (R^2) for regression $IPR = f(GDPpC)$

	2000 (N = 177)			2013 (N = 178)		
	MSE	MSE/N	R^2	MSE	MSE/N	R^2
Linear	18440	104.2	0.372	74357	410.8	0.509
Logartm	14177	80.10	0.517	35274	194.9	0.767
Monod1	14292	80.75	0.513	33760	186.5	0.777
Monod2	13690	77.34	0.534	32466	179.4	0.786
Logist	10946	61.84	0.627	35212	194.5	0.768
MaxLog	10261	57.97	0.651	33032	182.5	0.782
Planck	**9760**	**55.14**	**0.668**	**32011**	**176.9**	**0.789**

an inhibition lose relatively in strength and functions with a saturation win relatively in strength for this relationship.

Figure 4.14 shows the development of the best fit function for 2000 (blue) to 2013 (black). The Planck function achieves the best fit for the years between 2000 and 2013. This is again comparable with the relationship between the GDPpC and FTR. The results for the fit function show that the IPR has expanded between 2000 and 2007 for all GDPpC areas. The (Planck) fitting functions for the years 2008 and 2009 run in this respect slightly differently, because states with the biggest GDPpC have lower IPR values in contrast to the years before. However, this is not appropriate as the data show and is an artefact of the regression formation. This stems from the fact that countries in the medium GDPpC area between 2008 and 2009 tend to have a larger (relative) increase in their IPR than the countries in the top GDPpC area. The reasons for this are to be found in the beginning of the mobile Internet usage in countries with a medium GDPpC (see Sect. 4.1.5). One can observe the same phenomenon in 2013.

Figure 4.15 summarizes the data analysis of the relationship between the GDPpC and IPR. For the year 2000 it shows (blue) and 2013 (black) the point clouds, fitting functions as well as the average values (dashed lines) and their intersections (center) for the GDPpC and IPR. The arrow (red) indicates the shift of the center (C2000 → C2013) of the data between 2000 and 2013 with the corresponding x and y coordinates.

$$C2000 = (10919, 7.97) \rightarrow C2013 = (18484, 41.57)$$

In Fig. 4.15 is to be recognized once more that the IPR increased over all GDPpC areas, i.e., in all countries. The curve for 2000 (blue) runs clearly below the curve for 2013 (black) and there is no intersection of the curves. Furthermore, there is hardly a superposition of the two point clouds. The shift of focus is more upwards (towards IPR) than to the right (towards GDPpC) relatively regarded. This observation under-

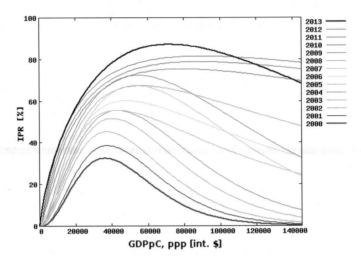

Fig. 4.14 Development of Internet penetration rate (IPR) as function of the GDP per capita (GDPpC), year 2000 (blue) until 2013 (black)

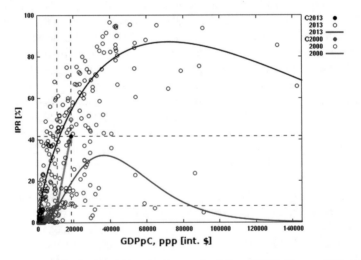

Fig. 4.15 Data, best fit-function and shift of center for IPR = f(GDPpC), year 2000 (blue) and 2013 (black)

lines the thesis that the catching up in Internet usage around the world runs clearly faster than catching up in regards to prosperity.

If one carries out a growth analysis from 2000 to 2013 for the GDP and the IPR in the corresponding GDPpC categories A, B, C, D and E (division, see Appendix A.1) the relationship between GDP growth and IPR growth can be specified. Figure 4.16 on the x-axis illustrates the respective GDPpC category and on the y-axis the growth rate for the GDPpC (blue) and the IPR (burgundy). The respective number refers to

Fig. 4.16 (Total) growth from 2000 to 2013 for GDPpC (blue) and IPR (red)

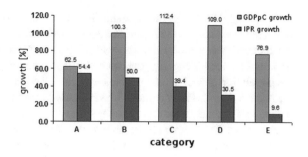

the overall growth between 2000 and 2013. Here, category A stands for the richest and category E for the poorest countries.

The highest average IPR growth of 54.4% took place in the highest GDPpC category A. However in this category the slightest GDPpC growth of 62.5% is given what, nevertheless, in terms of absolute growth is high. The reason is that the richest countries have a higher starting level than poorer countries. The findings are interesting in that as states are poorer, average IPR growth is lower. This means that in category B, the second strongest IPR growth takes place with 50%, followed by the category C with 39.4%, category D with 30.5% and category E with 9.6%. On the one hand, the catching-up in the usage of Internet is generally confirmed. But on the other hand, the findings for the global spread of Internet access show that still much is to be done, especially in developing countries.

In summary, the empirical relationship between the GDP and the IPR both in 2000 and in 2013 is positive. The Planck function is the best model for this context (within the considered class of functions) for the years 2000–2013. The course of the Planck function in 2000 is similar to the one of the relationship between the GDPpC and FTR, suggesting also the existing physical connection of Internet and fixed telephony (see Sect. 4.4.2). Up until 2013 almost all states had increased their IPR where on average the richest states showed a high increase and poorest states a lower increase. With the relative increase it is exactly the reverse.

4.1.4 Internet and Social Balance

In this section the influence of the social balance level or social inequality in the form of the distribution of income is examined according to the Equity Theory for the Internet penetration rate. The Equity Theory describes the social inequality level of countries mathematically by so-called Lorenz distributions or Lorenz curves of a particular type (see Sect. 2.3.2). Lorenz curves $F : [0, 1] \to [0, 1]$ in general describe which part $y = F(x) \in [0, 1]$ of the income the poorest groups $x\%$ of a society own. Extreme balance corresponds to the diagonal $y = x$ (equal distribution of income). Inequality means that F runs near to the zero axis for some time.

With regard to the questions arising here the following can be assumed: The higher the social inequality or the lower the social compensation level in a country, the lower the Internet penetration rate. The reasoning is as follows: First, rich countries are socially balanced. Secondly, the more balanced a country, the more the poor part of the population participates. Consequently, more people can afford Internet access in the balanced case. The argument, however, could then be wrong if access to Internet would be extremely expensive and therefore a higher degree of inequality would be necessary so that few would be able to use this special good. An example for this is the sport of polo. However, the Internet is not expensive in richer countries. The data situation for Equity Theory is limited and there is data for the equity factor for 30 countries from the years 2001 and 2009 in the work [206]. This data is used in the following for the investigation of the relation between the equity factor (ε) and the Internet penetration rate (IPR).

Figure 4.17 (below) shows the equity factor (ε) as an independent and the Internet penetration rate (IPR) as dependent variable for the year 2001. Accordingly, there is, as expected, a positive correlation between the size of the ε factor (growing ε means more social balance) and the IPR in of 2001. The ε factor rises with higher GDPpC as known from suitable studies [155, 206, 341, 510] and this has a positive affect here. In other words: The more balanced the distribution of income, the greater the IPR while a higher GDPpC is expected simultaneously. Equation (4.9) shows the linear regression model, the coefficient of determination (R^2) and the correlation coefficient (R). Accordingly, there is a moderate correlation.

$$IPR_{2001} = -29.8 + 108.9 \cdot \varepsilon_{2001}, R^2 = 0.343, R = 0.586 \qquad (4.9)$$

Figure 4.17 (top) shows this relationship for the year 2009. It is interesting that the relationship between the social inequality level and the IPR has strengthened between 2001 and 2009. Equation (4.10) shows the linear relationship where the correlation coefficient $R = 0.743$ also suggests a stronger correlation.

$$IPR_{2009} = -55.4 + 225.1 \cdot \varepsilon_{2001}, R^2 = 0.552, R = 0.743 \qquad (4.10)$$

Furthermore, the observed relation is checked for plausibility. Building on the knowledge that a level of high social balance promotes the use of the Internet of the whole population, one can consider the following: In particular, we suppose that with an IPR of x% in a country exactly the x% with higher income have access to the Internet. For example, if a country has an IPR of 70%, then according to the assumption, 70% of the higher income population in this country have Internet access and the 30% of lower income accordingly do not. Knowing the ε factor of a country, one is also able to compute the according Equity Lorenz curve and the Equity Lorenz density. Looking at the income y_i of the poorest person under the x% of higher incomes of the population, this has the value $\varepsilon \cdot F(x > 30\%)$ according to the differential equation defining the equity function (see Sect. 2.3.2). Under this policy, one can deduce the minimum income of the poorest person who can afford Internet access at the x% higher incomes.

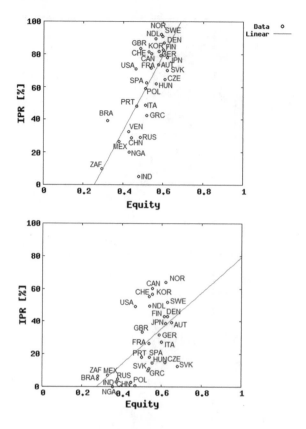

Fig. 4.17 Internet penetration rate (IPR) as a function of the Equity parameter, 2001 (below) and 2009 (above)

Figure 4.18 (bottom) shows the x-axis the income y_i of the poorest among those who can afford access to the Internet. This is in relation to the average income of the total population measured by means of GDP per capita (GDPpC) in 2001. The value $\frac{y_i}{GDPpC} = 2$, for example, says that in a country a person must have at least an income that is twice as high as the average GDPpC in order to afford access to the Internet. This is the case in Mexico and South Africa in 2001. In 2009 (above), the value for South Africa is about 1.5 meaning that a person (in relative terms) needed significantly less income than in 2001 to have Internet access.

The choice of (test) functions for the description of the relationship between the equity factor or the resulting minimum income in a country to afford Internet access requires other types of functions than considered in the previous sections. This is because the dispersion of the data is quite different here. For small values on the x-axis we get high values on the y-axis. If one moves (slightly) towards higher values on the x-axis, the y values fall rapidly.

In general, the data follow a curve which indicates an exponential decay, which is why such (test) functions are obvious here for a first modeling. To increase the accuracy of the modeling in certain circumstances, two more (test) function types

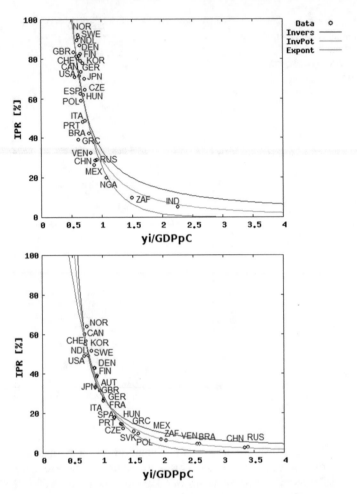

Fig. 4.18 Internet penetration rate (IPR) as a function of the relation of y_i to the GDPpC in 2001 (bottom) and 2009 (top)

(*InvPot, Invers*) are used in the investigation. The mathematical description of the respective functions is to be found in Sect. 2.4.8.

The result of this analysis states the following: The higher the value $\frac{y_i}{GDPpC}$ for a country, the lower the Internet penetration rate (IPR) There is obviously not a linear (negative) relationship given, but an exponentially decreasing function. The study concludes that a higher social equality level promotes the spread of Internet in a country. This result is to be seen positively as part of global sustainable development including the support of a higher social balance level and the propagation of ICT access.

Table 4.13 Descriptive statistic mobile Internet penetration rate (MBR) and GDP per capita (GDPpC) 2008 and 2012

	2008 (N = 29)						2012 (N = 83)					
	Sp	Min	Max	MV	MD	Sta	Sp	Min	Max	MV	MD	Sta
GDPpC	119878	3740	123618	32361	25984	24820	136874	709	137623	23804	18551	21205
MBR	73.78	0.01	73.79	9.18	3.42	16.6	106.4	0.02	106.6	33.06	28.25	29.98

4.1.5 Mobile Internet and Economic Performance

Since the 90s the Internet can be accessed with a mobile phone via the *Global System for Mobile Communications* (GSM), but only with very low bandwidth. In 2000 the mobile Internet was still a scarce commodity to the general public. In 2002, the Universal Mobile Telecommunications System (UMTS) was introduced with significantly higher speeds than GSM. The ITU collects data for mobile broadband use since 2007, but at this time this data existed just for a few countries. Since 2008 one can find data for more countries.

Therefore, the analysis of the relationship between the mobile Internet penetration rate (MBR) and the GDP per capita (GDPpC) start in 2008 and end in 2012, because there is no data yet for 2013 at the time of this analysis. In further it is assumed for the presumption that the MBR or the growth of the MBR has a similar development, such as mobile phone penetration (MPR) (see Sect. 4.1.2). Accordingly similar trends for the relationship between the GDP and the MBR are expected, as for the relationship between the GDPpC and the MPR.

In Table 4.13 the values for the span (Sp), the minimum (Min) and maximum (Max), the mean value (MV) and median (MD) and the standard deviation (Sta) are listed. The values result from the combination of data records of GDPpC and MBR for 29 states in 2008 and 83 countries in 2012.

In 2008 the MBR achieved a maximum value of nearly 74% (*Korea*), followed by *Norway* with about 60%. The mean value lies at about 9% and the median at almost 3.5%. The span is 74% and the standard deviation is nearly 17%. In 2012, the average is 33% with a median of about 28%. As the data shows, about 75% of all Internet users (see Sect. 4.1.3) have mobile access to the Internet within these 83 States.[16]

For the description of the empiric relation between the GDP per capita (GDPpC) and the mobile Internet penetration rate (MBR) linear and non-linear regression models are used by analogy with the previous segments. Here the MBR is the dependent variable and the GDPpC is the independent one. Figure 4.19 shows all tested fitting functions for this relationship for 2008 (below) and 2012 (above).

The result of the regression analysis is that in 2008 there still was no convincing explanatory relationship between the MBR and the GDPpC that was visible. The

---

[16]Reminder: The worldwide total Internet penetration rate in 2012 lies (fixed and mobile access) at about 40%.

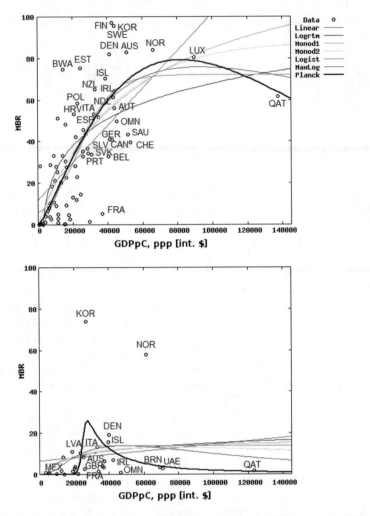

Fig. 4.19 Mobile Internet penetration rate (MBR) as a function of GDP per capita (GDPpC) in 2000 (bottom) and 2012 (top)

reasons for this are because there is only a small amount of available data and that the mobile Internet usage is still at the beginning of its dissemination. Another observation is that the MBR has increased significantly in many countries until 2012, and again the following basic statement (similar to the previous sections) applies: countries with low GDPpC have a slightly lower MBR. The higher the GDPpC, the higher the MBR in a country, whereas for very high values of GDPpC, the MBR decreases again. This again concerns countries where prosperity is marked by the possession of fossil fuels, such as the Gulf States.

Table 4.14 Parameters for the Planck function $MBR = f(GDPpC)$ in 2008 and 2012

t	a	b	c	d
2008	$-7.565 \cdot 10^{10}$	2.125	$3.358 \cdot 10^{11}$	$1.061 \cdot 10^{-3}$
2012	8.214	1.388	$1.726 \cdot 10^{5}$	$1.684 \cdot 10^{-5}$

Table 4.15 Mean squared errors (MSE) and coefficient of determination R^2 for the regression $MBR = f(GDPpC)$

	2008 (N = 29)			2012 (N = 83)		
	MSE	MSE/N	R^2	MSE	MSE/N	R^2
Linear	7599	262.0	0.014	43916	529.1	0.404
Logartm	7306	251.9	0.052	35867	432.1	0.513
Monod1	7297	251.6	0.054	35533	428.1	0.518
Monod2	7211	248.7	0.065	34801	419.3	0.528
Logist	6934	239.1	0.101	33336	401.6	0.548
MaxLog	7048	243.0	0.086	33745	406.6	0.542
Planck	**6615**	**228.1**	**0.142**	**33248**	**400.6**	**0.549**

In 2012, there is a positive relation between GDP and the MBR, which can be best described by a Planck function, see Eq. (4.11). The respective parameters a, b, c and d for the years 2008 and 2012: (see Table 4.14):

$$MBR_t = \frac{a \cdot GDPpC_t^b}{c \cdot \exp^{d \cdot GDPpC_t} - 1} \tag{4.11}$$

The tested functions and their mean squared error (MSE or MSE/N) as well as the coefficient of determination R^2 for the years 2008 and 2012 are listed in Table 4.15.

Figure 4.20 shows the evolution of the annual best fitting function from 2008 to 2012. It shows a pattern similar to the development of the previously analyzed access rates for mobile telephony. This finding largely confirms the assumption that the mobile Internet has taken and will further on take a similar development as the spread of mobile phone usage continues.

Figure 4.21 delivers a comprehensive presentation for the analysis of the functional relationship between the GDPpC and the MBR. This shows for 2008 (blue) and 2012 (black) the data, the best fitting function as well as the averages (dotted lines) and their intersections (center C2008 and C2012) for the GDPpC and the MBR. The arrow (red) indicates the shift in center (C2008 → C2012) of the point clouds between 2008 and 2012 with the corresponding x and y coordinates.

A growth analysis of the MBR in the GDPpC categories A till E can not be (usefully) carried out for reasons of data availability. For this, there is insufficient data to categorize, for example, in lower and lowest GDPpC. However, one can recognize

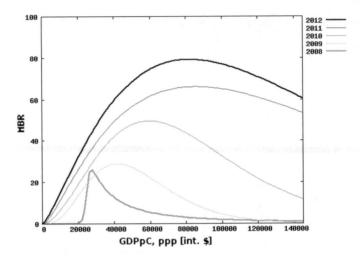

Fig. 4.20 Development of the mobile Internet penetration rate (MBR) as a function of GDP per Capita (GDPpC), years 2008 until 2012

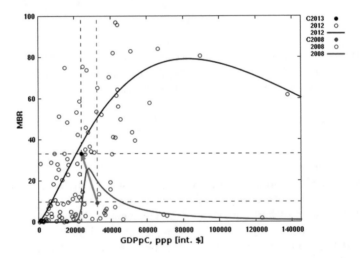

Fig. 4.21 Data, best fitting functions and shift in center for $MBR = f(GDPpC)$, years 2008 (blue) and 2012 (black)

at the fitting functions in Figs. 4.20 and 4.21 that the largest (absolute) growth rates for the MBR took place in the higher GDPpC areas. This is not surprising, as the adaptation of new technologies in this tendency firstly takes place in richer countries ("early adopters"). This is, unless, established solution systems, such as the banking sector or healthcare act as a "brake" for innovations (such as paying by mobile phone or telemedicine solutions).

For the relationship between the GDPpC and the MBR one can say in summary the following: First, data for the MBR is only available to a much lesser extent than is the case with the previously studied access rates of landline, mobile telephony and Internet. The use of mobile Internet is a consequence of the technological convergence of mobile telephony and Internet, so that its (global) deployment proceeds a similar development (by tendency) as the mobile phone. This possibility of communication is the hope to allow future Internet access for all.

4.1.6   Intermediate Result

The investigations concerning the empiric or functional relationship between the GDP per capita (GDPpC) in a country and the different access rates of fixed telephone, mobile phone, Internet and mobile Internet show a positive correlation from 2000 to 2013. The relation between the GDPpC and the access rates of fixed telephone (FTR) is well modelled in 2000 as well as in 2013 by the Planck function. The global FTR decreases over this period, although the absolute number of fixed lines slightly increase worldwide. The reason for this is on the one hand, the fact that the world population grows stronger than the number of fixed telephone connections. On the other hand, the existing technical possibility to bundle multiple fixed lines is a reason for this result. In addition, the increase in mobile phone access is another reason, because this strongly curtails the growth of the FTR. In the area of high GDPpC, for many states, one can observe a clear decrease of the FTR and in the area of low GDPpC in many states an almost steady level of the FTR over time. This is due to the strong population growth equal to the FTR and because these states are relatively fast in catching up with regard to mobile phone usage.

The relationship between the GDPpC and the mobile phone access rate (MPR) is best modelled in 2000 with the Planck function and in 2013, where all tested functions have a similarly good fit to the data, with a MaxLog function. For all GDPpC areas a nearly similar increase is to be observed. In 2013 many countries have a MPR of 100% which accordingly leads to a worldwide MPR of about 100%. Thus, one can say at that time that most people have mobile phone access. This example confirms the thesis that catching up in ICT usage runs much faster than in prosperity in general.

The result for the relation between the GDPpC and the Internet penetration rate (IPR) shows first that a strong positive linear correlation exists between the portion of a country in the world GDP and its portion in worldwide Internet users. Furthermore, in 2000 a very asymmetrical distribution of the Internet population could be ascertained in proportion to the distribution of the world population in countries. Concerning this a few rich states were represented very over proportionately in the Internet world. Until 2013, the circumstances were better balanced. In 2000, the IPR is best modelled as a function of GDPpC by a Planck function, which takes a similar shape, as in the view of the relationship between the GDPpC and FTR. The reason for this is the largely physical coupling of Internet access to the fixed telephone network in the year 2000. In 2013, again the Planck function shows the best fit, which now takes on a different functional form as for the relationship between GDPpC and FTR. It can be seen that in 2013, an Internet connection does not necessarily take place via the fixed network but in fact the mobile Internet usage increases significantly during the previous years. Furthermore, the IPR has increased over all GDPpC areas, even though we do have a need to catch up worldwide, for example when it comes to ensuring access to the Internet for all people. This is because between 2000 and 2013 the highest growth of IPR was among the richest and richer countries (category A and B) and the lowest growth of IPR among the poorer and poorest countries (category D and E). The hopes in this area are based on the mobile Internet access on the basis of technological convergence of mobile telephony and Internet.

The studies on the social inequality level of countries measured by the equity factor show the following: The more balanced a society, the higher its IPR. More economic balance corresponds to higher social balance and indirectly and directly to greater prosperity as well for the poor who are increasingly able to afford wireless Internet access. Also the following becomes clear: The more income a person must have in relation to the average income to afford an access to the Internet, the less is the IPR in a country.

A connection between the GDPpC and the mobile Internet penetration rate (MBR) is due to the short period of development only deducible since 2008. But also in 2008 the data is very small, because at this time mobile Internet access existed only in a few (rich) countries. In 2013, there is a positive relationship between GDPpC and MBR, which is modeled by the Planck function. The future of the Internet is no longer conceivable without mobile Internet access. It is very probable that smartphones will significantly promote the expansion of the Internet that in 2013 has a worldwide utilization rate of 30%.

4.2   Global Networking and Education Level 2000–2012

Education is a basic prerequisite for improving quality of life, overcoming poverty, achieving gender equality, reducing child mortality and stabilizing population growth. In addition it is also important for sustainable development and ultimately peace and democracy. One of the aims of the *Millenium Development Goals* (MDG) was to ensure primary education for children around the world until 2015.[17] Another goal is to promote global partnership between countries while taking advantage of information and communications technology. It is therefore key to promote the use of digital technologies in developing countries.

In becoming a worldwide information society, it is extremely important to offer the possibility to learn to read and write to as many people as possible. The "Education-

---

[17]Key figures of the MDGs illustrates that ensuring primary education for all children was not achieved worldwide. There are still substantial deficits. Besides, one has to ask for the quality obtained in each case to evaluate the success. Ensuring substantial education for all children worldwide remains a major challenge for the international community.

For-All-Global-Monitoring-Report" is a report ordered by UNESCO dealing with world education [476]. This action plan was initiated at the *World Education Forum* in April 2000 in Dakar/Senegal [468]. 164 countries passed a total of six aims for an education for all action plan. Aim 4 of the action plan is to halve the number of illiterate adults (about 880 million in 2000) till 2015 worldwide, especially among women. The progress is modest when looking at adult literacy. Currently, about 774 million adults worldwide are illiterate. The high proportion, including two thirds of women, remains unchanged for years [475]. The access of adults to basic and continuing education is to be ensured. UNESCO was also responsible for the *UN Literacy Decade* (2003–2012) [470]. Their aim is to create an environment worldwide for all people in promoting alphabetization and the acquisition of basic skills. The *World day of alphabetization* on the 8th of September [474] is a yearly reminder that it is a privilege in many countries to be able to read and write.

The following empirical analysis of the relationship between the considered technologies such as fixed telephone (FTR), mobile phone (MPR), Internet (IPR) and mobile Internet (MBR) and the level of education, as measured by the literacy rate (ALR) and the Education Index (EI) of a country, make a contribution to a better understanding of the importance of the MDGs. Here it becomes clear that education has a significant connection to digital networking. In the following it also becomes clear that GDP per capita (GDPpC) has an even greater impact on digital networking processes than education. However, a high GDP or GDPpC is also an important explanatory variable for the respective educational level and vice versa. In addition, education has a major impact on living standards. All these issues are correlated highly with one other where these manifold interactions are part of Sect. 4.3, in which the combination of factors will be considered as explanatory variables.

The education level of countries is determined in one step using the ALR. This indicator captures a minimum dimension of education. In the second stage, the educational level of countries is determined with the EI of the United Nations. Both indicators are analyzed for their empirical relation with the FTR, MPR, IPR and MBR.[18]

4.2.1 Fixed Telephony and Literacy

The literacy rate indicates the proportion of the population of a country or the world with the ability to read and write (see Sect. 2.3.3). The counterpart to this is the illiteracy rate, i.e., the proportion of the population who can not read and write. In a society that is dependent on the written language in addition to oral communication, a high literacy rate is a distinct advantage compared to illiteracy. The use of the Internet for example is restricted significantly for the illiterate because it works on

---

[18]At the time of the creation of this book there was no data on the education level of countries by 2013. So in deviation Sect. 4.1, the analysis (only) dates back to 2012. However, this has no impact on the investigations and the interpretations of the results.

Table 4.16 Descriptive statistics for fixed telephony rate (FTR) and adult literacy rate (ALR), 2000 and 2012

	2000 (N = 153)						2012 (N = 153)					
	Sp	Min	Max	MV	MD	Sta	Sp	Min	Max	MV	MD	Sta
ALR	90.25	9.54	99.79	81.16	89.08	21.20	68.48	31.37	99.85	86.18	93.80	16.71
FTR	73.05	0.02	73.07	18.91	10.32	20.58	62.05	0.02	62.07	17.93	14.95	16.59

a text-based nature mainly until today. The use of fixed and mobile phones as a comparison requires (almost) no literacy skills.[19]

Table 4.16 shows the data of adult literacy rate (ALR), fixed telephone (FTR), span (Sp), minimum (Min), maximum (Max), mean (MV), median (MD) and standard deviation (Sta) for the years 2000 and 2012.

The number N of the countries is 153 for both years. In 2000, the global average ALR is about 81.2%, while this rate increased to about 86.2% in 2012. Although in the same period the world population has increased by almost 16% from 6 to 7 billion people. Worldwide development programs relating to the MDGs and the above-mentioned campaigns contribute to this progress. The span of the ALR has decreased worldwide from about 90% to almost 70%, which is also a positive step in the direction of improving global education levels. The minimum value of about 9.5% (Niger) in 2000 increased to 31.4% (Niger) in 2012. A significant improvement, but still far from sufficient. The maximum value is 99.79% (Estonia) in 2000 increases slightly by 0.06–99.85% (Latvia) until 2012. The data confirms that most countries with a low ALR are located on the African continent. In Africa, the high population growth rate is a major challenge when it comes to increasing the total ALR. The Baltic countries are among those with the highest ALR. These high values are also a consequence of the high cultural importance of education in the previous Soviet Union. In both years the median value is with 89.1% in 2000 and 93.8% in 2012, slightly higher than the average. Most times it is the other way round, for example, in income distributions the average is (often significantly) higher than the median.

What is the reasoning? In many situations of data distributions there are great outliers upwards as very high income, that amount to multiple of the average income. In the present education data it is different. Almost all states are, with regard to the ALR, at a level greater than 50%, many greater than 80%. An ALR of exactly 100% is the upper limit. There are no outliers upwards, only those that are downwards. The median value is about 90% in 2000. The mean value is therefore determined by smaller ALR close to 50% and lower. This drives the mean value downwards in the direction of about 80%, which is less than the median. Looking closer at the available data, the following situation arises for 2000: From a total of 153 examined states 54 countries have an ALR which is lower than the average ALR. Accordingly, 99 states have an ALR, which is greater than the mean value. The so-called outliers are thus

---

[19]The use of the SMS service via the mobile phone requires reading and writing skills.

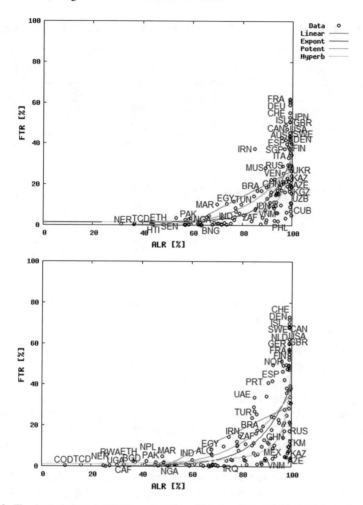

Fig. 4.22 Fixed telephone rate (FTR) as a function of Literacy Rate (ALR), 2000 (below) and 2012 (above)

the countries with a rather low ALR, which then pulls the mean value towards less ALR. Here, the data has downward outliers.[20]

The corresponding values for the FTR have already been discussed in Sect. 4.1.1 for the empirical relationship between the GDPpC and FTR. Slight changes to the descriptive statistics result from the different intersection of data of ALR and FTR compared to the average amount of data available for GDPpC and FTR, i.e., the number N of the countries varies.

Figure 4.22 shows the ALR as independent and FTR as dependent variables. The respective data points and tested fitting functions describe the relationship between

---

[20]For comparison: With the distribution of income one deals with outliers upwards.

Table 4.17 Mean squared error (MSE) and coefficient of determination (R^2) for regression $FTR = f(ALR)$, 2000 and 2012

	2000 (N = 153)			2012 (N = 153)		
	MSE	MSE/N	R^2	MSE	MSE/N	R^2
Linear	40939	267.6	0.364	24400	159.5	0.417
Expont	**31284**	**204.5**	**0.514**	**18842**	**123.2**	**0.550**
Power	32412	211.8	0.496	20003	130.7	0.522
Hyperb	37550	245.4	0.436	22232	145.3	0.469

Table 4.18 Parameters for exponential function $FTR = f(ALR)$

t	a	b	c	d
2000	0	0.011	0.083	0
2012	1.420	210.166	0.109	12.791

ALR and FTR for the year 2000 (bottom) and 2012 (top). From the graph it becomes clear that the data points are scattered on the ALR axis (on ALR = 100%) as well as on the FTR axis. This first optical impression of the data provides an indication of a possible class of (test) functions and the best possible adjustment based on the method of least squares (see Sect. 2.4.3). For reasons of completeness a linear regression model with a shift (red) is used first with a shift for the relationship that obviously, however, does not fit very well. Exponential growth processes are suitable for modeling connections in which almost "everything" happens at the end (here for large ALR), i.e., the FTR rapidly increases. A special feature of the exponential function is that the multiplication (for example, doubling, tripling, etc.) of an output happens within constant periods (for example, within the same period of time) of the input. Power functions model many physical processes (see Sect. 2.4.8), which is why a power function (green) is tested next. It describes the data well, but not as well as the exponential function. Finally, the relationship between the ALR and FTR is modelled by a hyperbolic function (yellow green).

For the tested class of functions, Table 4.17, show the values for the mean squared error (MSE or MSE/N) and the coefficient of determination (R^2), which were determined using the method of least squares.

The result show that the exponential function describes the connection between ALR and FTR according to Eq. (4.12) for 2000 and 2012 best of all. The corresponding parameters a, b, c and d Eq. (4.12) can be seen in Table 4.18 for 2000 and 2012.

$$FTR_t = a + b \cdot e^{c \cdot ALR_t - d} \tag{4.12}$$

In contrast to the previous Sect. 4.1, the years 2000–2012 and the best fitting function for the respective year are not compiled in a diagram. This is due to the very small changes in the fitting function from one year to the next, so that all fitting functions for the years 2000–2012 in one image would be untidy.

In principle, one can determine the following for each year: A (very) high ALR is a necessary condition for a high FTR, but by far not sufficient. Countries with a low ALR in 2000 are: *Bangladesh, Dem. Rep. of Congo, Nepal, Niger, Nigeria, Pakistan, Rwanda, Chad, Uganda* and the *Central Africa* region. Among these are mainly countries from Africa and highly populated Asian countries. However, there is also a group of states, which have a low FTR despite a very high ALR. This includes countries such as *Azerbaijan, Kazakhstan, Russia, Turkmenistan, Uzbekistan*, i.e., former USSR countries. Culturally, this should be justified with the former communist regime. At the time widespread education was one of the primary political aims of the USSR. The major disadvantage of the political system at that time was clearly the lack of innovation capability and economic performance that originates from missing incentives for companies and people to be competitive. This consequently had an impact on the availability of technology, its use and in particular on prosperity in general, which was relatively low.

Another group of countries (*Denmark, Germany, Finland, France, Great Britain, Canada, Iceland, Netherlands, Norway, Switzerland, USA*) have both high ALR and high FTR. These countries invest in overall literacy and education for their populations and have a high capacity for innovation and broad prosperity, which then is reflected accordingly in a high FTR.

Figure 4.23 shows a growth analysis of different ALR categories A to E. In category A are those countries with the highest ALRs, in category B countries with high ALRs, and so on until category E which contains the countries with the comparatively lowest ALR.[21] It can be seen very well that between 2000 and 2012, the ALR in category A has the lowest and in category E the highest growth. This is not surprising, since within most countries in category A, the literacy rate is almost 100%. Accordingly, the highest potential for growth lies within the countries of category E. Between 2000 and 2012 this growth was supported by worldwide development programs. In Fig. 4.23 an essential finding is the high correlation between the growth of ALR and FTR ($R = 0.759$), which leads to following conclusion: The higher the ALR-growth, the more the FTR has increased.

4.2.2 Fixed Telephone and Education Index

The Education Index (EI) is a measure of the education level of a country and is determined using the number of years of schooling that a 25-year-old has completed as well as the expected duration of a child's education at the age of its enrollment (see Sect. 2.3.3). The EI is part of the *Human Development Index* (HDI) of the United Nations that is an indicator of a country's prosperity in general. As mentioned above, the EI concerning the level of education of a country is a stronger measure than the

---

[21]The exact breakdown in different ALR categories can be found in Annex A.

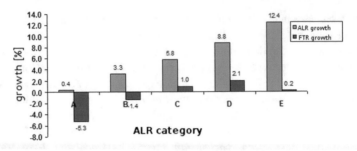

Fig. 4.23 (Total) growth from 2000–2012 for ALR (blue) and FTR (red)

Table 4.19 Descriptive statistic for fixed telephone rate (FTR) and Education Index (EI), 2000 and 2012

	2000 (N = 162)						2012 (N = 174)					
	Sp	Min	Max	MV	MD	Sta	Sp	Min	Max	MV	MD	Sta
EI	0.864	0.110	0.974	0.574	0.60	0.215	0.817	0.177	0.994	0.651	0.687	0.200
FTR	73.05	0.02	73.07	18.68	10.21	20.54	62.03	0.02	62.05	18.13	14.55	16.98

adult literacy rate (ALR). Accordingly, it is to be expected that lower values of EI can lead to higher FTR compared to the ALR.

Analogous to Sect. 4.2.1, Table 4.19 shows some values of descriptive statistics for the Education Index (EI) and the fixed telephone rate (FTR) for the years 2000 and 2012. The number N of studied countries is 162 in 2000 and 174 in 2012. The values of the FTR and the changes over time were discussed in detail in Sect. 4.1.1.

The maximum value for the EI of 0.97 was achieved in Australia in 2000. This value increases to 0.99 in 2012 and is achieved for *New Zealand*. It is no coincidence that these two countries are at a similarly high level in education. One reason for this is the special partnership with the UK and the US, which shows up in the membership of the *Commonwealth of Realm* as part of the *Commonwealth of Nations*.[22] Niger has the minimum value of 0.11 in 2000. This value increases to 0.18 and is given once more in Niger. Here one should remember the fact that Niger has also the lowest value of ALR (see segment Sect. 4.2.1). This example illustrates the importance of the ALR for the EI, because without broad literacy, a high level of education is not to be expected. However, widespread literacy does not lead directly to a high Education Index.

Between 2000 and 2012, the span of the EI has decreased from 0.864 to 0.817, the average has increased from 0.57 to 0.65 and the median from 0.6 to 0.69. The median is again higher than the arithmetic mean. The reasons for this have already been

---

[22]The Commonwealth of Nations is an organization of sovereign states, which support each other and cooperate in achieving international goals. It currently consists of 53 states, 16 of which have closer ties within the Commonwealth Realm of the United Kingdom and the British Crown. These include Australia and New Zealand [63].

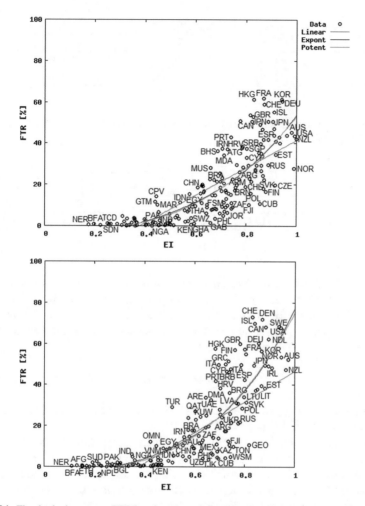

Fig. 4.24 Fixed telephone rate (FTR) as function of the Education Index (EI) for 2000 (bottom) and 2012 (top)

explained in the previous Sect. 4.2.1. These values point to the clear improvement of education levels worldwide, measured by EI, over the considered period of time.

Figure 4.24 shows the Education Index (EI) as an independent and the fixed telephone rate (FTR) as a dependent variable for the year 2000 (below) and 2012 (above). The corresponding data points and the tested fitting functions from the considered class of functions (see Sect. 2.4.8) can also be found. To guarantee a certain comparability to the relation between the ALR and FTR at this point, the same (test) functions are pulled up, similar to the previous segment Sect. 4.2.1 for the modeling. In particular a linear and exponential model as well as a family of power functions is used.

Table 4.20 Parameters for the power function $FTR = f(EI)$

t	a	b	c
2000	−1.709	76.701	3.005
2012	−2.291	55.356	2.804

Table 4.21 Mean squared error (MSE) and coefficient of determination (R^2) for the regression $FTR = f(EI)$, 2000 and 2012

	2000 (N = 162)			2012 (N = 174)		
	MSE	MSE/N	R^2	MSE	MSE/N	R^2
Linear	25792	159.2	0.620	18890	108.6	0.621
Expont	19557	120.7	0.712	16115	92.6	0.677
Potent	**19129**	**118.1**	**0.718**	**15868**	**91.2**	**0.682**

These (test) functions first of all point to a positive relationship between the EI and the FTR. The 3-parametric power function (green) according to Eq. (4.13) best fits the data, followed by the exponential (blue). The linear regression model (red) does not fit the data that well, but nevertheless it fits better here than for the relationship between ALR and FTR (see Sect. 4.2.1).

$$FTR_t = a + b \cdot EI_t^c \tag{4.13}$$

The corresponding parameters a, b and c of the power function in Eq. (4.13) for 2000 and 2012 are to be found in Table 4.20.

One can say that the best modeling of the functional relationship between the EI and the FTR has developed from a power function towards a (more) linear function between 2000 and 2012. Table 4.21 shows the values for the mean squared error (MSE or MSE/N) and the coefficient of determination (R^2) for the tested functions. The deviation between the coefficient of determination of two functions is lower in 2012 than in 2000. I.e., all the tested functions describe the data in 2012 similarly well.

An observation is that the countries in the upper area of the EI have a higher FTR in 2000 than in 2012. This is because states, in particular, *Denmark, Finland, Norway* and *Sweden*, but also *Australia, Great Britain, New Zealand, USA, Germany, France* and the *Netherlands* all lie in the upper area of the EI and have reduced their FTR. The reasons for this, related to the advent of mobile telephony, were dealt with in detail in connection with the consideration of FTR and economic performance (see Sect. 4.1.1). In the area of low EI, the FTR has slightly increased between 2000 and 2012.

Overall, the point cloud has shifted to the lower right between 2000 and 2012. While there are hardly any countries in 2000 with EI of at least 0.8 and an FTR clearly below 40%, in 2012 there are some countries that have an EI of more than

Table 4.22 Descriptive Statistics for Mobile Phone Rate (MPR) and Literacy Rate (ALR), 2000 and 2012

	2000 (N = 148)						2012 (N = 153)					
	Sp	Min	Max	MV	MD	Sta	Sp	Min	Max	MV	MD	Sta
ALR	90.3	9.54	99.79	81.99	89.79	20.87	68.48	31.37	99.85	86.05	93.80	16.91
MPR	76.39	0.02	76.41	17.52	5.64	23.37	182.43	4.98	187.41	102.55	106.17	40.46

0.8 and a FTR below 20% at the same time. A high level of education, as measured by the EI, is therefore (still) not a decisive factor in 2012, which usually indicates an inevitably high FTR.

4.2.3 Mobile Telephony and Literacy

To use the mobile phone, you have to be able to identify at least the digits 0 to 9.[23] This requires no skills for reading and writing of letters or words. The identification of numbers and corresponding basic operations such as adding numbers is part of a skill set displayed in the adult literacy rate (ALR). It is expected that the empirical relationship between the ALR and the mobile phone rate (MPR) is low, but should be lower in 2012 or functionally run significantly different than in 2000. The reason for this is the rapid spread of mobile telephony in this period across all countries (see Sect. 4.1.2).

Table 4.22 shows the descriptive statistics for the intersection of the data for ALR and MPR for the years 2000 and 2012.

The number N of the countries for which there are data to the ALR and MPR amounts to 148 in 2000 and 153 in 2012. Because the values of the descriptive statistics of the ALR are described in Sect. 4.2.1 in detail, it is renounced here as a new explanation of these values. The same is valid for the data of the MPR which are explained in Sect. 4.1.2. However, it should be mentioned again here that the median of the ALR is higher in both years than the arithmetic average and both values have grown similarly strong between 2000 and 2012. The reasons for this were explained in Sect. 4.2.1. It is valid for the MPR that the average is clearly higher in 2000 than the median, while the difference precipitates in 2012 more slightly. In 2000, many countries have rather a low MPR, which is why the few countries with a rather high MPR represent outliers and thus pull the average upwards so that the average is greater than the median. The fact that the difference between the arithmetic mean and the median in 2012 is smaller than previously implies that the MPR data have shifted towards an equal distribution, i.e., the poorer countries or countries with a lower education level catch up in the use of mobile phones.

---

[23] But there are many people who are able to read numbers but no words. Therefore there are different kinds of illiterates.

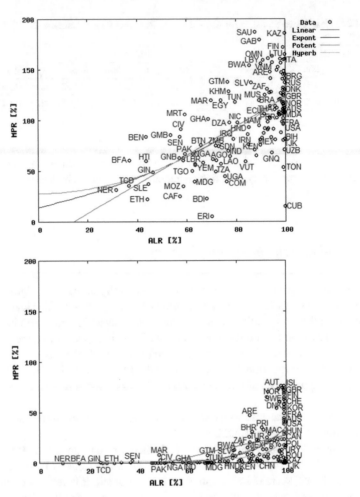

Fig. 4.25 Mobile phone rate (MPR) as a function of adult literacy rate (ALR), 2000 (bottom) and 2012 (top)

Figure 4.25 shows the adult literacy rate (ALR) as an independent and the mobile phone rate (MPR) as a dependent variable, illustrates the data points as well as regression functions tested for the relationship for 2000 (bottom) and 2012 (top). In both years there is a positive correlation between the ALR and MPR. In 2000 the data points are strewn along the ALR and MPR axis, which is why the exponential model and the power function describe the relation well. That means that only for high values of the ALR also high values of the MPR are reached. A similar pattern is to be observed for the relationship between ALR and FTR (see Sect. 4.2.1). It is probably the case that in 2000, for financial reasons, mobile telephony was linked to the fixed telephony, although the respective physical infrastructure was decoupled. The empirical relationship between fixed and mobile telephony is analyzed below in

Table 4.23 Mean squared error (MSE) and coefficient of determination (R^2) for the regression $MPR = f(ALR)$, 2000 and 2012

	2000 (N = 148)			2012 (N = 153)		
	MSE	MSE/N	R^2	MSE	MSE/N	R^2
Linear	61603	416.2	0.232	160139	1046.6	0.357
Expont	54214	366.3	0.324	159100	1039.8	0.361
Power	54311	366.9	0.323	158923	1038.7	0.361

Table 4.24 Parameters for the power function $MPR = f(ALR)$

t	a	b	c	d
2000	−0.638	0.019	0.0076	0
2012	0	$2.418 \cdot 10^{-9}$	5.089	–

Sect. 4.4.2 through the system theoretical approach to determine the interaction of the technologies themselves.

The relationship between the ALR and MPR changed by 2012 so that also for relatively low values of ALR, high values of MPR are reached. This change is clearly visible in the functional modeling of the relationship between ALR and MPR, because in 2012 the linear, exponential and power function have almost identical values for the mean squared error (MSE or MSE/N) and the coefficient of determination (R^2) (see Table 4.23). The catching up of developing countries, with the majority of illiterates, in mobile phone use decisively contributes to this positive development. The 100% mark of the MPR is achieved from an ALR of about 60%.

The 4-parametric exponential function according to Eq. (4.14) with the corresponding parameters a, b, c and d delivers the best fit for the correlation between ALR and MPR for 2000.

$$MPR_t = a + b \cdot e^{c \cdot ALR - d} \tag{4.14}$$

The 3-parametric power function according to Eq. (4.15) gives the best fit for 2012 in accordance with the method of least squares.

$$MPR_t = a + b \cdot ALR_t^c \tag{4.15}$$

Table 4.24 shows the parameter values for the corresponding years.

Figure 4.26 shows the result of the growth analysis of ALR categories A, B, C, D and E. The states with the highest ALR are in category A and the countries with the lowest ALR in category E (see Appendix A.3). The ALR growth was previously discussed already in Sect. 4.2.1.

For the relation between the ALR and MPR growth the following can be summarized: Between 2000 and 2012 a clear MPR growth across all ALR categories has taken place. The biggest MPR growth of about 100% is given for the categories B and

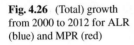

Fig. 4.26 (Total) growth from 2000 to 2012 for ALR (blue) and MPR (red)

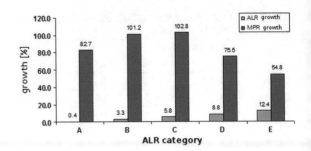

C, i.e., in the ambitious emerging countries the mobile telephony use has doubled. In states with the highest ALR (category A) the MPR has increased about 83%. States in the area of low ALR have improved their MPR about approx. 75%. In the lowest ALR category E a MPR growth of about 55% is given. In the category A and D a similar MPR growth is given and also in categories B and C an almost identical MPR growth took place. This indicates that between 2000 and 2012 the catching-up processes in the area of the mobile phone use was faster than the rise of the ALR. On the one hand education is associated with high costs and is a time-consuming process, which is closely coupled with cultural and social conditions (for example, child labor, role of women). Furthermore, the use of the mobile phones over the years has become easier and financially more affordable therefore a high ALR has become considerably less important.

4.2.4   Mobile Telephony Education Index

The presumption is that the use of mobile phones requires no level of higher education, as measured by the Education Index (EI). This is also evident if one looks at the results in Sect. 4.2.3 on the relationship between adult literacy rate (ALR) and mobile phone rate (MPR) involving the development in the mobile phone market in the considerations. These tend to show that the use of a mobile phone is made easier for the whole population and, in particular, for older populations through technical progress, e.g., through suitable user interfaces. The voice-enabled mobile phone may require no educational level for use once all the settings such as volume control, screen brightness or the management of the phone book are set. Newer generations of the mobile phone, so-called smartphones, have a clearly raised functional extent in comparison to older models which reduces usability. Indeed, simple phoning is also easy with smartphones.

The mobile telephony rate (MPR) provides access to the mobile network via a conventional mobile phone or via a smartphone. The latter begins its (global) dissemination from 2008 onwards. Table 4.25 shows the descriptive statistics for the intersection of the data for EI and MPR for the years 2000 and 2012.

Table 4.25 Descriptive statistic for mobile phone rate (MPR) and Education Index (EI), 2000 and 2012

| | 2000 (N = 157) | | | | | | 2012 (N = 174) | | | | | |
	Sp	Min	Max	MV	MD	Sta	Sp	Min	Max	MV	MD	Sta
EI	0.86	0.111	99.74	0.584	0.618	0.212	0.821	0.177	0.998	0.648	0.685	0.202
MPR	76.67	0.02	79.69	17.24	5.64	23.39	182.43	4.98	187.41	100.68	105.52	40.03

The number N of the countries for which there are data for EI and MPR amounts to 157 in the years 2000 and 174 in 2012. The data for the EI were further discussed in Sect. 4.2.2 above. The increase of the mean value is reducible to the successes of national and international education programmes. The median (Md) of the EI is greater than the arithmetic mean in both years and both values have a strong increase. The reasons for this situation were described in Sect. 4.2.1. The MPR values have been explained in the previous Sect. 4.2.3. Again: The average value in 2000 is significantly higher than the median value, but in 2012 the median value is slightly larger than the arithmetic mean.

Figure 4.27 shows the Education Index (EI) as the independent and the mobile phone rate (MPR) as the dependent variable. In addition, the data points are illustrated together with the regression function for the relationship for 2000 (bottom) and 2012 (top). From observation, a positive correlation is evident between the EI and the MPR for both years. In 2000 the MPR only reached appropriate values near to the mean value of about 17% for an EI of about minimal 0.5. Almost all countries that have an EI of less than 0.5, have a (very) low MPR, e.g., the data points lie almost on the EI axis. Furthermore we see that almost all countries with an EI of 0.8 and higher also have an above-average MPR. In 2000 the relation between the EI and the MPR is best modelled with a power function (green) (within the considered class of functions). The exponential regression model (blue) also describes the connection well. The linear regression function (red) does not adapt the data well. This means that high values of MPR are reached for high values of EI.

Up to 2012, the situation changed in a way that high MPR values can be reached for small EI values. The change between 2000 and 2012 is also to be observed in the functional modeling of the relation between EI and MPR. In 2012, the linear and exponential function have almost identical values for the mean squared error (MSE or MSE/n) and the coefficient of determination (R^2) (see Table 4.26). The power function adapts the data best of all and the power function (green) has a shape which is similar to a shifted logarithmic function.

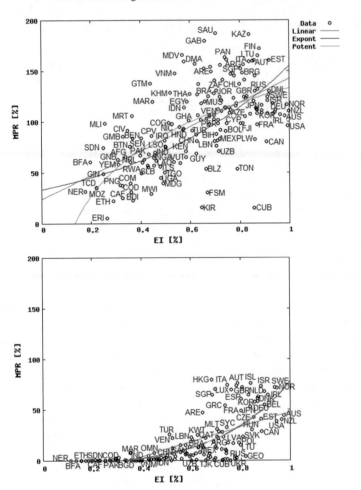

Fig. 4.27 Mobile phone rate (MPR) as a function of Education Index (EI) for 2000 (below) and 2012 (above)

Table 4.26 Mean squared error (MSE) and coefficient of determination (R^2) for the regression $MPR = f(EI)$, 2000 and 2012

	2000 (N = 157)			2012 (N = 174)		
	MSE	MSE/N	R^2	MSE	MSE/N	R^2
Linear	47958	305.5	0.438	177415	1019.6	0.360
Expont	40292	256.6	0.528	177456	1019.8	0.359
Potent	**39810**	**253.6**	**0.534**	**173036**	**994.5**	**0.376**

Table 4.27 Parameters for the power functions $MPR = f(EI)$

t	a	b	c
2000	−1.576	77.880	3.417
2012	14286.98	14153.86	0.0046

Equation (4.16) shows the 3-parametric power function with the appropriate parameters a, b and c, which gives the best fit according to the method of least squares for 2000 and 2012 (Table 4.27).

$$MPR_t = a + b \cdot EI_t^c \tag{4.16}$$

This positive development is contributed by the developing and emerging countries which consist of large populations of illiterate and uneducated groups. The 100% mark for the MPR is already achieved from an EI of about 0.25. It is crucial that the value of the coefficient of determination R^2 for all tested functions between the year 2000 and 2012 decreased, indicating a decoupling of the correlation between EI and MPR. From this, one can conclude that the catching-up processes have run faster in mobile phone usage, than catching up of developing and emerging countries in the area of education, although much improvement has been made. The empiric studies confirm the expectation that the use of the mobile phone assumes no more high educational level measured by the EI during the last years. The mobile phone is used by almost everyone, also the difference to the relationship between the ALR and MPR becomes clear. However, a very high level of ALR as a weaker measure of the educational standard of countries is also still necessary in 2012 (see Sect. 4.2.3).

4.2.5 Internet and Literacy

At the beginning of the Internet age, the network was accessible and used by highly-educated people only, since it can not be used by people with an average level of education, because it is too complicated for that. The *World Wide Web* (WWW) allows the use of the Internet for the wider public with which the Internet transforms in a sort of mass media. After more than 25 years it is still mainly text-based and requires reading and writing skills for its use. Meanwhile different uses and services are offered, as for example the possibility to phone and watch video contents, which partly requires alphabetization.

In this section the empiric as well as the functional relation is examined between the Internet penetration rate (IPR) and the adult literacy rate (ALR) of countries in a worldwide perspective from 2000 to 2012. Table 4.28 shows the data for the ALR and IPR for the years 2000 and 2012.

Table 4.28 Descriptive statistic Internet penetration rate (IPR) and adult literacy rate (ALR), 2000 and 2012

	2000 (N = 151)						2012 (N = 154)					
	Sp	Min	Max	MV	MD	Sta	Sp	Min	Max	MV	MD	Sta
ALR	90.25	9.54	99.79	81.28	89.30	21.29	68.49	31.37	99.86	85.97	93.94	16.98
IPR	51.98	0.02	52.00	8.29	1.78	13.39	95.41	0.8	96.21	40.05	38.68	28.89

The number N of countries for which there are data for ALR and IPR is 151 in 2000 and 154 in 2012. The data for ALR were explained substantially in Sects. 4.2.1 and 4.2.3. Although the IPR data were already examined in connection with the economic performance (see Sect. 4.1.3), these are described here with regard to the occurred catching-up processes by countries with low ALR once more compactly, because these stand now in connection with education. The mean value (MV) of the IPR in 2000 of about 8.3% is more than four times as large as the median (MD) with a value of nearly 1.8%. These two dimensions then have a similarly large value of about 40% in 2012. The reasons for this are the following. At the beginning of the *new economy* in 2000 only some countries with a very high ALR had a high IPR, i.e., these countries were statistical outliers who lead to a higher average than median. In the period from 2000 to 2012 many states in the area of the low ALR have raised their IPR, which is why both masses have approached.

Figure 4.28 shows the adult literacy rate (ALR) as an independent and the Internet penetration rate (IPR) of a country as a dependent variable, as well as the corresponding data points and tested regression functions for the correlation for 2000 (bottom) and 2012 (top).

From the observation it can be seen seen that in 2000 as well as in 2012 a positive relation exists between the ALR and the IPR, however, of another type: In 2000 high values of the IPR are reached (only) in the uppermost area of the ALR. The data points are strewn along the ALR axis and only from an ALR of at least 80% does the IPR rise recognizably. However, there are many countries that have a very high ALR, but whose IPR still remains at a low level, e.g., the group of ex-USSR countries (see Sect. 4.2.1). In other words: a high ALR in 2000 is a necessary condition for a high IPR, but by far not sufficient. This can be seen in the spread of the data along the IPR-axis for high IPR values. Therefore it is obvious that an exponential or even hyperbolic function in 2000 models the relationship between the ALR and IPR well.

In Table 4.29 the mean squared error (MSE or MSE/N) and the coefficient of determination (R^2) are to be found.

In 2000, the hyperbolic regression model (green) according to Eq. (4.17) with the parameter a reaches the best fit to the data, followed by the exponential function (blue) and the power function (green). The linear function (red) adapts the data most badly. The modeling with the help of a hyperbolic (test) function confirms the (worldwide) low extent in Internet use in the area of low ALR in 2000. This year signals the beginning of the new economy. Previously, the Internet was used by sophisticated states in which nearly the whole population was alphabetized.

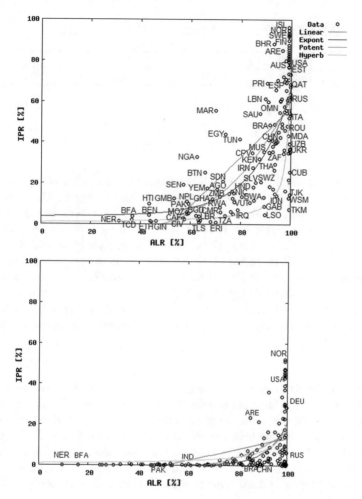

Fig. 4.28 Internet penetration rate (IPR) as a function of adult literacy rate (ALR), 2000 (below) and 2012 (above)

Table 4.29 Mean squared error (MSE) and coefficient of determination (R^2) for the regression $IPR = f(ALR)$, 2000 and 2012

	2000 (N = 151)			2012 (N = 154)		
	MSE	MSE/N	R^2	MSE	MSE/N	R^2
Linear	21700	143.7	0.194	63819	414.4	0.501
Expont	18657	123.6	0.307	**52578**	**341.4**	**0.588**
Potent	19165	126.9	0.288	54542	354.2	0.573
Hyperb	**18361**	**121.6**	**0.318**	120478	782.3	0.057

Table 4.30 Parameters for the power function $IPR = f(ALR)$

t	a	b	c	d
2000	0.0095	–	–	–
2012	3.829	0.273	0.069	1.424

$$IPR_t = \frac{1}{1 - a \cdot ALR_t} \tag{4.17}$$

The point cloud in 2012 shows another shape. The data indicates a clearly higher dispersion and are also not strewn any more so strongly among the ALR or IPR axis. This points to the fact that the connection does not have a hyperbolic shape between the ALR and IPR any more like in 2000. Table 4.29 even shows that the hyperbolic regression function fits the data worst in 2012 within the considered class of functions. The coefficient of determination (R^2) has a value of almost zero, indicating that the hyperbolic function models the connection similarly well or badly, as a constant function at a level of the average of the IPR. The 4-parametric exponential function according to Eq. (4.18) fits the data best of all, followed by the power function and the linear function. Table 4.30 shows the parameter values for the respective function in 2000 and 2012.

$$IPR_t = a + b \cdot \exp^{c \cdot ALR_t - d} \tag{4.18}$$

One could define a *new IPR* and interpret or evaluate the data in a different way under the assumption that only people who are able to read and write can use the Internet. If, for example, country i had an ALR of 80% and an IPR of 40%, the new IPR would be the relation between IPR and ALR, which is 0.5. In 2012, Morocco has an ALR and an IPR of 60% what indicates that this country has managed to offer Internet access to all people with reading and writing skills. Such consideration favors the developing and emerging countries and relativizes the high-level IPR in the rich states.

Figure 4.29 shows a growth analysis for different ALR categories A to E (see Appendix A.3).

For the relationship between ALR and IPR growth the following can be expressed: There is a high negative correlation between these values. An IPR growth between 2000 and 2012 has indeed taken place over all ALR levels, but with the biggest IPR growth in the highest ALR category and the lowest IPR growth in the lowest ALR category.

4.2.6   Internet and Education Index

The use of the Internet still requires basic literacy as was discussed previously in Sect. 4.2.5. But what is the relationship between the Internet penetration rate (IPR)

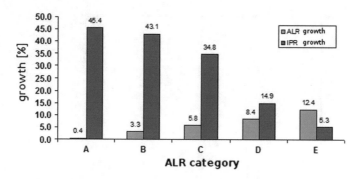

Fig. 4.29 (Total) growth from 2000 to 2012 for ALR (blue) and IPR (red)

Table 4.31 Descriptive statistic Internet Penetration Rate (IPR) and Education Index(EI), 2000 and 2012

	2000 (N = 161)						2012 (N = 174)					
	Sp	Min	Max	MV	MD	Sta	Sp	Min	Max	MV	MD	Sta
EI	0.863	0.111	0.974	0.579	0.605	0.213	0.812	0.177	0.989	0.6487	0.685	0.202
IPR	51.98	0.02	52.00	8.14	1.98	13.12	95.41	0.8	96.21	39.02	36.38	28.42

and the level of education as measured by the Education Index (EI)? A basic thesis is as follows: The higher the EI of countries, the higher the IPR should be. Another theory stemming from this book is that the importance of education levels of country's from 2000 to 2012, for the Internet penetration rate that decreases continuously and will most likely decrease in the future.

Table 4.31 shows the descriptive statistics for the intersection of the data for the Education Index (EI) and the Internet penetration rate (IPR) for the years 2000 and 2012. The number N of countries for which there are data for EI and IPR is 161 in 2000 and 174 in 2012. The values of the descriptive statistics for the EI have been widely discussed in Sects. 4.2.2 and 4.2.4 as well as for IPR in Sects. 4.1.3 and 4.2.5.

Figure 4.30 shows the Education Index (EI) as an independent and the Internet penetration rate (IPR) of a country as a dependent variable, the corresponding data points and for the correlation tested regression functions for 2000 (bottom) and 2012 (top). The first sight at the point cloud shows that in both years a positive correlation exists between the EI and the IPR. If one looks at the previous Sect. 4.2.5 in regarding the relationship between the education level, measured by the adult literacy rate (ALR), and the IPR, the first difference is to be observed here. While a high ALR is a necessary but not a sufficient condition for a high IPR, we have in the high range of EI also correspondingly high values for the IPR in 2000, but one can see this also in the fact that many data points are spread along the EI axis up to a value of about 0.6 and up from this EI value the IPR increases. A compression of the data along the IPR axis is here, in contrast to the correlation between ALR and IPR, in 2012 not present. One also notes this difference with the comparison of the

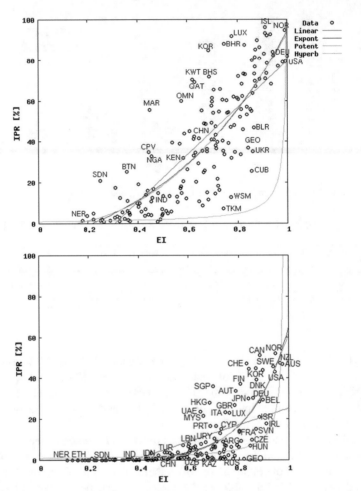

Fig. 4.30 Internet penetration rate (IPR) as function of Education Index (EI), year 2000 (bottom) and 2012 (top)

single regression models. The hyperbolic function (yellow green) delivers the worst adaptation for the relation between EI and IPR.[24] The relationship between EI and the IPR is modelled with the help of a power function (green) best of all, followed by the exponential function (blue) which is almost as good as the power function. The linear regression model delivers (red) a partly good adaptation (see Table 4.32).

Up to 2012 the dispersion of the point cloud changes what can be easily recognized. The power function according to Eq. (4.19) (green) in Fig. 4.30 gives in 2012, as well as in 2000, the best fit within the considered class of functions (Table 4.33).

---

[24]The hyperbolic function has the best fit for the relationship between ALR and IPR in 2000 (see Sect. 4.2.5).

Table 4.32 Mean squared error (MSE) and coefficient of determination (R^2) for regression $IPR = f(EI)$, 2000 and 2012

	2000 (N = 161)			2012 (N = 174)		
	MSE	MSE/N	R^2	MSE	MSE/N	R^2
Linear	15623	97.04	0.433	46135	265.14	0.670
Expont	8859	55.02	0.678	42011	241.44	0.699
Potent	**8818**	**54.77**	**0.680**	**41817**	**240.33**	**0.701**
Hyperb	19429	120.68	0.294	312670	1796.95	−0.168

Table 4.33 Parameters for $IPR = f(EI)$

t	a	b	c
2000	0.308	61.445	5.849
2012	−1.288	94.352	2.245

$$IPR_t = a + b \cdot EI_t^c \tag{4.19}$$

The exponential function again provides the second-best fit. The linear model approximates the data in 2012 significantly better than in 2000. Basically, all three functions deliver a similarly good modeling of the relation between the EI and the IPR. A high EI delivers a high value of the IPR and, however, already a middle EI level is sufficient to reach rather high values of the IPR. Countries with a low EI also tend to have a low IPR, although there are also countries that, despite a relatively low EI, have rather high values for their IPR. Catching up in Internet use has progressed worldwide for all countries faster than the development in the area of education. This fact suggests that global convergence processes in Internet use run significantly faster than the convergence in the context of education.

Of interest is the negative value of the coefficient of determination (R^2), which is obtained for the hyperbolic function in 2012 (see Table 4.32) and occurs in this book for the first time in this form. This is first counterintuitive, because one would think the R^2 to be a squared size, because of its symbolic description. This would then always have values greater than or equal to zero, which is not the case here. In memory would be expelled to Sect. 2.4.6 and to the definition. It does not concern a size in the square. From the definition that in the case in which the adaptation of a function to the data is worse than a horizontal line at the level of the average value, the size R^2 is negative. In this example one can interpret this as follows: The negative value of the R^2 says that the carried out modeling of the data with a certain function (here as hyperbolic assumed) is worse than the modeling of the data with a steady function which runs at the level of the average of the IPR.[25]

---

[25]This anomaly also occurs after repeated repetition of the optimization (minimization of the Mean Squared Error Sum) of the hyperbolic regression function with the help of the least squares method with the Levenberg–Marquardt algorithm.

4.2.7   Mobile Internet and Literacy

In this section the empirical and functional relationship between the mobile Internet penetration rate (MBR) and the adult literacy rate (ALR) of countries is examined. The analyses begin, aberrantly to the previous segments Sects. 4.2.1–4.2.6, in 2008, because between 2000 and 2008 there is lack of data on MBR due to the lack of widespread mobile Internet usage. In Sect. 4.2.3 it was shown that the functional relationship between the mobile phone rate (MPR) and the adult literacy rate (ALR) of countries can be well described in 2000 by an exponential function and in 2012 by an exponential and linear function. The catching up process of mobile phone usage in poor countries, in which relatively most illiterates live, is a reason for the fact that in average a MPR of 100% is already reached from an ALR of 60% in 2012.

A thesis represented in this book is that the technological convergence of mobile telephony and Internet to mobile Internet will accelerate the spread of Internet usage worldwide, similar to the spreading of mobile telephony. Today, from an educational-technical view, the use of the mobile phone hardly needs reading and writing abilities. This also raises hope for the quick expansion of mobile Internet use worldwide.[26]

4.2.8   Mobile Internet and Education Index

In Sect. 4.2.4 it was shown that the mobile phone rate (MPR) today has no connection to the education level, measured by the Education Index (EI). Everyone can call with their mobile phone, therefore for very low values of the EI high values of the MPR can also be observed. The Internet penetration rate (IPR) has a certain relation with the EI. As a result of technological convergence of mobile phone and Internet to mobile Internet, now the question arises, how the empirical or functional relationship between the mobile Internet penetration rate (MBR) and the EI looks like.

Table 4.34 shows the descriptive statistics for the intersection of the data for the Education Index (EI) and the mobile Internet penetration rate (MBR) for the years 2008 and 2012. The number N of the countries for which there are data for EI and the MBR is 29 in 2008 and 83 in 2012. The corresponding values for the EI differ significantly from the values shown in the previous Sects. 4.2.2, 4.2.4 and 4.2.6 as span (Sp), minimum (Min), maximum (Max), average (MV), median (MD) and standard deviation (Sta). This is due to the number N of observed states. Crucial are the values for the MBR in Table 4.34. The span (Sp) has increased between 2008 and 2012 from about 73.8% to about 105.6%. A value higher than 100% is possible because the MBR can reach values higher than 100% if there is, e.g., on average more than a mobile broadband connection per inhabitant. The average (MV) has increased from slightly above 9% in 2008 to 33% in 2012. Comparing this value with the average value for the conventional Internet penetration rate (IPR) results in

---

[26]Surely must be considered here that applications on smartphones and the user interface design must contribute substantially for this, to make the use of smartphones possible for illiterates.

Table 4.34 Descriptive statistic mobile Internet penetration rate (MBR) and Education Index (EI), 2000 and 2008

	2008 (N = 29)						2012 (N = 83)					
	Sp	Min	Max	MV	MD	Sta	Sp	Min	Max	MV	MD	Sta
EI	0.548	0.439	0.987	0.789	0.814	0.138	0.727	0.260	0.987	0.724	0.731	0.174
MBR	73.78	0.04	73.82	9.18	3.42	16.59	105.62	0.8	106.42	33.06	28.25	29.98

Table 4.35 Parameters for $MBR = f(EI)$

t	a	b	c	d
2008	1.435	0.002	11.997	2.191
2012	1.634	94.352	79.620	

a rough approximation that now three of four Internet connections trace back to the mobile Internet.

Figure 4.31 shows the Education Index (EI) as independent and the mobile Internet rate (MBR) as dependent variable, the data points as well as the regression functions tested for the relation for 2008 (left) and 2012 (right). From the first observation it is evident that in both years a positive connection exists between the EI and the MBR. The MBR in 2008 reaches an average level of 9% only from an EI of about 0.7. But a higher value for EI does not guarantee a high MBR (Table 4.35).

In 2008 the relation between EI and MBR is modeled best by an exponential function (blue) and a power function (green). The hyperbolic model fits the data similarly well. The linear regression function (red) does not fit the data well. The results suggests that only for high values of EI correspondingly high values for the MBR can be achieved in 2008. Up to 2012 this changes, so that also for EI values in the middle range high values of the MBR exist. In 2012 the exponential function (blue) and the power function (green), like in 2008, model the data best, followed by the linear function (red). An essential difference from 2012 to 2008 is that the hyperbolic function does not model the data well. The negative value of the coefficient of determination R^2 in Table 4.36 is a clue for the bad adaptation of the hyperbolic function to the data.[27] In 2012, the linear and exponential function have almost identical values for the mean squared error (MSE or MSE/N) and the coefficient of determination (R^2) (see Table 4.36).

$$MBR_t = a + b \cdot \exp^{c \cdot EI_t - d} \tag{4.20}$$

$$MBR_t = a + b \cdot EI^c \tag{4.21}$$

---

[27]The hyperbolic function models the data even worse than a constant function on the level of the mean value (horizontal line) of the MBR.

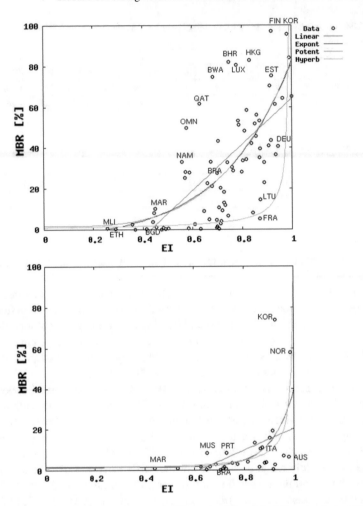

Fig. 4.31 Mobile Internet penetration rate (MBR) as function of the Education Index (EI), 2008 (bottom) and 2012 (top)

Table 4.36 Mean squared error (MSE) and coefficient of determination (R^2) for the regression $MBR = f(EI)$, year 2000 and 2012

	2008 (N = 29)			2012 (N = 83)		
	MSE	MSE/N	R^2	MSE	MSE/N	R^2
Linear	6206	214.0	0.195	42097	507.2	0.429
Expont	**5443**	**187.7**	**0.294**	**38594**	**465.0**	**0.476**
Power	**5444**	**187.7**	**0.294**	**38651**	**465.7**	**0.476**
Hyperb	5456	188.1	0.292	110711	1333.9	−0.083

4.2.9 Intermediate Result

The studies on the empirical and functional relationship between the level of education in a country, as measured by the adult literacy rate (ALR) and the Education Index (EI) and the different access rates of fixed telephony (FTR), mobile phone (MPR), Internet (IPR) and mobile Internet (MBR) show a positive correlation in 2000 and 2012, as well as for all years in between. The connection between the ALR and the FTR is well modeled in 2000 as well as in 2013 with an exponential function. For the relation between the ALR and the FTR a power function fits well in both years. A (very) high ALR is generally a necessary condition for a high technology penetration in a country, but by far still not a sufficient one. There is no country which shows a relatively high spreading of the communication technologies with a low ALR. This statement is valid in a weaker form for the EI as a stronger measure of the educational standard, i.e., there are isolated states which show moderately high values of the respective access rate with a relatively low EI. This situation is slightly different for the mobile telephony, since also for a relatively low ALR and EI, MPR values are high because the mobile phone is used by everyone, regardless of educational background.

4.3 Combination of Key Figures Between 2000 and 2012

The univariate regression method analyzes the relationship between (only) two variables.[28] If one uses instead multiple variables for predicting a size, one goes into the field of multivariate and multiple regression analysis (see Sect. 2.4.7). In this regression coefficients (here A, B, C and D) are used as weight variables that specify which weight the respective predictor enters the forecast. The aim is the use of a suitable linear combination (weighted sum) of the estimator with regard to the independent variables as a predictor. The weights are estimated in a multiple regression by the method of least squares in a way that results in the highest correlation between the estimated and the observed values with the determined linear combination of belonging to the independent variables estimators.

When considering several independent variables in general, the question arises, of which one affects the dependent variable the most or the least. The effect of each explanatory variable on the dependent variable measures (indirectly via the associated estimators) the respective regression coefficient. If the independent variables own different units, the dimensions of the regression coefficients cannot be compared (directly). Standardized regression coefficients compensate the effect of different scales. These result from the corresponding non-standardized coefficients by being multiplied by the standard deviation of the independent variable and divided by the

---

[28]The univariate regression analyses are necessary in this book to find the best (not linear) functions for the correlations which can be used then as a linear combination in the multivariate regression analysis.

standard deviation of the dependent variable. Thus all variables of the regression model receive a statistically uniform unit. The contact with the negative weights which can appear in the optimization process is also interesting. Comparing the dimension of the influence of the respective independent variables one has to use the absolute value and compare them with one another.

Until now, only univariate and non-linear regression analyses were used in this book. They were used for analysis of the relationships between the various access rates for fixed telephone, mobile phone, Internet and mobile Internet on the one hand and economic performance as well as the level of education of states on the other hand. Now the multiple linear regression analysis is used to determine the relationship between these access rates as the dependent variable and the combination of economic performance (per capita GDP) and education (adult literacy rate and Education Index) as explanatory variables. Because for the most part non-linear regression functions were used to model the certain relations (see Sects. 4.1 and 4.2), a transformation of the corresponding data is necessary with the help of the respective non-linear regression functions to be able to apply the multivariate linear regression model according to the approach introduced in Sect. 2.4.7. In the following the respective (specific) approach for modeling the relationship between economic performance, level of education and technology use are described. The suitable transformation functions that provide the best fit (in the univariate case) are described in the segments 4.1 and 4.2. Finally, the results of the multiple regression are interpreted.

4.3.1 Fixed Telephony, Economic Performance and Education Level

Considering the fixed telephone rate (FTR) as the dependent variable and the GDP per capita (GDPpC), the Education Index (EI) and the adult literacy rate (ALR) as the three independent, i.e., explanatory variables, the question is to what extent and with what weight the individual independent variables affect the FTR. Here the interaction is to be considered implicitly between the three explanatory variables. The approach is the following (see Sect. 2.4.7):

$$FTR = A \cdot f_1(GDPpC) + B \cdot f_2(EI) + C \cdot f_3(ALR) + D \qquad (4.22)$$

Accordingly the GDPpC with weight A, the EI with weight B and the ALR with weight C have influence on the FTR. The constant D has no (explanatory) meaning. It is a part of the optimum adaptation. The weight amounts (just) make a statement about the relation of the independent variables to each other with respect to their influence on the dependent variable. Based on the results in Sects. 4.1.1, 4.2.1 and 4.2.2, the assumption is that the GDPpC affect the FTR the most, followed by the FI and the ALR.

In Sect. 4.1.1, the relationship between GDPpC and FTR both in 2000 and in 2012 was modelled using a Planck function, which is now used as a transformation function f_1 for GDPpC data. It should be remembered that such transformation functions for the multiple linear regression, which result from the combination of non-linear regression functions (see Sect. 2.4.7), are needed. The relationship between FTR and the EI has been described both in 2000 and 2012 using a power function (see Sect. 4.2.2), which is now used as a transformation function f_2. Furthermore, the relationship between FTR and ALR in 2000 and 2012, was modelled by an exponential function (see Sect. 4.2.1), which now acts as a transformation function f_3.

For 2000 the regression weights of A, B, C and D arise after the realization of the multiple analysis:

$$\Rightarrow A = 0.517, B = 0.448, C = -0.218, D = -0.203$$

The result shows that the FTR most relates to the GDPpC and that the EI affects the FTR second most. This corresponds to the expectations. Besides, the FTR is determined by the GDPpC as well as by the EI similarly strong, which can point to a (possible) high correlation of economic performance and education level of a country, accepted in this present work. In proportion to these both explanatory variables the ALR has (clearly) less influence on the FTR. A negative regression weight, as is the case for C, can appear as a result of the optimization. For comparison purposes, the absolute value is used in this case. Considering the (proportional) influence of all independent variables, the following relationship is found: The GDPpC influences the FTR proportionately to 43.7%, the EI to 37.9% and the ALR to 18.4%. What could this result be related to?

On the one hand the implicitly accepted high correlation between GDPpC and EI can be responsible for this. Furthermore a vast or exhaustive fixed network infrastructure in a country is expensive. In Sect. 4.1.1 it was found out that in 2000 in states such as *Denmark, Germany, Finland, Great Britain, Switzerland, Sweden, Canada* and the *US* show a very high FTR (see Fig. 4.1). These countries also have very high levels of education (see Fig. 4.24). Within this group of states one can see the close interaction between GDPpC and EI. However, there are also states like *Bahrein, Brunei, Qatar, Kuwait, Oman, Saudi Arabia* and *United Arab Emirates* that in spite of a very high GDPpC show a relatively low FTR (see Fig. 4.1) and at the same time a mediocre education level (see Fig. 4.24). It was also found that a high FTR occurs only in countries that have a very high ALR, but that there are also states that, despite a very high ALR, only have a small FTR (see Fig. 4.22). The (high) education of the wide population is expensive, but it leads again to more prosperity and influences the build up of the fixed network infrastructure positively. Although a widespread literacy is not a sufficient condition for a correspondingly high level of prosperity and a solid fixed network infrastructure, which is why the ALR in 2000 has also significantly less impact on the FTR.

Table 4.37   Weight proportions GDPpC, EI and ALR on FTR, 2000 and 2012

FTR	GDPpC (%)	EI (%)	ALR (%)
2000	43.7	37.9	18.4
2012	44.6	41.9	13.5

In how far these relations have changed after twelve years of development towards a global information and knowledge society since the beginning of the *new economy-dynamics* in 2000? For 2012, the following regression weights A, B, C and D result:

$$\Rightarrow A = 0.423, B = 0.368, C = 0.137, D = -2.476$$

It turns out first that the FTR still is still strongly related to the GDPpC, once again followed by the EI. The ALR has again the smallest influence on the FTR. In this respect the basic order of impact strength has remained the same since 2000. In case of a consideration of all independent variables the following proportions of weight arise: The FTR is influenced by the GDPpC to 44.6%, by the EI to 41.9% and by the ALR to 13.5%. So these relations have changed up to 2012 also (only) slightly.

Table 4.37 gives an overview of the respective weight numbers of GDPpC, EI and ALR with regard to the influence on the FTR in the years 2000 and 2012. Essentially one can summarize the results as follows: The GDPpC and EI have both in 2000 and in 2012 a similarly strong influence on the FTR, but the GDPpC has a slightly greater impact than the EI. Compared to these, the ALR has in both years a (significantly) lower impact on the FTR, and it is even lower in 2012 than in 2000. The influence of GDPpC is roughly at a similarly high level for both years. The importance of EI has increased a bit relative. How could these results be explained, if one considers the results from Sects. 4.1.1, 4.2.1 and 4.2.2?

The FTR has slightly sunk between 2000 and 2012 worldwide as a result of the stronger global population growth in comparison to the (absolute) growth of the number in fixed telephone connections. The advent of the mobile telephony, which happens partly at the expense of the fixed telephone connections, is considered as another reason here. A decline in FTR can be observed in more wealthy countries, while the states (in the central GDPpC) area have expanded their fixed network infrastructure, thus increasing their FTR. These compensation phenomena could be a reason for the similarly high influence of the GDPpC on the FTR in both years. Similar to 2000, a high correlation between GDPpC and EI exist in 2012, especially in the years leading up to 2012 in oil states like *United Arab Emirates, Saudi Arabia* or *Oman* with their already high GDPpC how to raise their education level perceptibly. The education level has (slightly) improved at the same time worldwide. The rise of the EI was strongest within countries in the middle GDPpC range, these countries have also clearly improved her FTR. This is the reason for the slightly greater impact of EI on the FTR in 2012 compared the year 2000. Poorer countries have remained with their FTR at a constant level, although they have expanded their fixed network infrastructure. In these countries the growth in population has led to this compensatory effect with regard to the FTR. These poorer countries have considerably increased the ALR. The richer countries show an ALR of nearly 100% and cannot raise their ALR any further, i.e., they are already on the upper border. Thus, the global increase in the ALR between 2000 and 2012, especially in the poorer regions of the world, and the fall in global FTR from 2000 to 2012 due to world population growth, are primarily responsible for the lower correlation between the ALR and FTR after twelve years of development. It is clear that with a worldwide ALR of 100%, i.e., everyone is literate, there would no longer be a relation with the FTR.

The worldwide expansion of the (fixed) broadband infrastructure can effect the FTR. However, further expansion will still be a long process, especially if one thinks about the world population rising to 10 billion people up to 2050.

4.3.2 Mobile Telephony, Economic Performance and Education Level

Looking at the mobile phone rate (MPR) as the dependent variable and the GDP per capita (GDPpC), the Education Index (EI) and the adult literacy rate (ALR) as the (three) independent variables, the question arises at what extent each individual independent variable affect the MPR. Besides, the interaction between three explanatory variables is to be considered once more implicitly with. The approach is the following:

$$MPR = A \cdot f_1(GDPpC) + B \cdot f_2(EI) + C \cdot f_3(ALR) + D \qquad (4.23)$$

The MPR is obviously influenced by the GDPpC with the weight A, by the EI with the weight B and by the ALR with the weight C. As already explained in the previous Sect. 4.3.1, the constant D has no (explanatory) influence, which is why D is not further discussed. Based on the results in Sects. 4.1.2, 4.2.3 and 4.2.4, the presumption is that the MPR is related to the GDPpC most, followed by the EI and the ALR. As already observed above for the FTR, the meaning of the GDPpC will be the greatest in 2000, because in 2012 the mobile telephony is used across almost the entire world population therefore the other dimensions of influence become less important.

The relationship between the GDPpC and the MPR was modelled in 2000 using a Planck function, which will be used now as class of underlying transformation functions f_1 for the GDPpC data. In 2012, the transformation function f_1 is a MaxLog function (see Sect. 4.1.2). The relationship between MPR and EI has been described both in 2000 and 2012 using a power function (see Sect. 4.2.2), which is now used as a transformation function f_2 for the EI data. Furthermore, the relationship between MPR and ALR in 2000 and 2012 was modelled by an exponential function (see Sect. 4.2.3), which is therefore used herein as class of transformation functions f_3.

For 2000, the following regression weights A, B, C and D result:

$$\Rightarrow A = 0.737, B = 0.485, C = 0.291, D = -3.563$$

The result shows that the MPR is strongly related most to the GDPpC in 2000. It further shows that the EI affects the MPR the second strongest. Besides, the MPR is determined to a different extent by GDPpC and EI. In memory: The FTR was influenced equally by GDPpC and EI (see Sect. 4.3.1). In comparison to both these explanatory variables the ALR has a lower influence on the MPR. Considering the proportional influence of all independent variables, the following relationship is found: The GDPpC affects the MPR to 48.7%, the EI to 32.1% and the ALR to 19.2%.

How can this result be interpreted? In 2000, the mobile phone was used almost exclusively in the richer states. Thus, the GDPpC was the determining factor for the MPR worldwide. Furthermore, in these countries the ALR is consistently very high, i.e., in this group of states the ALR has no great importance for the height of the MPR for the purposes of a differentiation characteristic. The lower meaning of the EI concerning the MPR is determined above all by states like *Greece, Hong Kong, Italy, Korea, Portugal, Singapore, Spain* etc., that have a relatively high MPR, but a low EI in comparison to states like *Belgium, Denmark, Germany, Finland*, the *Netherlands, Norway* or *Sweden*. It can be assumed that the ALR in 2012 will have an even lower impact on the MPR, because firstly, the ALR has increased worldwide and on the other, the mobile phone is now used almost all over the world by many people. Moreover, in this connection it is be expected that the GDPpC and the EI will determine the MPR again at a similarly strong level.

For 2012, the following regression weights A, B, C, D result:

$$\Rightarrow A = 0.206, B = 0.184, C = -0.085, D = -2.412$$

It turns out, first, that all the regression coefficients have a lower value (absolute) than in 2000. The MPR is still influenced strongest by the GDPpC, followed by the EI, indeed, with the difference that now both explanatory variables have a similarly strong influence on the MPR. This result corresponds to the expectations. Looking at all the independent variables the following weight proportions result: The impact rate is 43.3% of the GDPpC, 38.7% of the EI and 17.9% of the ALR on the MPR. The latter value is even lower in 2012 than in 2000.

Table 4.38 gives an overview of the respective weight numbers for the independent variables concerning the influence on the MPR. Essentially one can summarize and explain the results as follows. The GDPpC has the strongest influence on the MPR in 2000 and 2012, but this influence decreases proportionally in this period. In comparison to the GDPpC, in both years the EI and the ALR have a lower influence on the MPR, with the ALR having the least. The EI increases its proportional influence during the considered period, while the connection with the ALR sinks a little bit. But what could be the reasons for this, if one takes into account the results from Sects. 4.1.2, 4.2.4 and 4.2.3?

Worldwide, the MPR has massively increased between 2000 and 2012, so that almost the entire world population nowadays uses the mobile phone. One conclusion

Table 4.38 Weight proportions GDPpC, EI and ALR with respect to the influence on the MPR, 2000 and 2012

MPR	GDPpC (%)	EI (%)	ALR (%)
2000	48.7	32.1	19.2
2012	43.4	38.7	17.9

for this is that the mobile phone is used by everybody no matter whether somebody is rich or poor, educated or uneducated, able to and write or not. In spite of this fact the analyses deliver a differentiated relationship between MPR, GDPpC, EI and ALR. This is due to the fact that the MPR is not limited upwards. There are many countries with a MPR significantly over 100%, because one person can usually have more than one access to a mobile phone. A MPR value over 100% is recognizable over all GDPpC areas. So it is not only given in rich states. Reasons for this were already explained in Sect. 4.1.2. Furthermore one can not observe any compensatory effects by the growth of the population in a country or in the whole world with regard to the increase of the MPR like this is the case with the FTR (see Sect. 4.3.1). Another reason could lie in the fact that the MPR shows a saturation effect in 2012 worldwide, but, however, there are still some countries in Africa with low prosperity and education levels (*Ethiopia, Burundi, Eritrea, Kiribati, Congo, Mozambique, Niger, Rwanda, Sierra Leone, Togo, Chad, Uganda, Central Africa* etc.) in which still less than half of the population has a mobile telephone.

4.3.3   Internet, Economic Performance and Education Level

If one looks at the Internet penetration rate (IPR) as the dependent variable and the GDP per capita (GDPpC), the Education Index (EI) and the adult literacy rate (ALR) as the independent variables, the question again is with which weight do the single independent variables influence the IPR. Here are implicitly the interactions between the three explanatory variables to consider. The approach is as follows:

$$IPR = A \cdot f_1(GDPpC) + B \cdot f_2(EI) + C \cdot f_3(ALR) + D \qquad (4.24)$$

According to this approach, the GDPpC affect the IPR by factor A, the EI by factor B and the ALR by factor C. The constant D has no (contentual) explanatory function of significant importance, which was already explained in previous sections. The weight levels (only) say something about the relation of the independent variables to each other with respect to the influence on the dependent variable. Based on the results in Sects. 4.1.3, 4.2.5 and 4.2.6 the presumption arises that the IPR is connected to the GDPpC strongest, followed by the EI and the ALR.

In Sect. 4.1.3, the relationship between the GDPpC and IPR both in 2000 and in 2012 was modelled using a Planck function, which is now used as a transformation function f_1 for GDPpC data, which were set as a basis for the multiple linear regression (see Sect. 2.4.7). The relationship between IPR and the EI has been described both in 2000 and 2012 using a power function (see Sect. 4.2.6). This class of functions now serves as a transformation function f_2. The relationship between IPR and ALR was modelled in 2000 with a hyperbolic function and in 2012 by an exponential function (see Sect. 4.2.5), which are the underlying transformation functions f_3 for the respective year.

For 2000 the following regression weights A, B, C and D result:

$$\Rightarrow A = 0.671, B = 0.584, C = -0.239, D = -0.222$$

The result shows that the IPR is most related to the GDPpC, followed by the EI and the ALR, what confirms the assumption described above. Here the GDPpC and the EI have a similar influence on the IPR. This points once more to a (possible) high correlation of economic performance and education level of a country. Compared to GDPpC and EI the ALR has (significantly) less impact on the IPR. A negative regression weight, as it is the case for C, is an "artefact" of the optimization. This was also explained above.[29] The negative regression coefficient C concerning the ALR means that with rising ALR values there is less of a relative probability for the MPR to accept high values. In comparison to the positive regression coefficients, as it is the case for the GDPpC and the EI, rising values for the GDPpC and the EI lead to rising probabilities for high MPR values. Considering the proportional influence of all independent variables, one finds the following relationship: The GDPpC affects the IPR to 44.9%, the EI to 39.1% and the ALR to 16.0%. What could this result be related to?

Although reading and writing abilities is a primary competence of Internet use, the cost of Internet access is of significant importance. Widespread literacy is a necessary condition for a correspondingly high level of prosperity in a country, but not a sufficient condition. A high level of prosperity in turn makes it more likely for individuals to afford Internet access. In 2000 an Internet connection is to be found almost only in rich countries and in these countries nearly the whole population is alphabetized. From this the low correlation between IPR and ALR results. Within this group of rich countries, in 2000, there were hardly any differences regarding the ALR, but differences regarding the education level in form of the EI what explains the strong meaning of the EI.

It is interesting furthermore that the weight proportions of GDPpC, EI and ALR show similar values like the weight proportions in the multiple analysis with regard to the FTR in 2000 (see Sect. 4.3.1). Based on the available data and the resulting differences in the number of countries that have been studied with respect to the relation of IPR and FTR. This result can be partly random. On the other hand, it points to a strong coupling between the fixed network infrastructure and the Internet

---

[29]In general the interpretation of the regression coefficients for not-linear regressions is not that easy, as it is the case for linear regression.

Table 4.39 Weight proportions GDPpC, EI and ALR regarding the influence on the IPR in 2000 and 2012

IPR	GDPpC (%)	EI (%)	ALR (%)
2000	44.9	39.1	16.0
2012	57.3	40.5	2.2

use in 2000, which was already discussed below (see Sects. 4.1.1 and 4.1.3) and is shown empirically below (see Sect. 4.4.2).

What is the situation after 12 years of development in 2012? The following regression weights A, B, C and D result:

$$\Rightarrow A = 0.624, B = 0.441, C = -0.023, D = -3.762$$

It becomes clear first of all that all the regression coefficients have reduced (according to absolute amount). This can be related to the used transformation functions. It also turns out that the IPR is still most related to the GDPpC followed by the EI. The ALR has the least impact on the FTR. Looking at all independent variables the following weight proportions result: The IPR is influenced by the GDPpC to 57.3%, by the EI to 40.5% and the ALR to 2.2%.

Table 4.39 gives an overview of the respective weight proportions of GDPpC, EI and ALR with regard to the influence on the IPR in 2000 and 2012. Essentially one can summarize the results as follows: The GDPpC in 2012 has a significant higher impact on the IPR than in 2000. The weight proportion of the EI is similarly high in both years. Compared to these explanatory variables the ALR in both years has a much lower influence on the IPR and additionally this influence sinks to a very low level in 2012.

What could these results could be related with, if one considers the results from Sects. 4.1.3, 4.2.5 and 4.2.6? The IPR has risen between 2000 and 2012 worldwide up to about 40%. In this time frame the biggest (absolute) growth concerning the IPR has taken place in the rich states (see Sects. 4.1.3), which explain the increase of the meaning of the GDPpC as an explanatory variable. In comparison, the slightest IPR growth took place in the poorest states (see Sect. 4.1.3). Within the group of rich states an accordingly high EI is given (see Sect. 4.2.6). Furthermore, these countries already have an ALR very close to the 100% limit (see Sect. 4.2.5), which consequently has almost no significance for the explanation of the IPR level. Now the IPR has increased worldwide, also as a result of access to the mobile Internet. How is it to be explained then that the weight of the ALR is so low in 2012 in comparison to the result concerning the FTR and MPR? Such a low value for the weight of ALR does not exist in connection with the investigations to the FTR (see Sect. 4.3.1) and MPR (see Sect. 4.3.2). With the results concerning the FTR, the low influence of the ALR is understandable (see Sect. 4.2.1), because the FTR has decreased worldwide a little bit and at the same time the ALR has clearly increased. With the low influence of the ALR on the MPR the argumentation is as follows: The MPR is worldwide in 2012 at the 100% level and those countries with an MPR over 100% are responsible for the ALR having a low influence over the MPR (see Sect. 4.2.3). An interpretation of the very low weight of the ALR regarding the IPR in 2012 is more difficult. The reason could lie in the fact that within the countries in which the Internet is widespread, i.e., a high IPR is given, there are (clearly) higher differences with regard to the EI than with regard to the ALR, since most countries have a very similar ALR. The result is understandable, because the ALR in states with Internet use were either very high before or had risen between 2000 and 2012 to a very high level (see Sect. 4.2.5). Finally, it should be noted that the weight proportions in 2012 concerning the influence on the IPR accept other values than concerning the influence on the FTR, while in 2000 for both adaptations similar weight proportions result. As already described above, the narrow coupling of the Internet access with the fixed network infrastructure is responsible for this in 2000, while up to 2012 the Internet access has partially decoupled from the fixed network infrastructure, namely by the success of the Mobile Internet (see Sect. 4.4.2).

4.3.4   Mobile Internet, Economic Performance and Education Level

It should be noted that results in this section should be considered with caution. A direct comparability to the investigations concerning the fixed telephone rate, mobile telephone rate and the Internet penetration rate is given only partly. This has to do with the fact that the investigations first began in 2008 and that there are much less data for mobile Internet rate (MBR) than for FTR, MPR and IPR.

Looking at the mobile Internet penetration rate (MBR) as the dependent variable and the GDP per Capita (GDPpC), the Education Index (EI) and the adult literacy rate (ALR) as independent variables, now it is examined with which weight the single independent variables influence the MBR. Here one has to consider implicitly the interactions between the three explanatory variables. Following approach is used:

$$MBR = A \cdot f_1(GDPpC) + B \cdot f_2(EI) + C \cdot f_3(ALR) + D \qquad (4.25)$$

According to this approach the GDPpC with the factor A, the EI with the factor B and the ALR with the factor C has influence on the MBR. Based on the results in Sects. 4.1.5, 4.2.7 and 4.2.8 the MBR is correlated strongest to the GDPpC, second strongest to the EI and lowest to the ALR. Furthermore it is to be supposed that the MBR is connected to all these explanatory variables in 2000 comparatively stronger than in 2012.

In Sect. 4.1.5, the relationship between GDPpC and MBR, both in 2008 and 2012 has been modeled using a Planck function, which is now used as a transformation function class f_1 for the GDPpC data that are required for multiple linear regression (see Sect. 2.4.7). The relationship between MBR and EI has been described both in 2008 and 2012 using a power function (see Sect. 4.2.8), which now serves as a of transformation function class f_2. The connection between MBR and ALR was modeled in the years 2008 and 2012 with an exponential function (see Sect. 4.2.7), which now represents the transformation function class f_3 for 2008 and 2012.

For 2008 the following regression weights A, B, C and D result:

$$\Rightarrow A = 0.587, B = 0.516, C = -0.262, D = -0.183$$

The result shows that the MBR is related to the GDPpC most, followed by EI and ALR, so the assumption from above is confirmed. The GDPpC and the EI effect the MBR similarly strong. Again this points to a close correlation between economic performance and the level of education. Compared to GDP and EI the ALR has less influence on the MBR. Considering the proportional influence of all independent variables, the following relationship is found: The GDPpC influences the MBR to 43.0%, the EI to 37.8% and the ALR to 19.2%. How could this result be explained?

Comparing the result here with the result for the Internet penetration rate, one notes that regarding the IPR in 2000 and the MBR in 2008 approximately the same weight proportions are given (see Sect. 4.3.3). Therefore, the same explanations are valid at this point similar to the analysis on the IPR. But how could the relationship be explained between IPR and MBR? The reason for this is to be found in the database. In 2008, MBR data where exclusively available for a few countries, in which Internet is widespread. What is the situation in 2012? The following regression weights A, B, C and D result:

$$\Rightarrow A = 0.243, B = 0.214, C = -0.115, D = -1.982$$

It firstly becomes clear that all regression coefficients show a smaller value (absolutely) than in 2000 what can be explained with the concrete (optimum) individual transformation function. The MBR is still influenced strongest by the GDPpC, followed by the EI. The ALR shows the lowest correlation to the MBR. This result corresponds to the expectations. Looking at all the independent variables the following weight proportions result for the year 2012: The MBR is influenced by the GDPpC to 42.5%, by the EI to 37.4% and the ALR to 21.1%.

Table 4.40 gives an overview for the respective weight proportions of the independent variables with regard to the influence on the MBR.

Table 4.40 Weight proportions GDPpC, EI and ALR regarding the influence on the MBR, 2008 and 2012

MBR	GDPpC (%)	EI (%)	ALR (%)
2008	43.0	37.8	19.2
2012	42.5	37.4	21.1

If one compares these results with that of the mobile phone rate (MPR), one can deduce that the weight proportions regarding MBR and MPR are similar in 2012 (see Sect. 4.3.2). This points to the fact that the access to the Internet is effected more and more by means of a mobile phone. Therefore the same explanations are valid at this point like in the analysis to the MPR.

4.3.5   Intermediate Result

The implementation of the multiple regression analysis to determine the relationship between the access rates to fixed telephone (FTR), mobile phone (MPR), Internet (IPR) and mobile Internet (MBR) as the dependent variables and the combination of economic potential (GDP per Capita) and education level (adult literacy rate and Education Index) as explanatory variables, yields the following results. First, it should be noted that all of these (four) data on access rates are most affected by the GDPpC, followed by the EI and the ALR. Furthermore the GDPpC, i.e., the prosperity level, and the EI, i.e., the education level, have a similarly strong influence on the respective access rates which also points to the explicitly existing (high) correlation between economic performance and education level. The ALR has significantly less impact on the individual access rates to technology compared to these two explanatory variables. This indicates that high literacy is a necessary but not a sufficient condition for a high level of prosperity, which then leads to greater use of technology in general.

It turns out essentially that in 2000 similar values for the influence of GDPpC, EI and ALR are observed on the fixed telephone rate and Internet penetration rate. This points to the fact that in 2000 the Internet access occurred almost exclusively over a fixed network connection. It was also found that the multiple analysis for mobile Internet usage in 2008 gives similar results as there are for the "conventional" Internet usage in the same year. In 2012 the investigation of mobile Internet delivers a similar result as for the mobile phone. In the trend, these results show that the access to the Internet occurs more and more via a mobile phone or smartphone and we experience worldwide the technological convergence of mobile telephony and Internet to the mobile Internet.

4.4   System Theoretical Approach for Determining the Interaction of Key Figures

The basic idea behind the underlying (system theoretical) approach to determine the (univariate and multivariate) interaction between the considered factors and technologies among themselves (FTR, MPR, IPR, MBR, GDPpC, EI and ALR) is the use of an equation system (see Eqs. (4.26)–(4.32)). This equation system covers the regressions used in this book in the form of a linear combination. If one wanted to determine from this system of equations, e.g., the relation between GDP per capita (GDPpC) and Internet penetration rate (IPR), then one has to put all other factors of influence in Eq. (4.26) back to zero. As a result, only the function $IPR_t = f_1(GDPpCt)$, which corresponds to the univariate regression model in Sect. 4.1.3, is analyzed. If one wants to open the equation for the multivariate regression model from Sect. 4.3.3 for the combination of factors from this equation system, one has again to put all factors except GDPpC, EI and ALR back to zero. Then Eq. (4.26) corresponds to the multivariate regression model $IPR_t = f_1(GDPpCt) + f_2(EI_t) + f_3(ALR_t)$ for the determination of the dependence of IPR on economic performance and the educational level. Accordingly one can derive from this equation system all other regression models which are used in this book. In this chapter in total 21 different regressions are carried out in the form of different univariate and multivariate approaches. The following equation system is in so far a chosen model of several options for determining interactions.

$$IPR_t = f_1(GDPpC_t) + f_2(EI_t) + f_3(ALR_t) + f_4(FTR_t)$$
$$+ f_5(MPR_t) + f_6(MBR_t) + f_7(IPR_{t-1}) \tag{4.26}$$

$$MBR_t = g_1(GDPpC_t) + g_2(EI_t) + g_3(ALR_t) + g_4(FTR_t)$$
$$+ g_5(MPR_t) + g_6(MBR_{t-1}) + g_7(IPR_{t-1}) \tag{4.27}$$

$$MPR_t = h_1(GDPpC_t) + h_2(EI_t) + h_3(ALR_t) + h_4(FTR_t)$$
$$+ h_5(MPR_{t-1}) + h_6(MBR_{t-1}) + h_7(IPR_{t-1}) \tag{4.28}$$

$$FTR_t = i_1(GDPpC_t) + i_2(EI_t) + i_3(ALR_t) + i_4(FTR_{t-1})$$
$$+ i_5(MPR_{t-1}) + i_6(MBR_{t-1}) + i_7(IPR_{t-1}) \tag{4.29}$$

$$ALR_t = j_1(GDPpC_t) + j_2(EI_t) + j_3(ALR_{t-1}) + j_4(FTR_{t-1})$$
$$+ j_5(MPR_{t-1}) + j_6(MBR_{t-1}) + j_7(IPR_{t-1}) \tag{4.30}$$

$$EI_t = k_1(GDPpC_t) + k_2(EI_{t-1}) + k_3(ALR_{t-1}) + k_4(FTR_{t-1})$$
$$+ k_5(MPR_{t-1}) + k_6(MBR_{t-1}) + k_7(IPR_{t-1}) \tag{4.31}$$

$$GDPpC_t = l_1(GDPpC_{t-1}) + l_2(EI_{t-1}) + l_3(ALR_{t-1}) + l_4(FTR_{t-1})$$
$$+ l_5(MPR_{t-1}) + l_6(MBR_{t-1}) + l_7(IPR_{t-1}) \tag{4.32}$$

The mutual interaction of the single variables can accept many forms. An interaction between the factors is defined here by a suitable time shift. An interaction originates, e.g., from the fact that the IPR in a country at time t is dependent in a certain way on the IPR at time $t - 1$. I.e., the interaction of a variable with itself is a possible form of interdependence which is to be considered.

It is also useful to consider the dependence or independence of a technology to another technology. For example, at the beginning of the Internet age you needed a landline access in order to connect to the Internet using a modem (see Sects. 4.1.1, 4.1.3 and 4.4.2). The situation is quite different with the relationship between fixed telephony and mobile phone (see Sects. 4.1.1, 4.1.5 and 4.4.2). One does not require any access to the fixed network to make calls via mobile communications. These two technologies are physically completely separate. But the access rates for mobile phone and fixed telephone may still have a certain dependence, since both access rates are primarily influenced by the level of prosperity.[30]

The design of an equation system of the considered type has to be constructed cycle-free. This is a basic condition so that the equation system can be dissolved generally. In principle, the design of the equation system leans upon the so-called multi-step process from numerical mathematics or the dynamic optimization in the Operations Research. This procedure uses in a kind of recursion the information from the before already calculated data values to calculate actual values (explicit procedures).

The following examine exemplary interactions between between the economic performance and education level (see Sect. 4.4.1) as well as the interactions between the information and communications technology (see Sect. 4.4.2).

4.4.1 Interaction Between Economic Performance and Education Level

In studies of the factors and their impact on technology access rates, a crucial result was that the economic performance, as measured by GDP per capita, and the level of education, as measured by the Education Index, have a similarly strong impact on the respective access rates (see Sect. 4.3). Even if the GDPpC is the most important factor for technology accesses, the EI has also a very determining influence here. The other assumption is that the GDPpC has a significant impact on the level of education in a country and vice versa. Therefore an interaction originates and is examined in the following.

---

[30]In examining the dependencies of technologies with each other, it can come in particular to a so-called *leap-frogging-effect* [436]. This effect can occur if some countries that are in arrears with an (older) technology compared to another country, from any time do not any longer invest in the old technology, but immediately in the (often less expensive and more efficient) new technology. Then as a result more people use the new technology than the old one. E.g., since 2010 in Africa more people use the mobile phone than the fixed network phone.

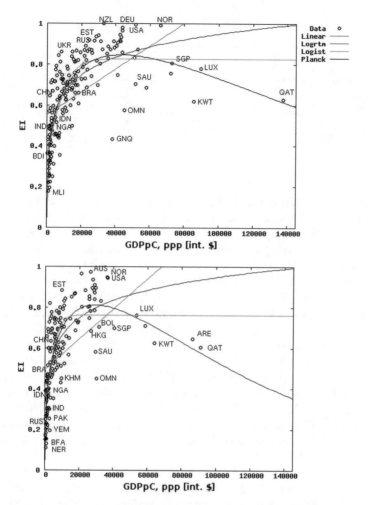

Fig. 4.32 Education Index (EI) as a function of GDP per capita (GDPpC), 2000 (bottom) and 2012 (top)

Relationship Between GDP per Capita and Education Index

This section describes the interaction between the economic performance (GDPpC) and the Education Index (EI). This investigation is carried out (only) for the years 2000 and 2012, since the changes within are not of particular interest. The presumption is that GDPpC and EI in 2000 and 2012 are in a close relation with each other.

To generate the empirical relationship between the GDPpC and EI, a (linear and non-linear) regression analysis is performed. Here, the EI is the dependent variable and the GDPpC the independent variable. Figure 4.32 shows the point cloud as well as the class of regression functions used for modeling the relation in 2000 (below) and in 2012 (above). The first observation is the following: There is a positive rela-

Table 4.41 Mean squared error (MSE) and coefficient of determination (R^2) for the regression $EI = f(GDPpC)$

	2000 (N = 168)			2013 (N = 178)		
	MSE	MSE/N	R^2	MSE	MSE/N	R^2
Linear	5.732	0.034	0.250	5.002	0.028	0.299
Logartm	2.918	0.017	0.618	2.578	0.015	0.639
Logist	2.816	0.017	0.631	2.396	0.014	0.664
Planck	**2.556**	**0.015**	**0.666**	**2.233**	**0.013**	**0.687**

tionship between GDPpC and EI. In 2000, there are many countries that are scattered along the EI-axis (for small values of GDPpC). This means, first, that there are many countries that have a very low value for the GDPpC, which is not surprising, since there are many relatively poor countries in the world, especially in Africa, but also in South-East Asia and Latin America. The interesting thing is that there are countries with low GDPpC and EI (*Afghanistan, Ethiopia, Bangladesh, Burkina Faso, Burundi, Dem. Rep. of Congo, Ivory Coast, Eritrea, India, Mozambique, Nepal, Nigeria, Senegal, Central Africa*), as well as countries with low GDPpC and moderate EI (*Egypt, Argentina, Armenia, Brazil, China, Georgia, Ghana, Indonesia, Iran, Kazakhstan, Kenya, Kyrgyzstan, Mexico, Moldova, Nigeria, Philippines, Russia, South Africa, Tajikistan, Thailand, Tunisia, Turkey, Ukraine, Uzbekistan, Vietnam*). On the other side there are also states which have a high GDPpC, but nevertheless a rather low to medium EI (*Bahrain, Brunei, Qatar, Kuwait, Oman, Portugal, Saudi Arabia, Singapore, United Arab Emirates*). The highest values for the EI are reached in 2000 as well as in 2012 in the high, but not highest, GDPpC area (*Australia, Denmark, Germany, Finland, France, Great Britain, Italy, Canada, Korea, New Zealand, Norway, Sweden, Spain* and the *USA*).

Table 4.41 lists the class of functions that were used to describe the relationship between EI and GDPpC in total. Furthermore the Sums of Mean Squared Errors (MSE bzw. MSE/N) and the coefficients of determination (R^2) are given. The result shows that for 2000 the linear modeling of the relationship between EI and GDPpC is not satisfying, while the logarithmic, logistic and Planck model describe this connection satisfying and similarly well. Nevertheless, the best adaptation has the Planck function. The reason for this lies once more with the very rich oil states which have a comparatively lower education level than the industrial states, which is why the curve for very high GDPpC values bends down.

For 2012 a similar course arises for the functions like in 2000. The values for the coefficient of determination (R^2) are similar. From this, one can derive that after a twelve-year development there is still a (strong) relation between GDPpC and EI. The difficulty in making progress in this area is also understood. This is valid for progress in the area of economic power as well as for the improvement of the educational standard.[31]

Thus, the assumption described above is confirmed. The results of the multivariate regression analysis (see Sect. 4.3) regarding the similarly strong influence of GDPpC and EI on the use of technology are strengthened or made plausible from this side.

4.4.2 Interaction Between ICTs

Another interesting question that arises in the context of technology use in general is the extent to which an (older) technology is replaced by another (new) technology. Answering or investigating this question can offer insights into the process of technology adaptation in countries. This also applies to the so-called *leap-frogging-effect*. This effect describes skipping or omitting stages in an innovation process. In the previous sections it was mentioned several times that access to the Internet in 2000 (actually) took place via the fixed network infrastructure, which is why the access rates for both technologies were seen and continue in a relationship with each other. The *leap-frogging-effect*, for which the mobile telephony is an example, can be observed in particular in developing countries. Furthermore, it was mentioned a few times that the physical infrastructure of fixed telephone and mobile telephone are decoupled. However, the access rates to both technologies could be coupled in a certain extent anyway, because of the effects of prosperity and education level.

Relationship Between Fixed Telephone and Mobile Phone

What is the empirical relationship between the access rates of these two technologies? In order to identify this relation, both a linear and non- linear regression analysis is used. The mobile phone penetration rate (MPR) is the explained (dependent) variable and the fixed-telephony rate (FTR), the explanatory (independent) variable. The dependency was chosen so that the newer technology (mobile phone) is a function of the older one (fixed telephone). Figure 4.33 shows for year 2000 (left) and 2013 (right), the data points and the class of functions that have been used to model this relationship.

The first observation is as follows: In 2000, there is a positive relationship between the MPR and FTR, where the more people in a country have access to the fixed telephone infrastructure, the more they have access to mobile phone infrastructure. This is the case, because the fixed network infrastructure was very well developed in rich countries in 2000 and these countries have made even massive investments in the communication via mobile phone. In the poorer countries of the world, both the fixed infrastructure was not well developed and the communication via the mobile phone was relatively expensive, which has meant that only few people could afford a mobile phone.

---

[31]Note: Achieving (worldwide) progress in the area of Literacy is much easier than to improve the education level of a single country.

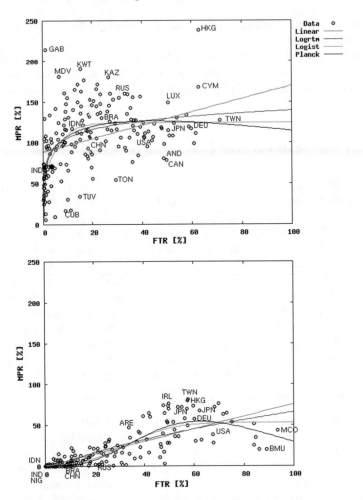

Fig. 4.33 Mobile telephone penetration (MPR) as function of the fixed telephony rate (FTR), year 2000 (bottom) and 2013 (top)

In 2013, there is also a positive relationship between the MPR and FTR. This relationship, however, is much weaker than in 2000, which is also visible on the wider distribution of the data points. The reason is that the global MPR is about 100% in 2013, because infrastructure investments were made in the mobile communication in the poorer countries and the mobile phone usage (clearly) has become cheaper.

Table 4.42 lists the class of functions that have been used to describe the relationship between the MPR and FTR, the corresponding mean squared error (MSE or MSE/n) and the coefficient of determination (R^2). The result shows that the linear modeling of the relationship between FTR and MPR is satisfactory in 2000, but this relation is described by the logarithmic, logistical and Planck's model even better. The Planck function has a better fit to the data than the logistical and logarithmic

Table 4.42 Mean squared error (MSE) and coefficient of determination (R^2) for regression $MPR = f(FTR)$

	2000 (N = 198)			2013 (N = 199)		
	MSE	MSE/N	R^2	MSE	MSE/N	R^2
Linear	36819	185.955	0.630	311592	1565.789	0.143
Logartm	35376	178.667	0.644	255780	1285.327	0.296
Logist	29227	147.611	0.706	250423	1258.407	0.311
Planck	**27709**	**139.944**	**0.721**	**251169**	**1262.156**	**0.309**

function. Of interest in this regard is that as a threshold for the FTR, the point cloud has a very high dispersion and a relationship between FTR and MPR is no longer observed. Figure 4.34 illustrates this finding. The linear relationship between FTR and the MPR shows in 2000 (chart below) for values FTR < 40% a high correlation. For values FTR > 40% is observed a weak correlation (chart above). Overall, the correlation (linear case) or the regressive context (nonlinear case) between FTR and MPR has fallen to a low level worldwide by 2013. One can therefore say that the fixed telephone use is largely decoupled from the use of mobile communications in 2013.

Relationship Between Fixed Telephony and Internet

In 2000, generally a fixed telephone line was required to establish a connection to the Internet via modem. Over time, Internet usage was made possible via a mobile phone. Such a connection today no longer requires access to the fixed network infrastructure. So the assumption is that a positive correlation exists for the empirical relationship between the fixed telephone rate and the Internet penetration rate, while this relationship should decouple slowly until 2013. This means that the correlation is not high, but still exists, because global Internet users can not all afford or want to use smartphones.

To answer this question, again a linear and a non-linear regression is carried out, where the Internet penetration rate (IPR) is the dependent and the fixed telephone rate (FTR) the independent variable. The relation is again chosen so that the new medium (Internet) is a function of the older technology (fixed telephone). Figure 4.35 shows for 2000 (below) and 2013 (above) the data point cloud and the class of functions that are used for the description of the relation. The result shows that the IPR and FTR are in a positive association in 2000, and in this case there is a high correlation. Thus, the presumption above and the results from Sects. 4.1.1, 4.1.3 and 4.3.3 are confirmed.

Table 4.43 lists the class of (test) functions that have been used to model the relationship. In principle, all (test) functions fit the data well, and again the Planck function fits best. The correlation between FTR and IPR decreases by 2013 to a more mid-range level, if one uses the linear modeling of the relationship. However, the logarithmic, logistic and Planck's function have in 2013 again a good match. In total, worldwide we still have a (close) coupling between the FTR and IPR, suggesting that

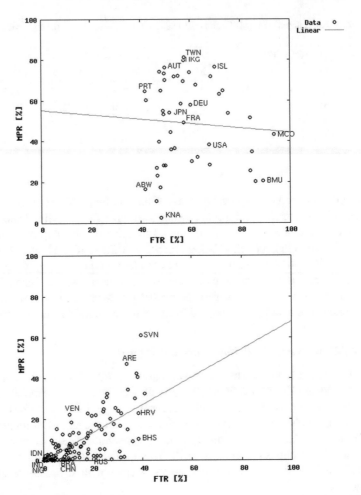

Fig. 4.34 Mobile telephone penetration (MPR) as function of the fixed-telephony rate (FTR), year 2000, FTR < 40% (bottom) and FTR > 40% (top)

the "fixed" Internet access is important in the poorest countries, because smartphones are (still) expensive. In richer countries the fixed telephone infrastructure remains important due to the expansion of broadband network.

Relationship Between Mobile Telephony and Internet

The Internet and mobile telephony are (physically) separate technologies, at least in year 2000, before the era of mobile Internet. Mobile Internet then emerged from the technological convergence of mobile phone and Internet. Nevertheless, based on previous findings, a relation between the MPR and the IPR is to be expected, as the value of both two factors is primarily based on the economic performance of a country. Thus, the presumption is that a positive correlation between the MPR and

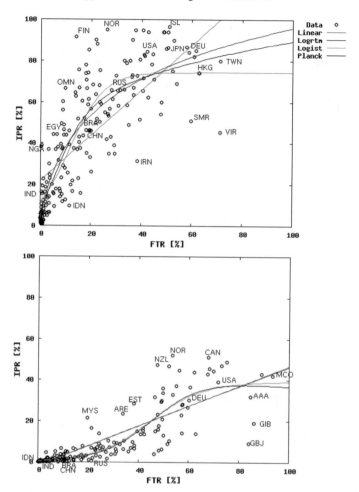

Fig. 4.35 Internet penetration rate (IPR) as function of the fixed telephony rate (FTR), year 2000 (bottom) and 2013 (top)

Table 4.43 Mean square error (MSE) and coefficient of determination (R^2) for regression $IPR = f(FTR)$

	2000 (N = 199)			2013 (N = 196)		
	MSE	MSE/N	R^2	MSE	MSE/N	R^2
Linear	10189	51.201	0.712	70141	357.86	0.580
Logartm	10208	51.297	0.711	43819	223.57	0.738
Logist	8525	42.839	0.759	45386	231.56	0.728
Planck	**8461**	**42.518**	**0.761**	**42918**	**218.97**	**0.743**

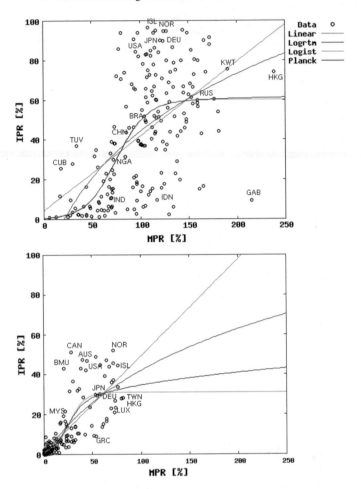

Fig. 4.36 Internet penetration rate (IPR) as function of the mobile telephony rate (MPR), year 2000 (bottom) and 2013 (top)

IPR exists in 2000 which decreases slightly by 2013, because the mobile phone is used by most people around the world in contrast to the Internet.

To explore this assumption, a regression analysis is carried out, in which the IPR is the explanatory (dependent) variable and the MPR the independent variable. Thus, the "new" technology depends on the "older" one. Figure 4.36 shows for 2000 (below) and 2013 (right) the data points and the treated class of (test–) functions that is used for modeling the relationship. The result is as follows: In 2000, there is a positive relationship between the MPR and IPR, i.e., the more people in a country have access to the mobile infrastructure, the more people have access to the Internet, running mainly on the fixed telephone infrastructure, as noted above. The correlation in the linear case is high and the logarithmic, logistical and Planck's function also

Table 4.44 Mean squared error (MSE) and correlation of determination (R^2) for the regression $IPR = f(MPR)$

	2000 (N = 192)			2013 (N = 199)		
	MSE	MSE/N	R^2	MSE	MSE/N	R^2
Linear	11789	61.401	0.663	120538	605.719	0.297
Logartm	11412	59.438	0.674	114175	573.744	0.335
Logist	11052	57.563	0.684	105574	530.523	0.385
Planck	**10971**	**57.141**	**0.686**	**105570**	**530.503**	**0.385**

model the relationship well. Table 4.44 lists the class of functions that have been used to describe the relationship between the MPR and IPR, and the corresponding mean squared error (MSE or MSE/n) and the coefficient of determination (R^2). In 2013, the correlation decreases in the linear case to a medium level, where the remaining models lose explanatory power, i.e., the relationship between the MPR and IPR decreases. Why is this? Interestingly in 2013, the logistic function modeled the data as well as the Planck function. The logistic function has its turning point at an MPR of about 70%. Looking now to the countries with an MPR of less than 70%, it turns out that no country has an IPR greater than 40%. The correlation between the MPR and IPR within this group of countries is vanishingly small ($R^2 = 0.08$). For countries with an MPR above the threshold of 70%, the whole span of the IPR values appears, from very low to the highest IPR for this group of countries. The correlation is even lower ($R^2 = 0.03$) than in the group with an MPR less than 70%.

In summary it can be said that the mobile phone has become widespread in the world in 2013 and is used by many people. If this technology is used by anyone, regardless of wealth or education, a (narrow) relation between the MPR and IPR cannot be expected.

4.4.3 Intermediate Result

The presented systems theory approach to determine the interaction of the factors considered covers the regressions used in this chapter as form of a linear combination. This was on the one hand, the relationship between economic performance and level of education, on the other hand examines the interaction between the information and communications technology. It becomes clear that a correlation between the level of wealth and the educational level of a country exist, both in 2000 and in 2012. Furthermore it is difficult to achieve sustainable progress in these two areas. The investigations on the interaction of the access rates of the technologies show that the fixed telephone rate and the mobile telephone rate are largely decoupled. The fixed-line rate and Internet penetration rate are (worldwide) coupled, both in 2000 and 2012. The empirical relationship between the mobile telephony rate and Internet penetration rate has a high correlation in 2000, which falls to a moderate level until 2013. It was also demonstrated that the more socially balanced (low inequality) a society is, the higher the Internet penetration rate is. However, the Internet penetration rate is less in a country, in which a person must have a (relatively) high individual-average income ratio to afford access to the Internet.

4.5   Conclusion

The studies on the empirical or functional relationship between the *economic potential* in a country in terms of GDP per capita (GDPpC), and the different access rates of fixed telephone (FTR), mobile phone (MPR), Internet (IPR) and mobile Internet (MBR) show a *positive relationship* in 2000 and 2013, including the years between. The respective relationships are modeled with the Planck function largely very good. For the global FTR is crucial that the world population is growing faster than the number of fixed telephone lines, which leads to a reduction of the FTR. In addition, the increase in the MPR is another reason, because it inhibits the growth of FTR. The MPR has an equal roughly increase for all GDPpC-categories between 2000-2013. Furthermore, many countries have an MPR of 100% or more in 2013, which also leads to a worldwide MPR of about 100%.

For Internet use it was initially found out that a strong linear positive correlation between how large a country's GDP as share of the world GDP is and country's share of global Internet users, both in 2000 as well 2013. In 2000, a strong *asymmetric distribution of the Internet population* was present in relation to the distribution of the world's population on countries. This situation is better *balanced* in 2013, due to the *catching-up on Internet usage in poorer and more populated states*. While in 2000 the Internet is largely used via the fixed telephone infrastructure, the *mobile access to the Internet has clearly spread* by 2013. It has been shown that the more balanced a society is in terms of income distribution, the higher the country's IPR. In addition, it was shown that the IPR in a country is lower, the higher income a person must have in relation to the average income in a country to afford access to the Internet can.

The studies regarding the relationship between the *level of education* in a country, as measured by the adult literacy rate (ALR) and the Education Index (EI), and the different access rates to ICT show a *positive relationship* for the years between 2000 and 2012. A (very) high ALR is generally a necessary condition for high ICT access rates in a country, but not a sufficient one. This statement applies to the EI in a weaker form, because there are countries which have a comparatively low EI values and high levels for the respective technology access rates. For mobile telephony, this is different. Here, even for relatively small values of both education indicators, the MPR has a high value.

The multiple regression analysis have confirmed that the investigated (four) access rates to ICT are most influenced by the GDPpC, followed by the EI and the ALR. Here, the GDPpC and EI have a similarly strong impact on the respective access rates. This also points to the *implicitly existing (high) interaction between economic performance and level of education* in a country. This interaction was also checked for plausibility with the featured systems theory approach to determine those interactions. In comparison, the ALR has a significantly lower impact on the ICT access rates. The investigations on the interaction of access rates to ICT among themselves show that the *FTR and the MPR are largely decoupled* from each other. This is expected, since both technologies use *separate physical infrastructure*. The FTR and IPR are (worldwide) coupled in 2000 and 2012, whereby this coupling is released somewhat due to the increase of mobile access to the Internet by 2012. *Overall, the studies conclude that the convergence regarding ICT usage proceeds worldwide faster than the convergence in terms of wealth and education in general*, so the initial formulated thesis of this book is confirmed.

Chapter 5
Global Networking and Worldview

The various worldviews of human beings describe a variety of different (cultural) phenomena that include the belief in a god, a world of deities or spirituality, such as the existence of a transcendent being or multiple transcendent beings. In this context it should be noted that different perspectives on the relationship between religion and culture are available in the world. On the one hand culture can be considered as part of religion and on the other hand religion is seen as part of culture. Religion generally always has had a significant influence on the action, thought and feelings of people and this probably will continue in the future. In the past, trade, wars and superior technologies favored some religions and made them world religions. The world's most famous religions *Christianity, Judaism, Islam, Buddhism, Hinduism* and *Taoism* are represented in various regions and continents and dominate to varying degrees (see Fig. 5.1). In addition, we have worldviews like *atheism, superstition*, or different forms of belief within the so-called *indigenous peoples*.[1]

Religious worldviews usually have a long tradition. They were often not peaceful and neither cooperative with each other despite the fact that religion in origin claims for a peaceful life. Often they postulated a monopoly on truth, which partially prevented technical development. In history, science and religion are partly controversial but also in a partly fruitful relationship. Technical superiority allowed religious communities to stand up their competition or oust other religions. Many religions shift people's aspiration for a better life to the future, in particular after death or in next life. This promotes tolerance, but hinders economic-technical momentum. The scientific knowledge of individuals like Einstein, Kepler, Newton, etc. were often associated

---

[1]Indigenous peoples are not the subject of investigations in this work, since they have no or only a very low level in technological development and there is very limited data available for a global analysis.

© Springer International Publishing AG 2018
H. Ünver, *Global Networking, Communication and Culture: Conflict or Convergence?*, Studies in Systems, Decision and Control 151,
https://doi.org/10.1007/978-3-319-76448-1_5

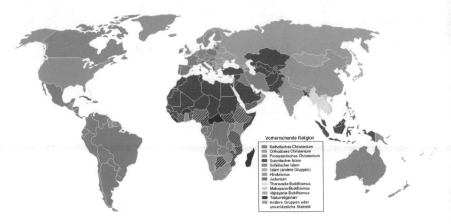

Fig. 5.1 Worldwide regional predominant large religions [498]

with religious motifs and reflections.[2] The fact that the Earth revolved around the sun like other planets, is a phenomenon no longer denied by humans today. But in the 16th century, there were two different astrological worldviews. First, the geocentric worldview in which the Earth is at the center of the universe. Second, the Copernican system in which not the earth but the sun is at the center of our solar system. Galileo favored this new worldview. He came therefore in conflict with the Church and its monopoly on truth. At the time, it was preached by parts of the Church that the earth is the center of the universe and it was not allowed to distribute a different worldview.

Certainly, the world religions have learned from history and adapted themselves in order to maintain and make progress in the modern world. What are the views of each major religion on the digital age of Internet and mobile telephony? How high are the user rates of the various information and communications technologies (ICT) such as the fixed telephone, the mobile phone, the Internet and the mobile Internet in the respective religious groups? What does modern Internet use mean for the Christian concept of man and what is the stance of the Islamic world on technological development? The same questions arise for the other major world religions, such as Buddhism and Hinduism, as well as for Judaism and others.

In Sect. 5.1 the empirical relation is discussed on the topics of global networking through the fixed telephone, mobile phone, the Internet, and the mobile Internet. This is initially done for *Christianity* and *Islam*, the *two largest world religions in 2010*.[3] This is followed by empirical analyzes of *Buddhism* and *Hinduism* (see Sect. 5.2).

---

[2]Quotations: *"God does not play dice."* and *"Subtle is the Lord, but malicious he is not."* Friedrich Dürrenmatt once said in a lecture on Einstein's 100th birthday: *"Einstein used so often to speak of God, that I almost suspect he was a theologian in disguise."* [435].

[3]Therefore, only the year 2010 is examined, because despite extensive research, a worldwide homogenous data basis to the proportion of the religious affiliation could be found only for this year.

Furthermore, *Judaism* and *other religions* as well as *people without religious confession* are considered (see Sect. 5.3). Finally, the analysis of the empirical relationship between the state of global networking and the so-called *Religion Diversity Index* (RDI) for 2010, which is a measurement method for the degree of religious diversity in a country (see Sect. 5.4).

The reason for this segmentation of the investigation lies in the comparison of the quantitative size of the respective population with a certain worldview, and not in theological considerations. It should be mentioned that *Christianity, Islam,* and *Judaism* as the *monotheistic* religions, more precisely the *Abrahamic religions*, have commonalities which have always played a great role as a connecting element alongside conflicts.

5.1 Global Networking: Christianity and Islam 2010

Christianity and Islam are the two largest world religions or worldviews. As mentioned above, they belong to the monotheistic religions, also known as Abrahamic religions.[4] Christianity is based on the Holy Bible,[5] and the Islamic world is committed to the revelations of the Qur'an.[6] The most important reference for further investigations is the comprehensive study of *Pew-Templeton Global Religious Futures* [333] of the *Pew Research Center* in Washington, D.C. from 2010 for over 200 countries. It is available only for the year 2010 and points out that approximately 2.2 billion Christians live on earth, which account for about a third of the world population, and is proportionately the largest religious group. Approximately half of the Christians are *Catholics*, about 37% of all Christians profess *Protestantism*, which in the widest sense includes the Anglican Church community and independent or non-confessional churches. Christian *Orthodox*, to be found predominantly in the Greek and Russian Church, have a share of about 12% of all Christians. People who belong to a different tradition of the Christian faith, such as, the *Mormons* or *Jehovah's Witnesses*, account for only about one per cent.

The Christian population is widely dispersed throughout the globe. The left pie chart in Fig. 5.2 shows the proportionate distribution of the Christian population in different regions of the world. The largest share of the Christian population lives in North and South America. In both regions, they make up roughly 36%, with twice as many Christians living in South America than North America. About 26% of all Christians are living in Europe. A significant share of about 12% of the Christian population lives in Asia and only 1% live in the Middle East and North Africa. About 40% of the African population is Christian. The majority of African Christians live in

---

[4]Christianity, Judaism, and Islam all refer to the prophet Abraham as the father of their faith [244].

[5]The *Bible* is a collection of scriptures, with the basic distinction in *Old* and *New Testament*. It is not just a book, but a multitude of books as a unit. This is true for both the *Old* and the *New Testament*.

[6]The Qur'an is the sacred book of Islam which, over the course of 22 years, has been handed over the revelations of God (*"Allah"*) to the Prophet Muhammad by angel Gabriel.

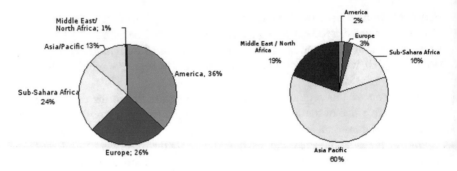

Fig. 5.2 Global distribution of Christian (left) and Muslim (right) population 2010

Eastern, Central and Southern Africa. The top 10 countries in which most Christians live are the *United States, Brazil, Mexico, Philippines, Nigeria, China, Germany, Ethiopia, Italy, and Great Britain.*

As of 2010, around 1.6 billion of the world population is *Muslim*, with the two main denominations of *Sunni* and *Shia*. Around 90% of all Muslims are Sunnis and 10% Shias. The right diagram in Fig. 5.2 illustrates the geographic distribution of the Islamic population in the different world regions. In 2010, 60% of all Muslims live in Asia. Thus, the Asian continent is home to almost two-thirds of Muslims. 19% of the Muslim population live in the Middle East and North Africa region, followed by the Sub–Sahara Africa region (16%). The lowest Muslim population share is in the Americas (3%) and Europe (4–5%). The Muslim population in Europe (28 countries in the EU, plus Norway and Switzerland) is estimated at about 26 million (5% of overall population) - up from 19.5 million (4%) in 2010. About 3.7 million Muslims came to Europe between 2010 and 2016 [340].

The thesis in this book states that convergence in the ICT sector is more rapid than, for instance in the field of economy, and it is much faster in terms of the convergence of income and wealth ratios. This also applies to the convergence in ICT usage in the large worldviews. In fact, the use of fixed telephone, mobile phone, Internet and mobile Internet is largely independent of the worldview. The same is the case in the usage of airplanes, cars, etc. The standard of living and education are decisive, not the worldview, as previously discussed in detail in Chap. 4.

Before analyzing the mobile phone and Internet as the central technologies of the digital age and its connection with the different worldviews, analysis of the fixed telephony and its interaction with the different worldviews is first carried out (analogous to Chap. 4).

5.1.1 Fixed Telephony

The empirical relationship between the fixed telephone rate (FTR) and the gross domestic product per inhabitant (GDPpC) in a country has been investigated in Chap. 4 (Sect. 4.1.1). The analysis results as follows: The higher the GDPpC a country has, the higher FTR can be assumed, whereupon in the area of the highest GDPpC the FTR is lower than in the medium GDPpC range. This is mainly due to the wealthy Gulf States, which owe their wealth to the abundant fossil energy sources. Considering that it can be expected that the FTR is higher in heavily Christianized states, i.e. in states with a high Christian population share compared to countries with a predominantly Muslim population, which is due to the high levels of wealth observed in Christian-dominant countries in comparison to predominantly Islamic countries.

This assumption is confirmed to a certain extent in the Fig. 5.3, in which the two diagrams illustrate how high the FTR is dependent on the share of the Christian (left) and Muslim (right) population in the overall population of a country in 2010. Countries are only considered, if they have at least a share of 1% of Christian or Muslim population.[7]

The dependent variable is the rate of persons using fixed telephone (FTR), the independent variable is the Christian (bottom) and Muslim (top) population share in a country. Both point clouds in the respective diagrams have a high variance, which at first sight suggests a rather small link between the FTR and the share of Christian or Islamic population. Furthermore, one can observe a condensation of the data in the marginal areas of the x–axis (0 and 100% edges).

The regression analysis results in following: The higher the share of the Christian population in a country, the more the FTR rises. The corresponding linear regression model, the coefficient of determination (R^2), and the correlation coefficient (R) are:

$$FTR = 17.001 + 0.0535 \cdot Share_{CHR}, R^2 = 0.0097, R = 0.098 \qquad (5.1)$$

In contrast, the FTR decreases the higher the proportion of the Muslim population in a country is. The linear regression model and the coefficient of determination (R^2) or the correlation coefficient (R) for the FTR as a function of the proportion of Muslims to the overall population are:

$$FTR = 27.198 - 0.211 \cdot Share_{ISL}, R^2 = 0.1758, R = 0.419 \qquad (5.2)$$

In essence, the observation is that, in 2010, the fixed telephone has a certain lead of users in the rather Christian countries, but there is generally no (significant) relation between the fixed telephone rate (FTR) and the share of Christian or Muslim population in a country. This can be seen numerically in the value of the coefficient of

---

[7]One reason for this definition lies in the data base that does not provide exact values for countries in which less than 1% of the population belongs to one or the other religion. With this 1% limit, the quality criterion accuracy and granularity is taken into account.

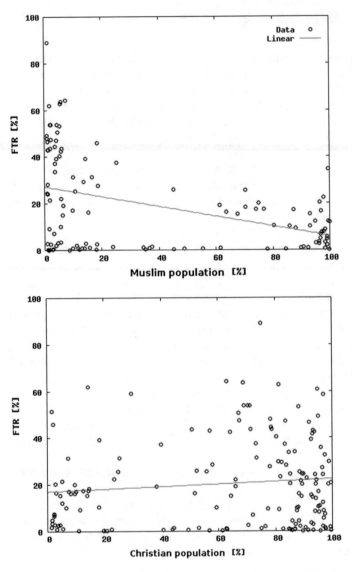

Fig. 5.3 Fixed telephone rate (FTR) as function of the Christian (bottom) and Muslim (top) population share in a country, year 2010

determination of $R_C^2 = 0.0097$ (Christian countries), where R is also the corresponding *Pearson*–correlation coefficient[8] that implies a rather small, almost not existing,

---

[8] According to Pearson, the correlation coefficient is a dimensionless measure of the strength of the linear relationship between two quantitative variables and is also referred to as the product–moment–correlation coefficient or the measure correlation coefficient.

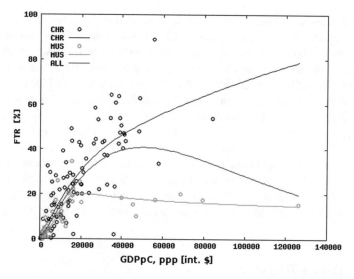

Fig. 5.4 Fixed telephone rate as function of the GDP per capita (GDPpC) for Christian (black) and Muslim (green) countries, year 2010

correlation. For Muslim countries, the coefficient of correlation is $R_M^2 = 0.18$, which implies a weak correlation, too. A R^2 or R of zero means that the regression line would exactly progress on a constant level of the mean value. In a concrete manner, this means that the global Christian population is (proportionately) more or less spread around the world, and thus has the world-wide average FTR in 2010 in this representation. To a certain extent this is also valid for the Islamic population.

In the following we will take a closer look at the reasons for the slightly higher mean value of FTR of Christian countries in comparison to Islamic countries. In Chap. 4 it was ascertained that the economic performance of a country is generally crucial for the (broad) use of technologies, which is why the FTR in the Christian and Islamic countries is now being analyzed as a function of the GDP per capita (GDPpC). Figure 5.4 illustrates this relation, in which the Christian countries (black), the Islamic countries (green) and the corresponding function with the best fit to each religious group (i.e. entire world population) are given.[9] The curve for the FTR in a global view (blue) was determined in Chap. 4, Sect. 4.1.1 and serves as a reference curve. One can see that the curve for the best adaptation of the Christian countries (black) is above the curve for the best adjustment of the Islamic countries (green). This confirms the result illustrated in Fig. 5.3. A further point is that the curve for the Christian countries (black) also runs above the curve for the whole world (blue), i.e. the fixed network infrastructure is better developed in more Christian countries

---

[9]The difficulty lies in a precise definition of when a country is rather Christian or rather Islamic. Due to the fact that these two worldviews are found, even if only slightly, in many countries, a 40% limit was chosen here. This means that countries whose population consists of at least 40% Christian or Islamic people are considered.

compared to the worldwide fixed network infrastructure. The curve for the Islamic countries (green) runs below the curve for the whole world. The reason for the (almost) horizontal curve from a value beyond 20,000 Intl\$. lies in the high level of wealth of the oil states that have a low FTR at the same time.[10] It is important to note that there are more Christian countries in the world than Islamic, which is why the curve for the whole world (blue) is dominated by the Christian countries (black). With regard to the present work, the result should be assessed as follows: fixed telephony is used in many Christian and Islamic countries. There is no clear, but rather small, link between the share of a country's Christian and Islamic population size and the use of fixed telephone in one country. The difference between Christian and Islamic countries lies in the economic performance. On average this is higher for rather Christian countries, which is why the fixed telephone network infrastructure is better developed and consequently more people have access to the fixed network.

5.1.2 Mobile Telephony

The result for the empirical relationship between the mobile phone rate (MPR) and the proportion of Christians or Muslims of a country's population is somewhat different from the result for the fixed telephone rate (FTR). From the communication point of view, the mobile phone is the ultimate device, which is why it is used by almost everyone. This applies to both the Christian and the Muslim populations.

Figure 5.5 illustrates the MPR as a dependent variable, i.e. the variable to be explained, and the Christian (bottom) or Muslim (top) proportion of a country as an independent variable, that is the explanatory one. The result is very simple to formulate. Whether the percentage of Christian or Muslim population in a country is high or low, the MPR has almost always a constant level of 100% in 2010, whereby, as expected, the straight line for the Christian population is on average slightly higher than the regression line for the Islamic population. The linear regression analysis gives the following equations and the coefficient of determination (R^2) or the correlation coefficient (R) for the MPR as a function of the proportion of Christian population in a country:

$$MPR = 100.07 - 0.1337 \cdot Share_{CHR}, R^2 = 0.0119, R = 0.109 \qquad (5.3)$$

The linear relationship between the MPR as a function of the proportion of Islamic population in a country is:

---

[10]It should be noted that the fixed landline telephony in the oil states is largely replaced by the mobile telephone, but in sum the number of connections to the fixed network and the mobile telephone are constantly growing. This situation and relationship to the use of technology has already been discussed several times in Chap. 4.

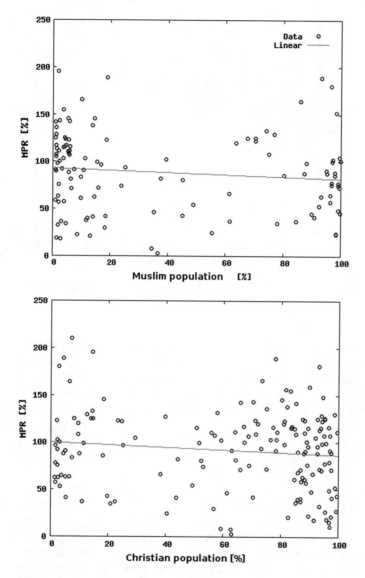

Fig. 5.5 Mobile phone rate (MPR) as function of the Christian (bottom) and Muslim (top) population share in a country, year 2010

$$MPR = 93.016 - 0.1127 \cdot Share_{ISL}, R^2 = 0.0111, R = 0.105 \qquad (5.4)$$

Consequently, the core observation is that there is hardly any link between the MPR and the share of a Christian and Islamic population in a country. This result can be checked against the previous Sect. 5.1.1 by looking at the existing relationship in the context of the economic performance.

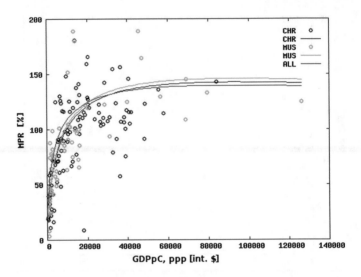

Fig. 5.6 Mobile phone rate (MPR) as function of the GDP per capita (GDPpC) for Christian (black) and Muslim (green) countries, year 2010

Looking at the result for the empirical relationship between the GDP per capita (GDPpC) and the MPR from Chap. 4, Sect. 4.1.2, the level of the GDPpC has a positive influence on the MPR. However, since the GDPpC in 2010 is no longer as important for the MPR as it was in 2000, there is hardly any relation between the MPR and the share of the Christian or Islamic population in a country. Since the mobile phone is now spread all over the world, this result of the regression analysis is plausible, that means a low correlation between the proportion of the Christian or Islamic population and the MPR. The mobile phone is a very useful and meanwhile so cheap device that it can be used by almost everyone.

Figure 5.6 illustrates the GDPpC as an explanatory variable and the MPR as a variable to be explained, with more Christian countries (black) and the more Islamic countries (green) being seen as data points. Figure 5.6 confirms the finding that the economic performance is not a decisive factor in relation to religious affiliation in 2010, which affects the MPR in a country. The curve with the best fit for the whole world (blue) is very similar to the curves for the Christian countries (black) and the Islamic countries (green), which are (not surprisingly) very similar to each other. In view of this book, the result should be assessed as follows: The mobile phone is used in all Christian and Islamic countries. There is no clear relation between a country's share of Christian or Islamic population size and mobile phone use. The difference between Christian and Islamic countries can be explained by the level of prosperity. However, since the level of prosperity or economic capacity have also (almost) no influence on the mobile phone usage rate of a country in 2010, it is no longer significant that the level of prosperity in more Christian countries is higher on average than in Islamic countries.

5.1.3 Internet

The Internet is a US invention and the majority of US citizens are of Christian faith. However, the Internet has spread throughout the world, albeit at different levels depending on the investigated countries and their level of economic performance and education (see Chap. 4). People of different worldviews use it for their own purposes. Looking at Fig. 5.7 and the percentage of Christian (bottom) or Muslim (top) populations as an independent variable and the Internet penetration rate (IPR) as a dependent variable, the result is that the empirical relation between the share of Christian population and the IPR remains (almost) at a constant level of 36% (= global IPR). This means that the linear fitting function is horizontally identical with the global average IPR value. The regression equations for the Internet penetration rate (IPR) as a function of the proportion of the Christian population in a country, the coefficient of determination (R^2), and the correlation coefficient (R) are:

$$IPR = 35.143 - 0.0038 \cdot Share_{CHR}, R^2 = 0.0002, R = 0.014 \qquad (5.5)$$

In essence, the finding here is that Internet use is widespread in the more Christian countries and there is hardly any connection between the IPR and the share of the Christian population in a country in 2010. One reason why the regression line is horizontal on the average of the global IPR is the wide spread of the Christian population in the world.

The regression equations for the Internet penetration rate (IPR) as a function of the share of the Islamic population of a country, the coefficient of determination (R^2), and the correlation coefficient (R) are:

$$IPR = 42.101 - 0.2351 \cdot Share_{ISL}, R^2 = 0.1019, R = -0.319 \qquad (5.6)$$

The data above show the result for the relation between the proportion of the Muslim population of a country and the Internet penetration rate (IPR). This relation is slightly different from the case of Christianity. The higher the proportion of Muslim population, the slightly lower the IPR. However, the coefficient of determination (R^2) and the correlation coefficient (R) are also small, which is why this relationship is at a rather low level.

In summary, it can be said that the IPR is somewhat higher in Christian countries compared to more Islamic countries as of 2010. How can this result be justified? Fig. 5.8 confirms the result for the more Christian and Islamic countries from the perspective of the level of economic performance, which is higher in the Christian countries on average than in the Islamic countries. It shows the IPR as a function of the GDPpC, whereby for the Christian countries (black) and the Islamic countries (green) the respective best fitting function are represented. It can be seen that the curve with the best adaptation for the Christian countries is above the curve with the best fitting for the Islamic countries. This again confirms that IPR is primarily dependent on GDPpC. It is interesting in this context that the curve for Christian

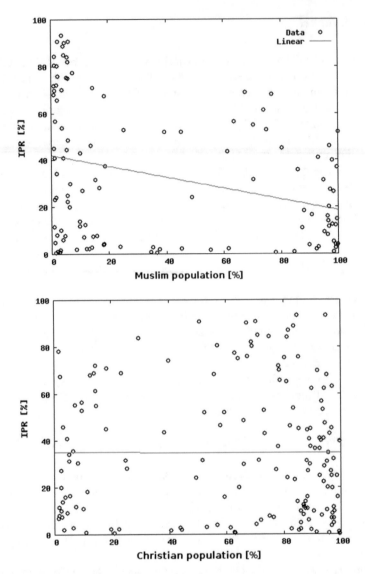

Fig. 5.7 Internet penetration rate (IPR) as function of the Christian (bottom) and Muslim (top) population share in a country, year 2010

countries (black) also runs above the curve for the whole world (blue). This means that Internet use is more widespread in predominantly Christian countries than in the whole world.

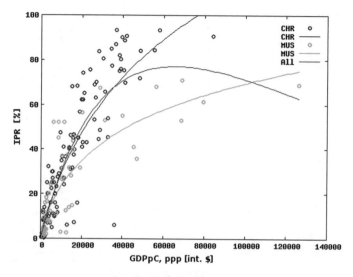

Fig. 5.8 Internet penetration rate (IPR) as function of the GDP per capita (GDPpC) for Christian (black) and Islamic (green) countries, year 2010

5.1.4 Mobile Internet

Mobile Internet access is often referred to as mobile broadband access. The mobile broadband user rate (MBR) is shown in Fig. 5.9 as a function of the proportion of Christian population in a country (bottom). At first glance, there is no relation between these two variables. A linear relationship is assumed again and a regression analysis is carried out. This results in following equation, the coefficient of determination R^2 and the correlation coefficient R for the mobile Internet penetration rate (MBR) as a function of the proportion of Christian Population in a country:

$$MBR = 25.841 + 0.0199 \cdot Share_{CHR}, R^2 = 0.0007, R = 0.026 \qquad (5.7)$$

The regression line in Fig. 5.9 (bottom) shows that there is no significant relation between the MBR and the proportions of Christians of a country's population. A correlation coefficient of almost zero confirms this result.

The linear relationship between the mobile broadband user rate (MBR) as a function of the proportion of Muslim population in a country is:

$$MBR = 29.961 - 0.2157 \cdot Share_{MUS}, R^2 = 0.0599, R = -0.245 \qquad (5.8)$$

The regression line in Fig. 5.9 (top) shows that there is a (slightly) negative relation between the MBR and the share of Islamic population in a country. This results in the following outcome: The global Christian population is somewhat more advanced in mobile Internet use than the global Muslim population. While higher levels of

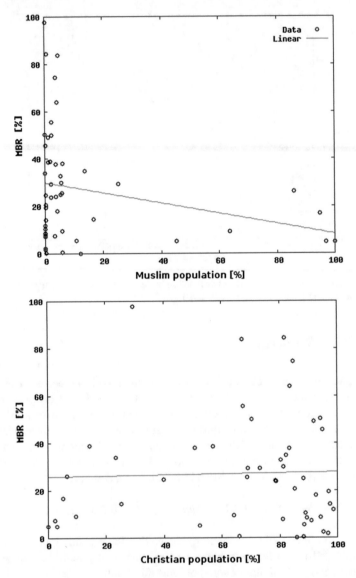

Fig. 5.9 Mobile Internet penetration rate (MBR) as function of the Christian (bottom) and Muslim (top) population share of a country, year 2010

the proportion of Christian population in a country are also linked with an (slightly) increasing mobile broadband user rate (MBR), the MBR is decreasing for higher levels of Muslims within a country's population. These tendencies are likely to be leveled up with the further spread of mobile Internet technology in the near future, as the conditions in general mobile phone use have already adapted. This will be the case due to the technological convergence of the mobile phone and the Internet to the mobile Internet.

5.1.5 Economic Performance

The analysis of access rates to the fixed telephone network, mobile phone, the Internet and the mobile Internet of the world's Christian and Islamic peoples has led to results that are plausibly explained with the economic performance of rather Christian and Islamic countries. Therefore we can also consider a clustering of the Christian and Islamic countries with a view to their prosperity level. The results described above are made plausible with this clustering in a further step. Subgroups are of particular relevance for the specific religious group. Figure 5.10 shows the economic performance (GDPpC) as a dependent variable and the share of a country's Christian (bottom) and Islamic population (top) as an independent variable. In a rough approximation, there are two levels of prosperity, which can be found within the worldwide Christian population or the Christian countries. One group consists of rich countries like the *USA, United Kingdom, Canada, Australia, New Zealand*, which are also closely linked culturally. Additionally this group comprises countries like *Germany, France, Austria, Belgium, Luxembourg, Netherlands, Italy, Sweden, Finland, Norway* and *Denmark*, which are culturally influenced by a common European idea. Within this first group are small as well as large states. What is common to all countries within this group is the fact that the proportion of Christian population on the one hand is not less than 50% and, on the other hand, not greater than 85%. There is room for diversity, openness and dynamism.

The second large group consists primarily of countries in Africa and Latin America. Among the African states, which have a moderate proportion of Christian populations (between 40 and 70%) and at the same time are very poor, are for example *Chad, Togo, Ivory Coast, Nigeria, Benin, Mozambique, Tanzania, Eritrea, Ethiopia* and *Cameroon*. There are also some countries in Africa with a relatively higher percentage of Christian population (>70%) who are also relatively poor. This group includes, for example *Congo, Liberia, Burundi, Zimbabwe, Rwanda, Kenya, Zambia* and *Ghana*.

The Latin American states are not as poor as the Africans on average, but they also belong to the poorer countries of the world. The share of the Christian population of states in Latin America is more than 80%. These countries include *Brazil, Argentina, Chile, Venezuela, Mexico, Colombia, Ecuador* and *Peru*.

Furthermore, states like *Russia, Belarus, Hungary* on the one, and *Spain, Greece, Portugal* on the other hand can be divided into two other small groups, which are neither rich nor poor. In the first small group (culturally Russian shaped) the proportion of Christians in the population is between 70 and 80%. In the second group (culturally European), this proportion has values between about 75 and 95%.

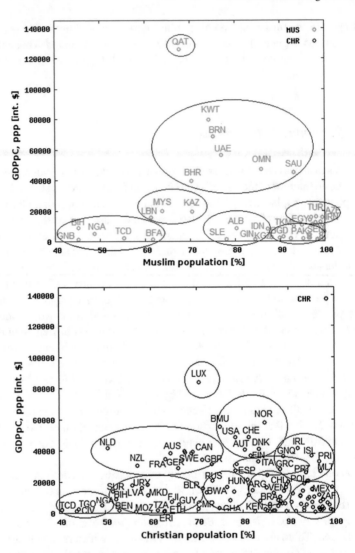

Fig. 5.10 GDP per capita (GDPpC) as function the Christian (bottom) and Muslim (top) population share in a country, year 2010

The study of the economic performance in predominantly Christian countries can be summarized as follows: There are very few countries (*Ireland, Equatorial Guinea* and *Iceland*) whose share in Christian population are very high (about 90% and more) and are rich at the same time, i.e. a *GDPpC* > 30, 000 (Intl$.). These countries are small. Within the group of rather rich states the share of the Christian population is not clearly extra high (<85%), but it is significant (>50%). This result is plausible from two perspectives. On the one hand, a low level of religious diversity can constrain the economic and technical dynamics of states. In a certain sense there is no modern state in such a situation. The factual reasoning is also that wealth attracts people from all cultures, even to attract a variety of talents and a stable population size with the help of migration processes, because the birth rate in rich countries is on average lower than in the poorer countries.[11]

The grouping of predominantly Islamic countries in Fig. 5.10 is somewhat simpler than the clustering of predominantly Christian countries. A rough classification into two main groups leads to the following cluster structure: In the first, rich, group of countries are the Gulf States, which have a high prosperity level, i.e. a $GDPpC >$ $40,000$(Intl\$.). These include *Bahrain, the United Arab Emirates, Oman, Saudi Arabia, Kuwait* and *Qatar*, the latter state having the highest GDPpC worldwide and therefore in Fig. 5.10 shown separately. These countries are small where their share of the Islamic population is high (about 70–90%), but is not completely dominant. These countries attract workers mainly from *India* or *Pakistan* and need this in the light of their political and demographic situation. The second larges group in the cluster of Islamic states has a GDPpC significantly below that of the Gulf States. If this group is subdivided into further subgroups, the following result: Especially the African states are very poor. These include both large and small states, with moderate to high levels of Islamic population such as *Niger, Burkina Faso, Gambia, Chad, Senegal, Nigeria* and *Sudan*. Countries in North Africa and the Middle East such as *Egypt, Turkey, Morocco, Tunisia, Algeria* and *Iran* have a very high number of Islamic populations and are at a moderate prosperity level. The Asian countries *Bangladesh, Pakistan, Afghanistan, Tajikistan, Kyrgyzstan, Turkmenistan, Uzbekistan* and *Indonesia* also have a high share of Islamic population, but tend to be among the poorer states in 2010.

5.1.6 Intermediate Result

The various investigations regarding the relationship between the proportion of Christian and Islamic populations in a country and the corresponding access rates such as fixed telephone rate (FTR), mobile phone rate (MPR), Internet penetration rate (IPR) and mobile Internet penetration rate (MBR) tend to lead to the same results. Firstly, this means that there is no decisive relation between the share of Christian and (only) a weak link between the share of Islamic population in a country and the corresponding access rates to ICT. The FTR is slightly higher for a higher proportion of Christian populations in a country, while the FTR slightly decreases for a higher proportion of Islamic population in a country. The result of the analyses of the relationship between the MPR with respect to these two religious groups is clear. There is no relation between the share of Christian or Islamic population in a country

---

[11]The reasons for this are primarily the role of women in the economically strong countries, such as, the longer education path and the embedding in the formalized work processes. In addition, in the poorer countries, many children are a kind of social or pension insurance.

and the MPR. The mobile phone is used by almost everyone in 2010. The analyses for the IPR and the MBR for the global Christian and Islamic population are essentially based on the same results. The IPR is somewhat higher in the predominantly Christian countries compared to predominantly Islamic countries. The correlation between the proportion of Christian populations in a country and the IPR or MBR is almost zero. For the Islamic countries, the correlation coefficient is at a negative level near zero. Overall, it can be said that Christian communities in the respective access rates are at a somewhat higher level than the Islamic communities. The reason for this is probably the (slightly higher) level of prosperity in average in Christian countries compared to Islamic countries, as has already been mentioned and shown.

5.2 Global Networking: Buddhism and Hinduism 2010

Buddhism and *Hinduism* are the two largest religions after Christianity and Islam. Buddhists live according to the teachings and tradition of Siddharta Gautama (*Buddha*), who lived in Northern India at the beginning of the 4th century. According to the teaching of *Buddha*, creatures that are not enlightened are trapped in an endless cycle of birth, life, death and rebirth until enlightenment allows the transition to the state of *Nirvana*. If the development or use of ICT would be regarded as suffering, Buddhists had to avoid the use of these technologies. But what is the empirical correlation in reality?

In 2010 there are about 488 million Buddhists in the world which makes up almost 7% of the world's population. The overwhelming majority of Buddhists of about 99% live in Asia, which is why a corresponding graphic for the worldwide distribution is renounced here. The only two other regions with more than 1 million Buddhists are Europe (1.3 million) and North America (3.9 million). In seven countries – *Cambodia, Thailand, Burma (Myanmar), Bhutan, Sri Lanka, Laos and Mongolia* – Buddhism represents the majority of the population (see Fig. 5.1).

With about 1 billion people of Hindu faith, there are more than twice as many Hindus as Buddhists. These form the third largest religious group in the world. Hindus, like Buddhists, are geographically not widespread (see Fig. 5.1) and are concentrated in Asia, where they form a majority in four states: *Bali* (90%), *Nepal* (81%), *India* (80%) and *Mauritius* (56%) [333]. At the same time, almost 97% of all Hindus live in these four countries. In *Bhutan* (25%), *Bangladesh* (12%), *Indonesia* (2%) and *Pakistan* (1–2%) Hindus live as a minority. In the religious sciences one finds the terms of Hindu tradition or Hindu religions, since the Hinduism is based on a very long tradition (until approx. 1750 BC) and comprises several structures of religious traditions. Hinduism today marks the whole complex of beliefs and institutions that have developed since the writing of the ancient (sacred) scriptures, the Vedas. Unified within Hinduism is the recognition of the Vedas as the source of revelation, the caste system, and the faith in life, rebirth and liberation.

5.2.1 Fixed Telephony

The worldwide Buddhist and Hindu populations show a similar development in the rate of persons having a fixed telephone connection (FTR), which can be seen in Fig. 5.11. The FTR is the dependent variable, the proportion of Buddhist (bottom) and Hindu (top) in a country's population is the independent variable. The FTR of the worldwide population of Buddhists is 24.8% on average. There are six countries (*Cambodia, Thailand, Bhutan, Sri Lanka, Laos* and *Mongolia*) in which more than 50% of the population are Buddhists. The countries in which the proportion of the Buddhist population is less than 40% and that lie above the regression line (see Fig. 5.11) are *Hong Kong, Korea, Japan, USA, Australia, New Zealand* and *Singapore*. The percentage of Buddhists in the US, Australia and New Zealand is less than 3%. Among these countries, *Japan* and *Singapore* have the largest share of Buddhists with 36 and 34%, respectively. A total of 10 out of 23 countries are located above the regression line therefore another 13 states are below the regression line. The largest Hindu communities live in *Nepal* (80.7%), *India* (79.5%) and *Mauritius* (56.4%).

The regression analysis shows the following: The higher the share of Buddhists, the lower the fixed telephone rate (FTR) of a country. The equation of the regression line and the coefficient of determination (R^2) are as follows, where the correlation coefficient (R) has a negative value and indicates a mean correlation. The regression line and the correlation of determination for the link between the Buddhist share and the FTR are:

$$FTR = 32.74 - 0.28 * Share_{BUD}, R^2 = 0.205, R = -0.452 \qquad (5.9)$$

The regression line and the correlation of determination for the link between the Hindu share and the FTR are:

$$FTR = 24.46 - 0.24 * Share_{HIN}, R^2 = 0.123, R = -0.351 \qquad (5.10)$$

There are 8 out of a total of 26 states above and therefore 18 states below the regression line. The four countries in which Hindus have the highest FTR on average are *Canada* (53.9%), *USA* (53.8%), *Australia* (47.4%) and *New Zealand* (43.1%). Buddhists make up 1.4% of *Canada's* population, 1.3% in *USA*, 1.4% in *Australia* and 2.1% in *New Zealand*. The comparison between the Buddhist and the Hindu world populations and fixed telephone rate (FTR) in the respective countries comes to the conclusion that the higher the share of one of the two religious groups, the lower is the FTR.

In the following, we will look at reasons why the FTR in predominantly Buddhist and Hindu countries decreases the higher the population share of the two religious groups in a country is. Again, an explanation from the view of the economic performance is sought, which is why the FTR in the Buddhist and Hindu states is examined as a function of GDP per capita (GDPpC). Figure 5.12 illustrates this correlation, in

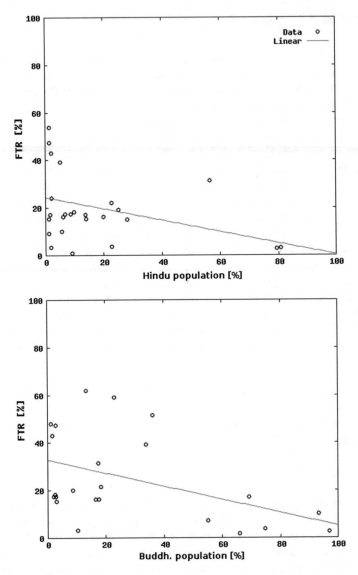

Fig. 5.11 Fixed telephone rate (FTR) as function of the Buddhist (bottom) and Hindu (top) population share in a country, year 2010

which the Hindu countries (green) and the Buddhist countries (black) and the corresponding function with the best fit to each religious group are shown.[12]

---

[12]As with the definition of rather Christian and Islamic countries above, there is also a difficulty in defining exactly when a country should be described as rather Buddhist or rather Hindu. Due to the fact that these two worldviews are found with a noticeable share (only) in a few countries,

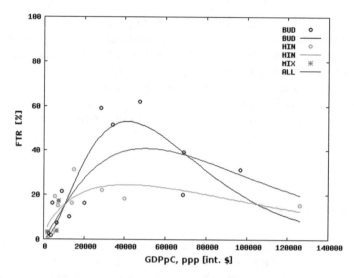

Fig. 5.12 Fixed telephone rate (FTR) as function of the GDP per capita (GDPpC) for Buddhist (black) and Hindu (green) countries, year 2010

Firstly, it should be noted that the results of the optimization should be taken cautiously, since the number of data points is small. The following results show: the curve for the FTR in a worldwide context (blue) was determined in Chap. 4, Sect. 4.1.1 and is used as reference curve. One can see that the curve with the best fit to the Buddhist countries (black) progresses above the reference curve until approximately 70,000 Intl$. This is partly an artifact of the optimization process. This is mainly due to countries such as *Korea, Japan, Singapore* and *Hong Kong* all having a (significantly) higher FTR than the global FTR and therefore pull the curve upwards. Countries such as *China, Brunei Darussalam, Sri Lanka, Malaysia* and *Vietnam* are in the middle of the global FTR. *Thailand, Mongolia, Bhutan, Nepal, Cambodia* and *Laos* have an FTR that is well below the global FTR. The last group of states also have the highest proportion of Buddhists in their country.

The curve with the best adaptation to the Hindu countries (green) progresses essentially below the reference curve (blue) for the world. One reason for this is that *Qatar* and *Bahrain* both have very a very high GDP per capita, but comparatively low FTR. In *Qatar* about 14% and in *Bahrain* 10% of the population are Hindus who live there as migrants. Many Hindus also live in the rest of the Gulf States, although the proportion of Hindu migrants is less than 10% and is therefore not included here. Another reason for the low curve of the more Hindu states (green) is the very low FTR in countries like *Bangladesh, India, Nepal* and *Bhutan*.

---

a 10% limit was chosen. That means, all countries whose population is comprised of at least 10% Buddhists or Hindus are considered. As reminder: Among Christians and Muslims, we had set this limit to 40%. An equal value for the Buddhists and Hindus would include too few countries.

In the context of this book, this result should be assessed as follows: There is a negative relation between a country's share of Buddhist or Hindu population and the use of fixed telephone connections, i.e. the FTR decreases with an increasing share of one or both religious groups. This is more true for Hindu countries than for Buddhist countries. The reason for this difference lies in the level of wealth. On average it is lower in the rather Hindu countries compared to the Buddhist countries. This is probably a major reason why the fixed network infrastructure in Hindu countries is, on average, less developed than in Buddhist countries and consequently fewer people have access to the fixed network.

5.2.2   Mobile Telephony

The result for the relation between the mobile phone rate (MPR) and the Buddhist and Hindu population share in a country is, to a certain extent, comparable with the result for the fixed telephone network. Figure 5.13 (bottom) illustrates following: the higher the Buddhist population share in a country, the lower the MPR. The same conclusion can be drawn for the Hindu population. The result for the Buddhists and their connection with the MPR is however only conditionally comparable with the result for the fixed telephone rate. This is because the average MPR with 107% and the minimum FTR with 34.3% in *Nepal* is considerably greater than the corresponding values for the FTR. Within the Hindu community the average MPR value is 103.2%. The minimum value with 34.3% is found again in *Nepal*. The regression line and the coefficient of determination (R^2) for the relation between the Buddhist population share and the MPR are:

$$MPR = 123.15 - 0.5378 \cdot Share_{BUD}, R^2 = 0.1619, R = -0.402 \qquad (5.11)$$

For the Hindu population the regression line and the coefficient of determination (R^2) are as follows:

$$MPR = 116.18 - 0.7936 \cdot Share_{HIN}, R^2 = 0.2232, R = -0.472 \qquad (5.12)$$

In summary, one can say that the mobile phone connection has spread widely within the Buddhist as well as the Hindu community. However, there is still a tendency to see that the MPR is lower if the country's share of Buddhist or Hindu populations is higher. Another result is that the correlation between the proportion of Buddhist population and the MPR is less than the correlation between the proportion of Hindu population and the MPR. However, the correlation between the proportion of Buddhist population and the FTR is higher than the correlation between the proportion of Hindu population and the FTR.

What can be a reason for this? The mobile phone has spread over the whole world, because it is a very useful device and the access to the mobile network has

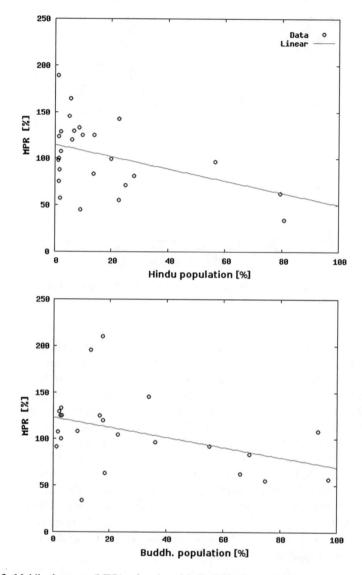

Fig. 5.13 Mobile phone rate (MPR) as function of the Buddhist (bottom) and Hindu (top) population share in a country, year 2010

become inexpensive. Looking at the MPR in relation to the economic performance as measured by the GDP per capita (GDPpC) in the Buddhist and Hindu countries, the results described above become clear. Figure 5.14 illustrates the MPR as variable to be explained and the GDPpC as an explaining variable, with the data points of the predominantly Buddhist (black) and predominantly Hindu (green) countries. At the same time the best adaptation curve for the two religious groups is shown in the respective color. In addition, the curve for the entire world (blue) is again used as a

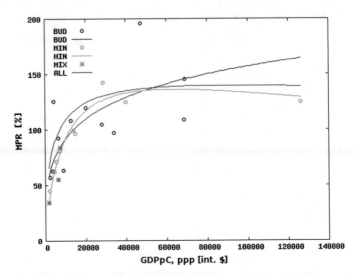

Fig. 5.14 Mobile phone rate (MPR) as function of the GDP per capita (GDPpC) for Buddhist (black) and Hindu (green) countries, year 2010

reference curve. The key findings are the following: The mobile phone is used in all Buddhist and Hindu countries. There is a slight negative link between a country's share of Buddhist or Hindu populations and the MPR. Both the best adaptation curve for the Buddhist (black) and Hindu (green) population are slightly below the reference curve (blue) and progress very similar. A single country (*Hong Kong*) with a MPR of nearly 200% pulls the curve for the Buddhist countries in the direction of larger GDPpC. Between these two religious groups, there is a difference in the average level of wealth. As mentioned in the previous Sect. 5.2.1, it is higher in predominantly Buddhist states compared to Hindu ones. In Chap. 4, Sect. 4.1.2 it was found that the GDP per capita is not, or is only slightly related to the MPR, even though the trend remains that MPRs are rising with rising GDP per capita.[13] Therefore, the curves of both religions are similar to each other and to the reference curve for the whole world. For this reason, it does not matter in the context that the Buddhists are slightly richer than the Hindus.

5.2.3 Internet

Looking at a country's Buddhist population as an independent variable and the Internet penetration rate (IPR) as a dependent variable (see Fig. 5.15) results in the

---

[13] As reminder: this effect is primarily due to the fact that the MPR is far above the 100% limit in many countries.

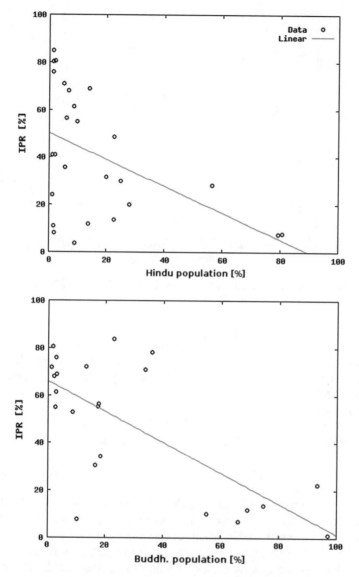

Fig. 5.15 Internet penetration rate (IPR) as function of the Buddhist (bottom) and Hindu (top) population share in a country, year 2010

following regression line and the following coefficient of determination (R^2) for the relationship between the proportion of Buddhist population and the IPR:

$$IPR = 66.11 - 0.646 \cdot Share_{BUD}, R^2 = 0.507, R = -0.712 \qquad (5.13)$$

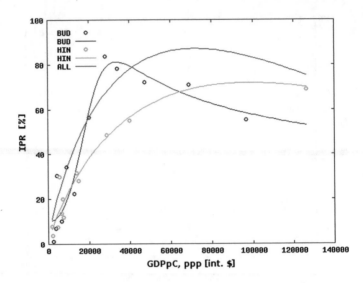

Fig. 5.16 Internet penetration rate (IPR) as function of the GDP per Capita (GDPpC) for Buddhist (black) and Hindu (green) countries, year 2010

This yields to the following: The higher the proportion of Buddhist populations in a country is, the lower the IPR in this country. The correlation coefficient R indicates a mean relationship. For the link between the Hindu population share and the IPR (see Fig. 5.15 right) results in the following regression line and the coefficient of determination (R^2):

$$IPR = 50.22 - 0.561 \cdot Share_{HIN}, R^2 = 0.222, R = -0.471 \qquad (5.14)$$

The higher the Hindu population share in a country, the lower its IPR. There is a correlation on a medium level.

In summary, both the Hindu and Buddhist communities are behind the Christian and Islamic countries with regard to Internet usage. Overall, the IPR in 2010 is slightly higher in most Buddhist countries compared to Hindu ones. How can this be explained? Fig. 5.16 confirms the outcome for the rather Buddhist and Hindu countries from the perspective of the level of wealth, which is slightly higher in Buddhist countries compared to the Hindu countries. It shows the IPR as a function of the GDP per capita, whereby for the Buddhist countries (black) and the Hindu countries (green) the data points as well as the respective best adaptation function are represented. Up to a value of about 65,000 (Intl$.) the curve with the best fit to the Buddhist countries progresses above the curve with the best fit to the Hindu countries. This again confirms that the IPR depends primarily on GDP per capita, because the average GDP per capita or the level of wealth is generally higher in more Buddhist than Hindu countries. It is interesting in this context that the curve for the Hindu countries (green) is below the reference curve for the whole world (blue). This means that Internet use is less common in the more Hindu countries in comparison with the rest of the world. Over a large range of GDP per capita the same applies to the Buddhist countries. Their curve (black) progresses below the reference curve (blue). Two countries within the Buddhist group form the exceptions here. Relative to their GDP per capita, *Korea* and *Japan* are better positioned than the rest of the world and are therefore above the reference curve. *Hong Kong* and *Singapore* are at a high level of IPR, although the IPR should be even higher if regarded in correlation with the GDP per capita. *China* and *Malaysia* lie exactly on the reference curve. Within the group of the Hindu countries, only *Guayana* is relative to its GDP per capita better than the reference curve.

Within these two religious groups, it is worth taking a separate look at the IPR in *China* and *India* which have by far the largest population in the world. As of 2010, about 1.35 billion people are living in *China*. *China* has an IPR of about 34%, which is roughly equal to the world average, and is therefore not surprisingly, in Fig. 5.16, on the reference curve.[14] As a result, *China* has the largest Internet population in 2010, with approximately 460 million Internet users.[15] As of 2010, about 1.3 billion people are living in *India*. With an IPR of almost 8%, the number of Internet users is thus just over 100 million people.[16] This makes *India* far behind *China*, which is due to the lower economic performance of *India*.[17]

As of 2010, in terms of Internet usage or IPR, the Buddhist countries are more progressed than the Hindu countries on average. This results mainly from the slightly higher prosperity level in the rather Buddhist countries than in the rather Hindu countries.

5.2.4 Mobile Internet

Figure 5.17 (bottom) illustrates the rate of persons using mobile Internet as a function of the Buddhist population share in a country. Compared to Christianity and Islam (see Sect. 5.1.4) there are less data points. At first glance, there is a minor negative correlation between both variables. Performing a linear regression analysis yields to following equation and a coefficient of correlation (R^2) or a correlation coefficient (R)

---

[14]In 2014, the IPR is about 46% in *China*.

[15]As reminder: Despite having the largest number of Internet users, *China* is still underproportion- ally represented when regarding this number in the context of the large population size (see Chap. 4, Sect. 4.1.3).

[16]In 2014 *India* has an IPR of about 20%.

[17]Note: The reasons for the lower economic performance and economic growth of *India* compared to *China* are historically also due to the fact that *India* has long been under the colonial rule of the British Empire. Politically, the reasons could also be that India has a democratic form of government and has longer and more time-oriented decision-making structures compared to China. With regard to religion, the Indian people are more religious than the Chinese. As described at the beginning of this chapter, this can lead people toward their future aspirations (after death), which can ultimately lead to the fact that the economic and technical dynamics are less pronounced.

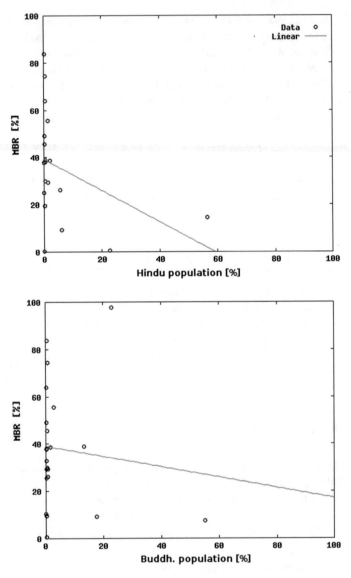

Fig. 5.17 Mobile Internet penetration rate (MBR) as function of the Buddhist (bottom) and Hindu (top) population share in a country, year 2010

for the mobile Internet penetration rate (MPR) as function of the Buddhist population share in a country.

$$MBR = 39.093 - 0.2194 \cdot Share_{BUD}, \, R^2 = 0.0123, \, R = -0.111 \qquad (5.15)$$

The linear correlation between the mobile Internet penetration rate as a function of the Hindu population share in a country is:

$$MBR = 39.239 - 0.6618 \cdot Share_{HIN}, R^2 = 0.1504, R = -0.389 \qquad (5.16)$$

Both the regression line and the coefficient of correlation indicate a medium negative correlation between the mobile Internet penetration rate and the Buddhist or Hindu population share in a country. The difference between these two religious groups lie in the regression line for the Hindu population in a country, which has a steeper negative rise compared to the one for the Buddhist population. The flat incline of the Buddhist line is a result of the adaption function, since the country, in this case *Korea*, has a share of almost 23% and a very high MBR of 97.7% and therefore is a positive outlier. The correlation coefficient $R = -0.111$ for the Buddhist religion is on a very low level. For the Hindu religion lies the correlation coefficient $R = -0.389$ on a medium level. As key observation, we should note that both religion have a minor negative correlation with the MBR. Summarized: the higher the share of Buddhist or Hindu population in country is, the lower is the MBR tendentially.

5.2.5   Economic Performance

The examination of the access rates to fixed telephone, mobile phone, Internet and mobile Internet in Buddhist or Hindu countries leads to results that can be plausible explained with the economic performance of these countries. For this reason we will have a closer look on a cluster of Buddhist and Hindu countries with focus on their economic performance. With this clustering, the former results can be additionally verified. It also gives a good impression of which subgroups have a special relevance to each religious group.

Figure 5.10 shows the GDP per capita (GDPpC) as depending and the Buddhist (bottom) or Hindu (top) population share of a country as independent variable. The arranged grouping of the Buddhist countries is following: The very rich (small) countries *Macao, Singapore* and *Hong Kong* ($GDPpercapita > 40,000$Intl\$.) have a minor Buddhist population share ($<40\%$). Countries like *Korea, Malaysia* and *Japan* have a medium to high economic performance ($20,000 < GDPpercapita < 40,000$Intl\$.) and furthermore a minor Buddhist population share ($<40\%$). Countries such as *China, Vietnam* and *Nepal* are poorer ($GDPpercapita < 10,000$Intl\$.) and have a very small Buddhist population share ($<20\%$). Countries with a medium to high Buddhist population share (50–80%) are all poor ($GDPpercapita < 10.000$Intl\$.). These countries consist of *Mongolia, Laos, Sri Lanka* and *Bhutan*. There are only two countries with a very high Buddhist population share ($>90\%$), which also belong to the poor countries ($GDPpercapita < 10.000$Intl\$.). There is no country in world where a medium to very high Buddhist population share comes with a high economic performance.

The following group arrangement can be made for the Hindu countries: analoge to the Buddhist countries there is no single country that has a high Hindu population share and a high economic performance. *India* and *Nepal* have by far the highest Buddhist population share (>80%) but also belong to the group of the poor countries (*GDPpercapita* < 10, 000Intl$.). *Mauritius* is the only country (although globally only of minor relevance) with a medium Buddhist population share (50–60%) and a medium level of wealth (10, 000 < *GDPpercapita* < 20, 000Intl$.). The remaining countries can be arranged in a single large group with a rather small Hindu population share (<30%). Within this group there are very small resource rich countries like Qatar, Bahrain and Trinidad & Tobago with a small percentage of Hindus (10–20%) livening there as migrant labor force and a high level of wealth.[18] In the other subgroup are the countries where (proportionately) few Hindus live and which also have a low level of wealth (<10, 000Intl$.). Countries belonging this group are *Bhutan, Fiji, Suriname, Sri Lanka, Guyana*, which are all very small states. *Bangladesh* is by far the largest state within this group. In general, the largest number of Hindus (absolute) in the world lives in *India*, which is one of the rather poor countries (Fig. 5.18).

5.2.6 Intermediate Result

The examination of the worldwide correlation between the Buddhist and Hindu population share and the corresponding access rates to fixed telephone (FTR), mobile phone (MPR), Internet (IPR) and mobile Internet (MBR) almost yield the same result. There is a minor negative correlation between the Buddhist and Hindu population share in a country and the corresponding access rates, i.e. they decline with a higher share of these religious groups. The result for the analysis of the correlation with the Internet penetration rate (IPR) is clear. For the IPR the correlation is more pronounced than the correlation for the other access rates. It can be concluded that the usage of Internet within these two religious groups is on a considerably lower level than is the case for the larger religious groups of Christianity and Islam.

5.3 Global Networking: Judaism and Other Worldviews 2010

Judaism, as well as *Christianity* and *Islam*, are *monotheistic religions* in which Judaism dates back more than 3,000 years, making it the oldest among the Abrahamic religions. According to Jewish tradition, Abraham is regarded as the founder of monotheism, the belief in the existence of one personal and transcendent God. The

---

[18]The special position of the gulf states *Qatar* and *Bahrain* was discussed in Sect. 5.1.5 in the context of the rather Islamic countries.

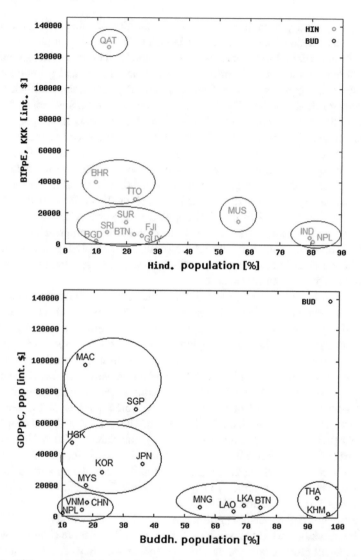

Fig. 5.18 GDP per capita (GDPpC) as function of the Buddhist (bottom) and Hindu (top) population share in a country, year 2010

Jewish faith is based on the *Holy Torah* – the Hebrew Bible – which is considered to be the *Book of the books* and consists of the five books of *Moses* including 613 commandments (248 positive commandments and 365 negative commandments). The meaning of the Hebrew term *Thora* is doctrine, teaching, instruction and law.

Around 14 million Jews were living around the world as of 2010 which makes up 0.2% of world's population. This number is calculated from all persons who – based on national surveys and censuses – denote themselves as Jews when asked for their religious affinity. It might be higher if the definition of being Jewish is extended to all persons whose grandparents are of Jewish faith. Certainly, it would be less if the definition is limited to the unbroken succession in the female line, so that Jews from different regions of the world share the same cultural and genetic heritage. Furthermore, the number of Jews from different denominations worldwide is not always clearly determinable. The Jewish community in the US is mostly affiliated with one of the four branches: Orthodox, Reform, Conservative and Reconstructionist. In Israel and elsewhere around the world it is distinguished between Charedim and ultra-orthodox, modern-orthodox and less traditional forms of Judaism.

Jews are concentrated primarily in North America (44%) and the Middle East (41%). The remainder of the global Jewish population is found in Europe (10%), Latin America (3%), Asia and the Pacific (1–2%) and sub-Saharan Africa (1%). Jews make up roughly 2% of the total population in North America and a similar proportion in the Middle East. In the remaining regions, they comprise less than 1% of the overall population. Judaism is the religion of the highest geographic concentration worldwide, i.e. mainly concentrated in two countries (*USA* and *Israel*), forming the majority in only one country (Israel).

People in other and no religion are not affiliated to any religious confession or are atheists, followers of folk religion or traditional rituals, and other faiths that do not belong to any of the listed religions. In 2010, there were about 1.1 billion people who class themselves as non-denominational. This corresponds to one out of six persons (16%) worldwide. Thus, the group of non-religious make up the third largest ideology after Christianity and Islam and is comparable in size to the number of Catholics worldwide. Non-religiousness encompasses atheism, agnosticism and persons who are neither atheists nor agnostics. Three-quarters of the religiously unaffiliated (76%) live in the Asia-Pacific region. The rest is distributed over Europe (12%), North America (5%), Latin America (4%), Sub-Sahara (2%), Middle East and North Africa (<1%). According to the *"Global Religious Futures"* project [333], there are six countries where the religiously unaffiliated make up a majority of the population: the *Czech Republic* (76%), *North Korea* (71%), *Estonia* (60%), *Japan* (57%), *Hong Kong* (56%) and *China* (52%).

It is estimated that about 400 million people, or 6% of the world population, practice various folk or traditional religions. These are faiths that are observed within a particular small group of people or a tribe, respectively. They often have no formal creeds or sacred texts. Examples of folk religions include African traditional religions, Chinese folk religions, Native American religions and Australian aboriginal religions. Folk religionists are most prevalent in the Asia-Pacific region, where nine-in-ten of the world's folk religionists (90%) reside. The remainder is concentrated in sub-Sahara (7%) and Latin America (2–3%).

The Baha'i faith, Taoism, Jainism, Shintoism, Sikhism, Tenrikyo, Wicca, Zoroastrianism account for 60 million persons worldwide and are classified into the "other" group. The vast majority (89%) of this group lives in the Asia-Pacific region.

5.3.1 Fixed Telephony

As already mentioned, the Jewish religion is heavily concentrated in one single country – *Israel*. Figure 5.19 (bottom) shows the correlation between the share of Jews in relation to the overall population and the share of persons having a fixed telephone subscription (FTR).

$$FTR = 39.256 + 0.0898 \cdot Share_{JUD}, R^2 = 0.0046, R = 0.068 \qquad (5.17)$$

The distribution of the data is of particular interest. With just one exception, all countries are concentrated close to the value 0 of the abscissa.[19]

An examination of the connection between the share of *other religions* to the overall population and the FTR results, under the presumption of a linear regression model, is shown in the following equation:

$$FTR = 16.576 + 0.3335 \cdot Share_{AND}, R^2 = 0.085, R = 0.292 \qquad (5.18)$$

In the following, we want to analyze why the FTR in Jewish countries is slightly higher compared to the rest of the world. Furthermore, an interesting question arose is why the FTR in countries with a large number of people affiliated with no religion is higher than the one of Jewish influenced countries and the rest of the world. For this reason, the FTR of both groups is examined in relation to the Gross Domestic Product per capita (GDPpC). Figure 5.20 visualizes this correlation. It shows the countries with Jewish share (black), countries with other religions (green) as well as the corresponding curve that has the best fit to the data.[20] A key finding is following correlation: The curve for the FTR worldwide (blue) was determined in Chap. 4, Sect. 4.1.1 and is used as reference curve. As it can be seen in the figure, the curve with the best fit of the countries with Jewish populations (black) lies slightly above the reference curve for the whole world (blue). Based on the data, this result is not surprising since the Jewish population is, although small in number, spread all over the world. The curve with the best fit of the countries with other religions (green) lie considerably above the reference curve for the whole world (blue). A reason for this is as the share of this group increase, the more prosperous a country becomes.

---

[19]Due to this high concentration, one can look on the data without Israel. Basically, it does not make big of a difference, if the share of Jews of a population of a country and the FTR is regarded with or without Israel. In both cases, there is, with a correlation coefficient close to zero, almost no correlation.

[20]The difficulty lies in the definition of when a country is considered rather Jewish. Compared to Christianity, Islam, Buddhism and Hinduism, there is no country beside Israel, which can be considered Jewish in a narrow sense. The Jewish population lives (almost) all over the world in an extremely small minority. Only in *Israel* (75.6%), the *United States* (1.8%) and *Canada* (about 1%) they make up more than 1% of the total population. For that reason, countries are only considered if they have a Jewish population of at least 0.1%. A 10%-limit was defined for the group of countries with other religions.

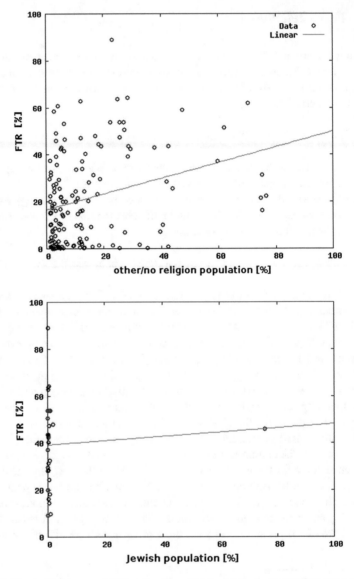

Fig. 5.19 Fixed telephone rate (FTR) as function of the Jewish (bottom) and "other" (top) religious population share in a country, year 2010

5.3.2 Mobile Telephony

Figure 5.21 (bottom) illustrates the result of the regression analysis for the correlation between the rate of people owning a mobile phone (MPR) and the share of the Jewish population. The following linear regression model describes this correlation. The

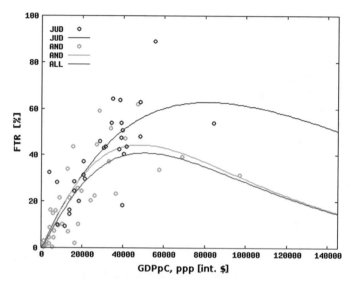

Fig. 5.20 Fixed telephone rate (FTR) as function of the GDP per capita (GDPpC) for countries with Jewish (black) and other religious (green) population share, year 2010

correlation coefficient is approximately zero, which implies a very weak correlation.

$$MPR = 116.3 + 0.065 \cdot Share_{JUD}, R^2 = 0.0012, R = 0.035 \qquad (5.19)$$

The graph of the MPR is an almost horizontal line, in which *Israel* lies on the regression line.[21]

The graph on the top in Fig. 5.21 visualizes the correlation between the share of other religions in a country as an independent variable and the MPR as a dependent variable. It implies a slightly positive correlation with a correlation coefficient of $R = 0.241$.

$$MPR = 83.838 + 0.6062 * Share_{AND}, R^2 = 0.058, R = 0.241 \qquad (5.20)$$

What would be a possible reason for this correlation? As it already described, the worldwide spread of mobile phones is based on prices that became affordable for all. Analyzing the MPR in correlation with the economic performance, measured by the Gross Domestic Product per capita (GDPpC), in countries that have a Jewish population share and persons who are affiliated with *other religions* helps to find an explanation. Figure 5.22 shows the MPR as dependent and the GDPpC as independent variable, in which countries with Jewish population (black) and countries with parts of the population affiliated to *other religions* (green) are shown as data

---

[21]Regarding this correlation without *Israel* implies a moderate negative correlation between the share of Jewish population and the MPR.

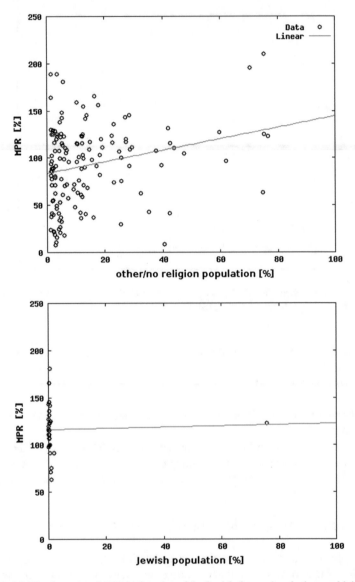

Fig. 5.21 Mobile phone rate (MPR) as function of the Jewish (bottom) and other worldviews (top) population share in a country, year 2010

points. Simultaneously, the curve with the best fit for both religious groups can be seen in the corresponding colors. The blue curve is the reference curve for the whole world. The result is as expected: The adaption curve for countries with Jewish populations (black) has a very similar progression compared to the curve for the whole world (blue). This again is not surprising, since Jews are spread all over the world,

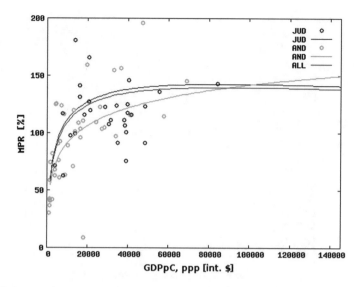

Fig. 5.22 Mobile phone rate (MPR) as function of the GDP per capita (GDPpC) for countries with Jewish (black) and other worldviews (green) population share, year 2010

although only living as a small minority in most countries. The adaption curve, which represents countries with a population share of other religions (green), is below the reference curve (blue) and below the curve for the countries with a Jewish minority. The reason for this is that numerous poor countries in the world, mainly located in Africa and Southeast Asia, have a share of persons without religious denomination. The MPR is to an appreciable extent independent of the economic performance of a country, since an already small GDPpC is sufficient to reach a 100% level for the MPR.

5.3.3 Internet

An examination of the Jewish population share as an independent variable and rate of persons using the Internet (IPR) as dependent variable (see Fig. 5.23 bottom), results in following regression line and a coefficient of determination (R^2) between the Jewish population share and the IPR:

$$IPR = 62.371 + 0.0659 \cdot Share_{JUD}, R^2 = 0.0016, R = 0.04 \qquad (5.21)$$

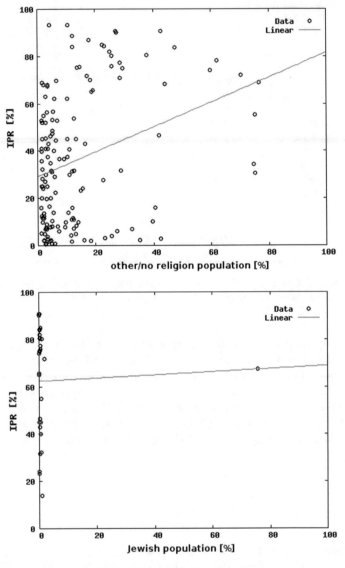

Fig. 5.23 Internet penetration rate (IPR) as function of the Jewish (bottom) and other religious (top) population share in a country, year 2010

It turns out that there is a positive relation, but with only a small correlation. Following equation describes this correlation if the *State of Israel* is excluded[22] from the analysis:

---

[22]The reason for excluding the State of Israel from the analysis is due to the statistic reasons that have been explained above.

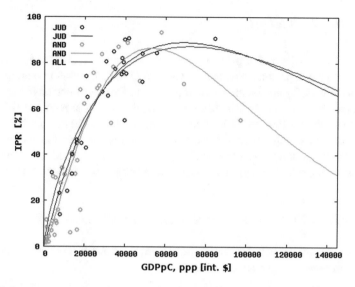

Fig. 5.24 Internet penetration rate (IPR) as function of the GDP per capita (GDPpC) for countries with Jewish (black) and other worldviews (top) population share in a country, year 2010

$$IPR = 63.484 - 2.8597 \cdot Share_{JUD_{ohneISR}}, R^2 = 0.0021, R = -0.046 \qquad (5.22)$$

The IPR in a country increases if the share of persons affiliated with *other religions* increases. This connection is positive, although having only a small correlation coefficient (R).

$$IPR = 29.042 + 0.5255 \cdot Share_{AND}, R^2 = 0.0985, R = 0.314 \qquad (5.23)$$

How can this result be interpreted? Fig. 5.24 illustrates the IPR as function of the GDPpC, in which countries with Jewish population share (black) and countries with persons affiliated with *other ideologies* (green) are shown as data points and their corresponding adaption curve with the best fit to the data points. The result shows the following: The best adaption curve for countries with Jewish population share (black) has almost the same progression as the reference curve (blue). This has already been ascertained for fixed telephone (see Sect. 5.3.1) and mobile phone (see Sect. 5.3.2), from the worldwide broad distribution of the Jewish community, living mostly as a small minority. Above a large GDPpC range, the curve for *other worldviews* progresses very similar compared to the reference curve. Only in the very-high GDPpC range this curve is pulled down by two countries (*Luxembourg* and *Singapore*). However, from a worldwide perspective these countries are rather insignificant due to them classified as small city state. Therefore, it can be said that the Jewish community and the population share affiliated with *other religions* is, regarding the usage of Internet, on a very similar level as the rest of the world.

5.3.4 Mobile Internet

Figure 5.25 shows the rate of persons using mobile Internet (MBR) as function over the share of Jewish persons in a country. Compared with the other two monotheistic religions Christianity and Islam and their relation to the rate of persons using mobile Internet (see Sect. 5.1.4), the data has a heavy concentration around x = 0. *Israel* is the only Jewish majority, while Jews form ethnic minorities in every other country. Performing a linear regression analysis results in following equation and the coefficient of determination (R^2) and a correlation coefficient (R) respectively for the rate of persons using mobile Internet as a function over the Jewish population share in a country.

$$MBR = 29.17 - 0.049 \cdot Share_{JUD}, R^2 = 0.0016, R = -0.04 \qquad (5.24)$$

The linear correlation between the mobile Internet penetration rate as a function of the share of *other religions* in a country is:

$$MBR = 13.158 + 0.5561 \cdot Share_{AND}, R^2 = 0.1607, R = 0.401 \qquad (5.25)$$

The regression analysis shows that in 2010 there is no recognizable relation between the rate of persons using mobile Internet and the share of Jewish population in a country. There is a positive connection between the share of persons using mobile broadband and share of persons affiliated with *other worldviews* in a country, which can be quantified being on a medium correlation level with a $R = 0.401$.

5.3.5 Economic Performance

The rate of persons having access to fixed telephone, mobile phone, Internet and mobile Internet was examined in countries with – even though minor – Jewish share of population and in countries with a population share affiliated with *other worldviews*. The results can be mostly explained with the level of prosperity of the examined countries. For this reason, we will arrange groups of countries classified by their economic performance (GDPpC).

Figure 5.26 illustrates the GDPpC as dependent and the Jewish (bottom) and *other worldviews* (top) share of population in a country as independent variable. *Israel*, with a Jewish population share of about 75%, was omitted from the picture because otherwise the (minor) differences in the remaining countries could not be seen. This shows in which (rich) countries the majority of Jews live. The group arrangement of countries with a Jewish minority is as follows: *Israel* has the highest percentage of Jewish population – following the *US* and *Canada*.[23] For that reason, latter two rich countries are grouped together. There is another group that also comprises rich

---

[23]Remember: Israel is also the only country in the world where a majority of citizens are Jewish.

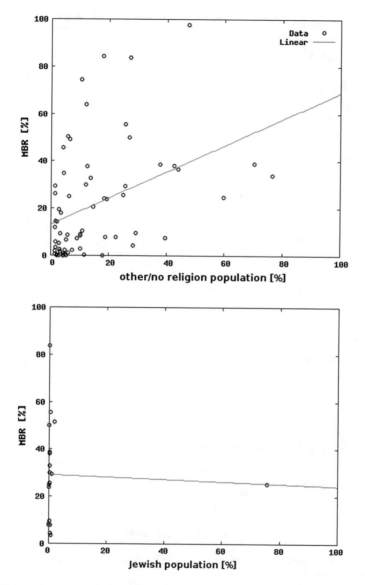

Fig. 5.25 Mobile Internet rate (MBR) as function of the Jewish (bottom) and other worldviews (top) population share in a country, year 2010

countries but with a minor Jewish population share (<1%). These countries are for instance *Germany, France, UK, Australia, New Zealand, Sweden, Luxembourg, Switzerland* and *Bahrain*. The last group also has a small Jewish population share and comprises rather, but not extremely, poorer countries (*Argentina, Chile, Russia, Ukraine* and *South Africa*). Basically, it can be said that the Jewish population, with

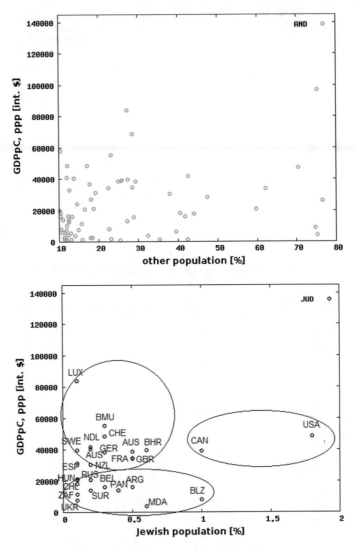

Fig. 5.26 GDP per capita as function of the Jewish (bottom) and other worldviews (top) population share in a country, year 2010

a perceptible share as a minority, cannot be found in any of the world's poorest countries – since they live in predominantly rich countries.

5.3.6   *Intermediate Result*

The worldwide analysis of the relation between the share of Jewish population, share of persons affiliated with *other religions* and also the non-religious population in a country with the rate of persons using fixed telephony (FTR), mobile phones (MPR), Internet (IPR) and mobile Internet (MBR) implies no correlation between these rates and the Jewish populations share. This religious group, with *Israel* being the only Jewish majority, is spread all over the world in small groups, the distribution of which is the basis of the corresponding analysis. The result of the analysis between the population share of *other religions* in a country and the respective access rates is clear. There is a positive correlation for all four rates of access, but only at a low to medium level.

5.4 Global Networking and Religion Diversity 2010

The *Religion Diversity Index* (RDI) (see Sect. 2.3.4) is an index based on the percentage of each country's population that belongs to one of the eight major religious groups. These groups are Christianity, Islam, Judaism, Buddhism, Hinduism and religiously unaffiliated, adherents of folk or traditional religions, and those belonging to the remaining religious groups. It has a range of 0 to 10, with the index being 0 if all of a country's population belong to one religion. An index of 10 means that all of the eight major religious groups are equally represented in a country, therefore with population share of 12.5% each. The RDI varies in different regions of the world. Among the six regions analyzed – North America, Latin America, Europe, Middle East- North Africa, sub-Saharan Africa and Asia-Pacific – the Asia-Pacific region has the highest level of religious diversity (highest RDI). Europe and North America have a moderate RDI, the Latin America-Caribbean and Middle East-North Africa regions have a low RDI.

5.4.1 Fixed Telephony

Figure 5.27 illustrates the Religion Diversity Index (RDI) as an independent variable and the rate of persons owning a fixed telephone connection (FTR) as the dependent variable. At first view, there is no obvious correlation between those two variables, whereby the point cloud has a large variance. The linear regression analysis results in following the equation:

$$FTR = 13.21 + 1.8918 \cdot RDI, R^2 = 0.0531, R = 0.231 \tag{5.26}$$

The result implies a minor positive correlation between the FTR and the RDI. The higher the level of religious diversity in a country is, the higher the FTR.

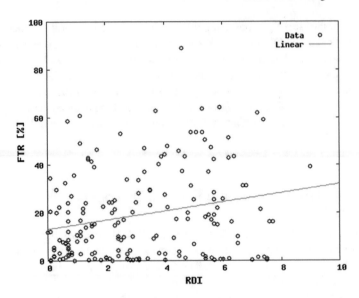

Fig. 5.27 Fixed telephone rate (FTR) as function of the Religion Diversity Index (RDI), year 2010

5.4.2 Mobile Telephony

Sections 5.1.2, 5.2.2 and 5.3.2 conclude that the usage of mobile phone is very common within major religions. Therefore, a similar result regarding the rate of persons using mobile phones (MPR) in a composition of different religions seems to be obvious. Initially, Fig. 5.28 does not make a correlation directly obvious.

Assuming a linear regression model, the following equation is given for the correlation between the MPR as the dependent and the RDI as the independent variable.

$$MPR = 81.335 + 2.6267 \cdot RDI, R^2 = 0.0198, R = 0.141 \qquad (5.27)$$

The result implies a minor positive correlation between the RDI and the FTR. Former examinations for the MPR are confirmed here. The usage of mobile phones is very common both in individual religions and in multireligious societies in 2010.

5.4.3 Internet and Religion Diversity

Figure 5.29 illustrates the Religion Diversity Index (RDI) of a country as the independent and the rate of persons using mobile Internet rate (MBR) as the dependent variable.

Assuming a linear correlation results in following regression line and following coefficient of determination (R^2) for the correlation between IPR and the RDI:

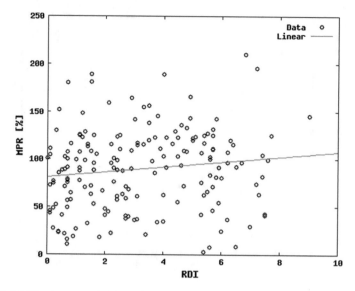

Fig. 5.28 Mobile phone rate (MPR) as function of the Religion Diversity Index (RDI), year 2010

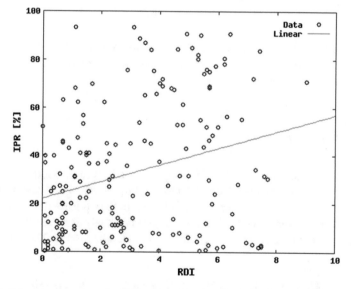

Fig. 5.29 Internet penetration rate (IPR) as function of the Religion Diversity Index (RDI), year 2010

$$IPR = 22.084 + 3.4838 \cdot RDI, R^2 = 0.0816, R = 0.286 \qquad (5.28)$$

Although a low linear growing relationship between the RDI and the IPR is recognizable, the great heterogeneity of the data, which hardly permits a correlation, is

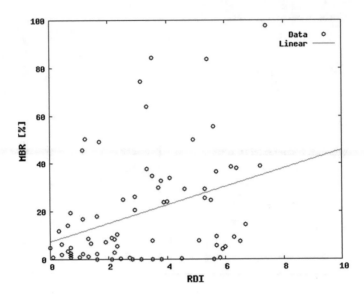

Fig. 5.30 Mobile Internet penetration rate (MBR) as function of the Religion Diversity Index (RDI), year 2010

dominant, which manifests itself in a very small degree of correlation coefficient R. The higher the religious diversity in a country, which corresponds to a high RDI, the higher the IPR in a country.

5.4.4 Mobile Internet

Figure 5.29 illustrates the Religion Diversity Index (RDI) of a country as the independent and the rate of persons using Internet (IPR) as the dependent variable. At first glance, there is a minor correlation between those two variables, whereby the point cloud has a large variance. The linear regression analysis results in the following equation and the corresponding coefficient of determination (R^2) and correlation coefficient (R):

$$MBR = 7.389 + 3.8307 \cdot RDI, R^2 = 0.1221, R = 0.349 \qquad (5.29)$$

The result of the regression analysis confirms the first impression, that there is (only) a minor positive relationship between the MBR and the RDI. The higher the level of the composition of different religions in a country, the higher the MBR (Fig. 5.30).

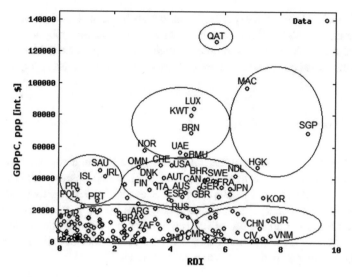

Fig. 5.31 GDP per capita (GDPpC) as function of the Religion Diversity Index (RDI), year 2010

5.4.5   Economic Performance

In this section a further plausibility check is made for the results regarding the link between the different access rates to fixed telephone, the mobile phone, Internet, mobile Internet and the Religion Diversity Index (RDI). The assumption is that a certain degree of RDI, i.e. a certain religious-cultural diversity, points to a higher GDP per capita (GDPpC). The reason is probably that high diversity is tendentially associated with increased economic activity.

Figure 5.31 illustrates the correlation between the GDPpC (dependent variable) and the RDI (independent variable).

The assumption is confirmed by looking at the grouping in Fig. 5.31. In the center of this figure is the group of rich and rather large states such as the *US, Canada, Australia, UK, Germany, France, Italy, Spain, Japan* and the Scandinavian countries, *Denmark, Finland, Sweden* and *Norway*. All of them have a certain religious diversity. Some of the world's richest countries are *Qatar, Kuwait, United Arab Emirates, Luxembourg, Macao, Singapore* and *Hong Kong*. All of these countries are very small but have a high religious diversity. Then there is a number of European countries with a moderate level of economic performance, such as *Ireland, Poland and Portugal*, where the religious diversity is rather small. In the area of low to moderate levels of prosperity, there are countries with very low religious diversity (*Turkey, Iran, Egypt, Indonesia, Philippines, Pakistan, Bangladesh, Brazil, Argentina, South Africa*, etc.) but also high religious diversity (*China, Vietnam, Nigeria* and *Ivory Coast*). It can be concluded that there are no rich states with a small religious diversity. A moderate degree of religious diversity seems to be important for both small and large states to allow a high level of prosperity. It could also be seen the other way with the conclusion that both small and large rich countries must involve migrants as workers into their economic processes in order to maintain their prosperity level. This then leads to the cultural diversity in these countries and modern states emerge.

5.4.6 Intermediate Result

The worldwide examination of the link between the RDI in a country and the corresponding access rates to fixed telephone (FTR), mobile phone (MPR), Internet (IPR), and mobile Internet (MBR) points that the RDI has a positive relationship with the different access rates. A high diversity of religious groups in a country thus has a positive effect on the use of information and communications technology in 2010. The reason for this is that a certain degree of religious or cultural diversity in a country and its prosperity level are positively related.

5.5 Conclusion

Summarized, the empirical analyses of the relationship between the major world worldviews *Christianity, Islam, Judaism, Buddhism, Hinduism* and the *group other*, which consist of people without affiliation to a religion, atheists and folk religion, draw no significant correlation between the share of a respective religious group in a country and the different access rates to fixed telephone, mobile phone, Internet and mobile Internet. In general, the access rate of the respective technology is slightly higher with a higher share of the Christian population in a country. For the Islamic population, there is a minor negative correlation. Within the Buddhist and Hindu world population the different access rates show a slight decrease, the higher the percentage of these groups in a country. The percentage of Jewish populations in a country is not clearly linked to the respective access rates, however, the correlation for the share of persons affiliated with other religions is positive. Furthermore, the diversity of religious groups in a country has a positive impact on the level of all access rates. This confirms the thesis that the religions, in a global perspective, have little or no relation to the different access rates to information and communications technology, and that *convergence in the use of ICT among the major religions in 2010 has advanced quite far even if absolute convergence is not reached yet.*

Part III
Power Issues and Reflections on Possible Future Developments

Chapter 6
Global Networking, Power and Control Issues

It is not surprising that in the sphere of such a mighty system like international (mobile) communication and Internet usage, massive political, economic, ideological, military power and control issues are based. In looking at the origin of the Internet, the US takes on a specific role. A *conflictual* subject in the area of power and control issues in the context of *cultural convergence* was the position of the US as the supervisory nation of the Internet Corporation for Assigned Names and Numbers (ICANN). Until October 2016, its role was to manage the central technical Internet functions such as the management of top-level domains and IP addresses worldwide. In March 2014 the US government, as a reaction to global critics, announced its termination of this position as a part of the current IANA-contract between itself and ICANN. The IANA-transition is completed now. Since ICANN's founding in 1998, the multi-stakeholder community controls and manages the Internet's critical resources.

After a brief overview of the history and current state of the Internet Governance Ecosystem, in this chapter, power and control issues are considered from a technical, economic and political perspective. Discussion includes the bargaining positions of states and cultures on the processes of changing or maintaining mechanisms in Internet governance with a special emphasis on the responsibility of the *International Telecommunications Union* (ITU) as an UN body. Finally, change processes of power structures are described in this context in attaining global information and the creation of a knowledge society. In addition, a comparison is made with former social and cultural transformation processes, i.e. the industrial revolution.

© Springer International Publishing AG 2018
H. Ünver, *Global Networking, Communication and Culture: Conflict or Convergence?*, Studies in Systems, Decision and Control 151,
https://doi.org/10.1007/978-3-319-76448-1_6

6.1 Internet Governance Ecosystem

In this section, the so-called *Internet Governance Ecosystem* (IGE) [58] is described with the major actors and their relationships with one another. A unique description for the IGE does not exist. This concept is an approach to capture the complex Internet world with the actors that influence the technical, economic and political development of the Internet. In this ecosystem, the *multi-stakeholder model* is of central importance where organizations and bodies (technical community, governments, businesses, civil society) have a specific responsibility for different areas of the Internet.

6.1.1 Historical Development

Traditionally, out of the genesis of the Internet the US has taken a leading role on the global Internet than any other country or culture. The rest of the world, e.g. the EU Commission, require more say. In Brussels in 2014, the European Commission called for an Internet power structure based on *fundamental freedoms* and *human rights* [101]. A similar position is addressed by the EU at the IGF in Brazil in November 2015, whereby the focus was, inter alia, in the areas of *net neutrality* and *zero rating*. With a look at the past, a key challenge for the future becomes clear here. What do the major global cultural systems understand by the concepts of fundamental freedoms and human rights? In order to answer this question, reaching a consensus requires an understanding of the position of others and a type of global and cultural empathy, as it should be in other international cases. This kind of empathy is also a key prerequisite for a harmonious way of humanity towards a well-functioning human–technology–system, the so–called *global superorganism* (see Chap. 7).

For example, the EU calls for a model that is based on an interaction between various stakeholders in accordance with the multi-stakeholder model. The last international conferences concerning this, the *Internet Governance Forum* (IGF), took place at the beginning of September 2014 in Istanbul [186] and in mid November 2015 in Brazil [187]. Meanwhile, IGF 2016 took place in Mexico and IGF 2017 in Geneva. At the core of these conferences were debates around the question of who has the oversight on key resources in maintaining and operating the (global) Internet. The main topic was the exact arrangement of the multi-stakeholder model in controlling the Internet, which is intended to secure a beneficial interaction of governments, private sector, technical community and (world) civil society. Specifically, it was about issues such as freedom of expression, broadband expansion, network neutrality and zero rating, data protection and privacy as well as more participation via Internet in all cultures of the world, which in turn have partially very different and partly similar ideas and visions. In principle there are three different thematic areas (technology, economy, politics) which are under discussions and concern Internet governance.

First, this is the technical domain, which is the basis for the two overlying areas of economy and politics. Till this day, the management of core (technical) functions of the Internet was primarily in the hands of the US due to the state–militarily promoted development of the Internet in the 1960 s as a computer network for the American science and research. At the beginning of the Internet age, it was managed by scientists and IT engineers by a certain form of the self-regulation principle, which includes the possibility to quickly link many computers. Meanwhile many organizations are involved in the technical development of the Internet. In particular the IETF, ICANN, IANA, IEEE, W3C, ITU-T etc. belong to these (see Sect. 6.1.2).

Since 1997, almost every country on earth raised (political) demands for an internationalization of the central Internet functions. The US government responded with a transfer of its tasks to a privately organized company (ICANN) that consists of various stakeholders, particularly including the user community. However, this company should be seated in the US and remains obliged by contract to the supervising (US) government agencies (in particular the US Department of Commerce). The demands for a globalized control of critical Internet resources therefore remained.

The emergence of Internet and WWW (economically) in the 1990 s led to a rapid expansion of usage and functions. An interest arose in a co-creation of the new communication medium due to the growing IT economy and the number of Internet users as a whole economy. As the Internet has become central for economic prosperity, it has quickly gained widespread usage in *China, Russia, India*, the *Arab world* and *Latin America*. First, the challenge for the IT industry consisted in providing suitable hardware and software for Internet access for a population. Meanwhile, the access has been widely expanded (see Chap. 4), but will be (worldwide) extended even further. Furthermore, already existing network access is constantly getting faster. In this context, a central question is whether the existing so-called net neutrality can be maintained in the future. Phones and mobile phones capable of Internet also allow an access to the Internet and are becoming increasingly important. Currently, another interesting question (from power and control point of view) is to which extent the (data-driven) knowledge regarding the behavior of Internet users is a crucial economic factor. In this area, the digital global players who currently have the final say operate in global markets with significant cultural differences such as the free market economy versus the state-regulated markets.

Increasingly, governments worldwide are involved in the co-creation of the Internet in a political sense. The spread of *political contents*, fake news as dangerous form, on the Internet was not seen with much pleasure by global powers and entailed much reactions. At the political level, the Internet is used by people and systems in different cultures which overlay partly geographic or political space and state formations. If it is a matter then, e.g. of pursuing an *election campaign* or aspects such as *freedom of speech*, it quickly escalates to clashes of interest between cultures. *Regarding freedom of expression, there are still significant differences between the world's cultures, i.e. it is rather a conflict than a convergence in this field*. These differences manifest themselves in both with regard to global management processes on the Internet as well as Internet use as a tool of formulating or creating opinion.

Following the *World Summit on Information Society* (WSIS) in 2003 (Geneva) and follow-up event in 2005 (Tunis), economic and political opportunities and the potential risks of the Internet became more clear. The design of Internet Governance has become an important political issue, but also a matter of power. Currently, worldwide there are mainly two different cultural perspectives on the direction that Internet Governance should take going forward. On the one hand, this is the position taken up by the *US, EU* and *Japan*, which all have a desire to protect the status quo of self-regulation through the multi-stakeholder approach. Here we see a similar perspective or convergence of positions across cultural borders [255]. This is the position of economically strong countries particularly rich countries with a far developed civil society and widespread rights and freedoms enjoyed by all citizens.

On the other hand, there are countries like *Brazil, Russia, India, China* (BRIC) and *South Africa*, as well as *Iran* and *Saudi Arabia*. All of which are in intensive development processes, have a difficult (global) situation of competition and have poorly mobilized civil societies.[1] In these countries non-state actors have rather less power in relation to state institutions. These countries are from quite different cultural systems, however, with regard to Internet Governance, share similar mindsets and interests. They want to organize the Internet governance primarily through an international organization within the UN system such as the International Telecommunication Union (ITU) or at least to transfer more responsibility to the ITU in this matter.

6.1.2   Technical Power of Control

The Internet was originally an innovation of a technical type in an academic space. This is why the technical matters still widely follow the academic habits [66]. This book focus on the main actors in this subject area due to the variety of organizations involved in the technical management of the Internet

The *Internet Society* (ISOC) is the parent organization of the *Internet Engineering Task Force* (IETF), the *Internet Architecture Board* (IAB), the *Internet Engineering Steering Group* (IESG) and the *Internet Research Task Force* (IRTF). The ISOC is generally responsible for the maintenance and further development of the Internet infrastructure. One of the main tasks of ISOC is the publication of the so-called *Request for Comments* (RFCs), which represent Internet standards through the IETF[2]

The *Internet Engineering Task Force* (IETF) is primarily concerned with *"the matter of protocol agreements, especially their changes and adaptations with which*

---

[1]The upheaval and transformation processes in the Arab world show the power that can be achieved by mobilizing large segments of the population, whereby intentions and results of corresponding processes can differ widely [408].

[2]Note: Although all Internet standards are documented in RFCs, but not all RFCs are Internet standards. RFC2026 says the following to that matter: *RFCs cover a wide range of topics in addition to Internet Standards, from early discussion of new research concepts to status memos about the Internet* [90].

the current and future use of the net is determined" [66]. The IETF is responsible for the specification of protocols as for example the *Internet Protocol* in versions 4 (IPv4) and 6 (IPv6). The *Hypertext Transfer Protocol* (HTTP) was also developed in the rows of the IETF. The IETF encourages contributions from the whole world, particularly as the work takes place widely online through mailing lists as well as in an open process of development. Therefore all interested and technically competent people can contribute through a so–called a working group (WG).[3] Although most of the work takes place online at the IETF, there are meetings three times per year.

The work of the IETF follow the principle of *rough concesus* with which no formal voting is required. The motto of Dave Clark[4]: *"We reject kings, presidents, and voting. We believe in rough consensus and running code"* is interesting in this context. The notion of not voting is because there is no membership but only participants in the IETF. Therefore a formal basis of a group of authorized voters is missing. The broad agreement is often determined by hand signs or by loud humming [183]. No mandatory vote counting is conducted. Conflicts are resolved in discussion groups (mostly online). Final decisions can be made only on the basis of mailing lists.

With regard to the cultural issues which are the center of this present work, the approach of the IETF includes potential conflict situations. This is especially the case from the perspective of authoritarian states where decisions are primarily made behind closed doors. However, even from the point of view of democratic states, a rough consensus is unusual and politically not feasible. One can designate this culture as culture of a pure (technical) pragmatism or as *technocracy*.

The *Internet Architecture Board* (IAB) maintains an overview of the standardization activities of the IETF. The IAB is primarily concerned with the (strategic) development of the Internet. It furthermore checks the management of the allocation of protocol parametre values by the *Internet Assigned Numbers Authority* (IANA). In November 2014 the IANA implemented protocol standards regarding encryption methods and to activate them by default. This issue is gaining more and more importance as a result of the recent news surrounding the NSA [177].

The *Internet Assigned Numbers Authority* (IANA) [178] is a department of the *Internet Corporation for Assigned Names and Numbers* (ICANN) [180]. It is responsible for the coordination and function of (global) numbers as well as names on the Internet (see Sect. 3.2.1). ICANN executes the functions of IANA in the name of the *National Telecommunications and Information Administration* (NTIA), an authority of the US *Department of Commerce*. The IANA functions can be divided essentially into three areas [181]. This includes the management of the *Domain Name System* (DNS), such as the registration and processing of *top-level domains* (TLD) including .com, .org or .net, in the root zone file, which are to be authorized in the *root servers* [395]. The IANA conducts updates on the root zone database, publishes these and lists contact information for all TLD operators [179].

---

[3]The IEFT-WG are divided in different subjects, such as routing, transport, security, etc. and are all directed by *Area Directors*, which are also part of the *Engineering Steering Group* (IESG).

[4]American computer scientist and Internet pioneer, who was *Chief Protocol Architect* of the *Internet Activities Board* between 1981 and 1989, which later became the *Internet Architecture Board*.

Furthermore, the IANA is responsible for the allocation of IPv4 and IPv6. Here the IP address blocks are allocated to the five *Regional Internet Registries*[5] (RIR), which then allocate their addresses to the corresponding *Local Internet Registries* (LIR). These Local Internet Registries are usually the providers for the specific region or country (e.g. DENIC for Germany). The reviews and assignment of unique values for various parameters (e.g. port number or protocol numbers), used in various Internet protocols, is another function of the IANA. This function is provided in consultation with the IETF.

A stock of global IPv6 addresses, which is larger than the amount of currently available IPv4 addresses, is necessary to ensure the sustainable and long-term development of the global and open Internet. Global efforts to increase broadband availability in urban and rural areas are to be observed. Many new smartphones and network-capable devices have appeared on the market and the number of Internet users is constantly increasing (see Chap. 4). About 90% of the users own computer operating systems that are able to handle IPv6. This means that most households and (small) companies are waiting for their service providers to offer IPv6 connections.

By which factor is the address space in the transition from IPv4 to IPv6 extended? It is extended from 2^{32} to 2^{128}, which corresponds to a factor of 2^{97}. One can imagine providing every square centimeter of the Earth's surface with an IPv6 address. The earth has about 510 million km^2, which is exactly $5.100.000.000.000.000.000$ cm^2. This figure is roughly equivalent to 1.96^{64} (just under 2^{64}).[6] One can therefore state that it is possible to provide every square centimeter of the earth with about 2^{64} IPv6 addresses, which is more than sufficient from today's perspective.

The functions of the IANA are comparable to those of the notary [224]. During the creation of ICANN in 1998, the US government announced its intentions of dispensing this notary function. A declaration released March 2014 cedes the ICANN under specific conditions. This notarial function is the determining control position of the Internet, from the point of view of many countries. For many governments, the myth is that the functions of IANA and ICANN are of purely technical nature. The daily maintenance and management of the DNS system is undoubtedly a technical matter, but the policy behind ICANN relates to economic and public interests [350].

Another important organization for Internet standards is the *World Wide Web Consortium* (W3C) [518] whose founders include Tim Berner Lee. The W3C operates in accordance with the mission to develop relevant protocols and guidelines for the Web. This is achieved primarily through the creation and publication of web standards. With reference to these web standards, hardware and software manufacturers can offer their products and programs. Most web browsers like *Mozilla Firefox, Internet Explorer* or *Google Chrome* follow several W3C standards, which can interpret HTML and CSS code. If the different browsers comply with the W3C standards,

---

[5]The five RIRs consist of: 1. Réseaux IP Européens Network Coordination Centre (RIPE NCC), 2. American Registry for Internet Numbers (ARIN), 3. Asia-Pacific Network Information Centre (APNIC), 4. Latin American and Caribbean Internet Addresses Registry (LACNIC), 5. African Network Information Centre (AfriNIC).

[6]$0.04^{64} = (2 - 1.96)^{64}$ is a very big number but the aim of this estimation is to draw an idea and mental image for the amount of available IPv6 addresses.

the web pages can be displayed consistently across different browsers. In addition to HTML and CSS standards, the W3C also offers standards for web graphics (e.g. PNG images). The W3C also develops standards for Web applications, Web scripting and dynamic content. Furthermore the W3C provides recommendations for issues including privacy and website safety and security.

The *Institute of Electrical and Electronics Engineers Standards Association* (IEEE-SA) is also a vital organization in the areas of Internet standards, primarily responsible for standards of hardware (e.g. ethernet, WiFi, bluetooth and fiberglass connections) [182]. The development process of standards by the IEEE-SA is open for members as well as non–members of this organization. The membership allows a narrower integration and participation, e.g. by the takeover of a leading role in a working group. The IEEE-SA is led by the *Board of Governors* (BOG), which is elected by the members. The standard Committee monitors the development process of standards. Its members are elected by the IEEE-SA-members.

The ITU-T is the standardization sector of the ITU and is concerned with the development of global standards and norms in the telecommunications area, which are called *Recommendations* [195]. The recommendation process of ITU-T standards is carried out on the one hand, in accordance with the *Traditional Approval Process* (TAP) and on the other hand, after the *Alternative Approval Process* (AAP). Which procedure is used depends on the adopted recommendation. In particular, the duration and the formal character play a role here. The AAP was introduced in 2001 so that standards with purely technical character can be put on the market, which are urgently demanded by the industry. The prevailing majority of standards are released in this manner. These standards, which may have a regulation effect, need to go through the traditional approval process (TAP). The TAP process finds use if it's foreseeable that a recommendation can have regulatory or political impact. The adoption takes place exclusively by a corresponding positioning of the Member States. The following examples provide an insight into which standards are monitored by which process:

- Traditional Approval Process (TAP) (3–4 years): *E.108 Recommendation ITU-T describes the requirements for a disaster relief mobile message service. In the aftermath of a disaster, communication facilities are often overloaded due to many users attempting to contact friends or relatives to determine the safety of people who may have been affected by disasters. As a result, communication attempts often fail. The intent of a disaster messaging service is to allow an alternate method to communicate safety status information. Two approaches are presented. The first is a text-based messaging system, and the second is a voice-based messaging system.*
- Traditional Approval Process (TAP) (3–4 years): *X.1341 (ex X.cmail) The objective of this Recommendation is to define the certified mail transfer protocol (CMTP) and certified post office protocol (CPOP) in order to foster the exchanges of electronic certified mails in the world in a secure way by providing confidentiality, identification of the correspondents, integrity and non-repudiation.*
- Alternative Approval Process (AAP) (1–2 months): *J.1005 This draft new Recommendation specifies an architecture and requirements to digital right management (DRM) system for cable television content delivery service including multiple*

device viewing experiences. It is anticipated that architecture and requirements identified in this Recommendation can be applied to the DRM service that covers protected IP-type contents (IP VOD, IP linear, etc.) delivery from content provider or cable operator to end terminal devices (PC, tablet, smart phones, etc.) via cable television network.

6.1.3   Economic Power of Control

There is massive interest in the future development of the Internet in the field of economics (e.g. the issue of *net neutrality*). Should the Internet be a commodity, such as electricity grids, or the access and the transportation speed should be dependent on user profiles and possible financial expenses? With this question primarily related to *net neutrality* which pursues the principle that network and Internet providers have to equally prioritize all applications, services end users, i.e. strive for equal treatment of all users. The providers neither may block the access to the net for certain user's groups nor throttle the speed dependent on users. Furthermore, no user may be particularly treated, while providers offer incentives to these new users such as faster access. The *US* has taken a pioneering role in the enforcement of network neutrality. The regulatory authority of the *United States*, the *Federal Communications Commission* (FCC) has positioned itself against a two-tier society in the Internet in 2015. The US regulation authority has presented a set of rules which ensures equal treatment of end users on the Internet and mobile Internet [107]. But now the open Internet is in danger. On December 2017 the FCC announced its plan to kill net neutrality and let ISPs to run the Internet with throttling censorship and fees.

The so-called zero-rating is another issue of dispute that was extensively discussed at the UN conference in Brazil in November 2015. This concept corresponds to a quasi service that provides large Internet companies such as *Google, Facebook, Twitter* and *Wikipedia* the opportunity to offer their services free of charge. For this purpose, those companies have entered into separate agreements with mobile phone companies in many developing countries like the *Philippines, South Africa, Nepal* and *Kosovo*, where, for example, *Wikipedia Zero* is available for 30 countries where users do not have to complete the appropriate mobile phone plan. Another version of these contracts ensures that the services can be used by the respective companies, without removing the traffic for the contract services from the posted data volume. This ability to access certain services, however, has little to do with a real Internet connection. Such a connection does not allow access to the global (open) Internet. The principle of *net neutrality* is not the case here, because this principle may not favor any website or service of the access provider. They must treat all services and data equal and derive the same quality through their data network.

Meanwhile companies in the Internet world reach gigantic sizes. There is a risk that the digital global players, most notably *Amazon, Apple, Cisco, eBay, Google, Facebook, Microsoft, Twitter* or *Yahoo* are trying to control and direct the consumption patterns of their users even more than before. Here, the corporate dominance is clearly from the side of the *US*, but competitors from *China* including *Alibaba, Huawei, Baidu, Lenovo,* and *Xiaomi* are catching up. Below are some companies and their technical, economic and legal control mechanisms exemplified.

As online retailers *Amazon* and *Alibaba* use several management mechanisms for customer retention and for the monopolization of online trading. Firstly, Amazon controls its customers by a technical nature, for example the Kindle reader for electronic books accepts only a certain (from Amazon created) format. From Amazon one can only buy electronic books (eBooks) for the house brand Kindle, which accordingly can not be read by other devices (eReaders). This is the supply side of control. On the demand side, it is a mirror image so that if someone buys an ebook from another source than Amazon, these ebooks are not readable on Kindle devices. Additionally Amazon manages its customers with the terms of service by law. It says that the related eBooks via the Amazon Portal are not purchased directly, but rather are made available for the user permanently. With this contractual approach, Amazon established itself into a kind of global library. If a customer should violate the terms of service, for example, by developing or using a program to convert other formats to Kindle formats or even vice versa, Amazon may remove all the books from the Kindle library of the client. The purchase price will not be refunded. In a sense, Amazon also determines the economic conditions on the market. All prices of books and other products worldwide are observed with a software. The aim is to offer the cheapest product on Amazon where the retailer is not permitted to offer a lower price on other platforms except for Amazon. Today, different prices for one and the same good are available in the area of air transport services. At this point big data could be of increasing importance [109, 405].

Google handles worldwide requests with its search engine. The company has an impact on search results to favor services or advertisers. Google dominates the online advertising market. In addition, it leads the smartphone business with its Android operating system. With Google+ it also penetrates the market for social platforms such as Gmail as e-mail service provider and Youtube as a video platform. In addition it has its own browser, Google Chrome. Finally, Google can link the data from the various services together and thus obtains a fairly precise picture of each individual user.

Facebook members are participants, customers, employees and suppliers at the same time. This is especially the case in social networks, but also in the networked world where a breakthrough innovation can result in future digital business models. In a sense, the legwork of unsalaried employees is the next level of low-wage jobs that are now being accepted, even at zero cost. This is a clear competitive advantage of companies in the digital world as compared to companies in the real economy. Facebook is the parent company of *WhatsApp*. The term digital expropriation can designate the effects of the most popular messenger app with about 800 million users as of April 2015 and about 1.5 billion users as of December 2017. With its general terms and conditions, the company can use text content, images and selfies of users for free, and even resell without their consent.[7] For example, WhatsApp can use an image of a user as an advertisement for commercial purposes, which was originally served for another user. Thus, text and image rights are held by WhatsApp.[8] The terms and conditions states [9]:

> *"The copyright remains .... however WhatsApp has the unrestricted right of use of the data to copy, and advertise in any media formats."*

This becomes particularly problematic when a user has sent pictures of which he himself did not have the rights (e.g. downloaded from the Internet). If this is detected by the originator of the sent image, the user can be warned for this and be held liable in circumstances where costly charges arise.

The economic power of control in the digital world is heavily dominated by *US* companies. The dominant companies are using similar methods, algorithms and patterns in different markets for the monopolization of their markets. Increasingly, however, companies from *China* compete with these *US* corporations. Although the Chinese and American companies come from very different cultural backgrounds, they use similar processes and control instruments for the monopolization of markets, even if the market in the US is based on the ideas of free markets and that one in *China* is controlled by the government.

6.1.4   Political Power of Control

Currently, the political power of control largely depend on national regulations and is primarily concerned with the issues surrounding *cybersecurity*, *privacy*, *censorship*, *freedom of speech* and how to deal with *intellectual property rights*. The world is still far from having a general data privacy agreement for the Internet. Here there are clear cultural differences. Due to the cross-border architecture of the Internet, a national or supranational (such as in the EU) data protection agreement is not effective enough when it comes to establishing and maintaining a high level of data protection for (world) citizens. Today, a global agreement on data protection is more wishful thinking due to the strong global differences and conflicts regarding values and visions in the political field of the Internet. But if a global agreement on data protection can ever be achieved, then this will most likely occur on the application of relevant laws in the UN framework, for example, in the area of Universal Human Rights. Article 12 of the Universal Declaration of Human Rights states [459]

---

[7] The terms and conditions are hardly noticed in the digital economy. About half of all Internet users in Germany accepted the terms and conditions mostly unread [447]. The terms and conditions are likely to read even less in countries with a lower level of education than Germany.

[8] Starting in April 2016, WhatsApp uses end-to-end encryption for all communication. Therefore, WhatsApp and Facebook can only use metadata but do not have access to the actual communication.

[9] WhatsApp made his terms and conditions only in English available. The service has to offer this in German since May 2014, as judged by the district court Berlin [246]. It can not be expected from the German customers to understand the terms and conditions in English.

"No one shall be subjected to arbitrary interference with his privacy, family, home or correspondence, nor to attacks upon his honour and reputation. Everyone has the right to the protection of the law against such interference or attacks."

In this context, the United Nations Commission on Human Rights in its resolution 68/167 requested a report for the protection of the right to privacy and promoting the right to privacy with regard to national and extraterritorial monitoring mechanisms. Furthermore, it was reported on the surveillance of digital communication and the collection of personal data, including monitoring on a large scale and mass surveillance. It should be emphasized that other human rights are affected by mass surveillance or by monitoring the digital communications and the collection of personal data. These include, e.g., the right to freedom of expression and the right to search, receive and send information and the right to peaceful assembly and association. Other rights, such as the right to health can also be affected by digital surveillance practices, for example, where an individual does not communicate sensitive health information due to fear that his or her anonymity can be compromised. The declaration of human rights at the national level include similar provisions.

It has been claimed by some organizations and individuals that in some cases the exchange of information by electronic means is part of a conscious compromise by which individuals give out information about themselves and their relationships voluntarily in return for access to digital info, content and services. However, doubts arise in this case about the extent to which users are really aware of what data they share or post, with whom they share this data and for which purpose the data is really used in a certain period of time. According to a report to the US president [105], the reality of *Big Data* is that it is very difficult to ensure the anonymity of data once it is all collected. While there are many research projects that endeavor to conceal personal information, but they are not truly successful, because the efforts and investments in gaining large amount of info and data are much greater than the efforts made to improve privacy and security.

The focus on controlling the collection and storage of personal data is important, but is no longer sufficient for the protection of personal privacy as big data analytics brings up unexpected results from the data. Similarly, it was claimed that the pure interception or collection of big data does not constitute an invasion of privacy. But this argument is not convincing. Aggregating data or metadata may generally give insights into a person's behavior, social relationships, private preference and sexual identity. As the European Court has discovered, meta data at an aggregated level can allow very precise conclusions to be made about the private lives of individuals [99]. It thus follows that the (mere) existence of a mass surveillance program represents an invasion of privacy of people. The state has the burden of proof that such mass surveillance programs are not arbitrary and not illegal, which raises the question of what arbitrary and illegal really means in the digital age.

6.2   Bargaining Positions of Some States and Cultures

At the core of this book is the debate on who or what should have supervisory function over the critical resources for the operation of the global Internet, such as the root servers. The current *multi-stakeholder approach* combines different bodies from industry, science, politics and civil society, who all are or feel responsible for the management and development of the Internet. The announcement of the US government in March 2014 to to transfer the supervision of the IANA functions of ICANN under certain conditions to an international (new) organization provided the need for a worldwide discussion. In October 2016, the IANA stewardship transition has been completed. Looking back on the recent years worldwide commitment resulting in the successful conclusion of the almost 20 years long stewardship of the US government over the IANA functions. Therefore this section deals with the bargaining positions of large countries and cultures in the context of the multi-stakeholder model. However, it should be noted that towards the end of the publication process of this book net neutrality has become increasingly important because the Republican US government has announced its intention to kill net neutrality.

6.2.1   United States of America

The United States is the heavyweight nation when it comes to retaining the status quo in the area of *Internet Governance*, whereby status quo does not mean to have standstill in development. Starting from the status quo on the topic as a fall-back position, they advocate the multi-stakeholder model, a bottom-up process which includes political organizations, technical experts, governments, companies, civil society and individual users. From the standpoint of the United States this position is in their own interest and reasonable, because in all these areas, they are in a strong position or at least adequately represented when compared to other nations, particularly developing countries.

The strategy of the US for defending their position can be driven by the goal to be able to further customize the global Internet and thereby assert its own interests and also the interests of the leading US companies in this field (see Sect. 6.1.3). Here, Washington wants to make the multi-stakeholder approach more transparent to impede attacks or assaults from multilateral institutions that want to achieve a different form of Internet governance. Today, the challenge for the United States is to prevent an extension of the ITU's mandate in the direction of more participation in Internet governance, although many countries (e.g. the BRIC-states) mainly support this extension.

As one of the strongest arguments in favor of the multi-stakeholder model, the United States pointed out that they want to give up the coordination of IP address assignment by ICANN after the expiry of the last contract by end of 2015 and transfer its functions to another institution. On *October 1 2016*, the contract between ICANN and NTIA to perform the *IANA functions officially expired*. This point in time marks the *transition of the coordination and management* of the Internet's crucial resources to the private-sector, a process that started 1998. This is of *great importance for the cultural convergence* and is argued in the direction of the multi-stakeholder model in so far that the influence of the US over the Internet would decrease. The question then arises whether the IP address assignment by ICANN is used as an instrument of power (see Sect. 6.1.2).

The US further explains the success of the multi-stakeholder model where the Internet has expanded from a network dedicated to American science and research to international mass media. This is a success story where this model yielded suitable technical decisions, primarily by the IETF. It is an advantage that the IETF was largely protected against political influence from outside.[10] The larger involvement of international organizations such as the ITU, which are based on majority decisions between states, would not be a suitable solution for this. The IETF is an open organization for all individuals and other institutions with technical expertise, whereby decisions are based on the principle of a general or rough consensus, but requires a running code as a demonstration of effective implementation. There is no formal voting (see Sect. 6.1.2).

The US is generally opposed to an extension of the ITU's mandate in Internet Governance. They argue that the ITU should focus on its (technical) know-how and explicit (technical) mandate, especially since the work on issues such as *cyber security* and *data protection* are more political and social issues as opposed to technical. Washington wants to address these issues outside a UN framework, similar to trade issues (WTO) [488] and issues of intellectual property rights (WIPO) [93] as well as questions on the global financial system [426]. The ITU should continue to focus on issues such as how global access to information and communications technology can be expanded, since they have achieved a lot in the past. The focus can be placed on global investments for communications networks.

From the perspective of the United States, the ITU can play a role in supporting developing countries with international events and forums dealing with issues of cyber security and privacy. For example, the ITU can help developing countries to participate in the *Messaging, Malware and Mobile Anti-Abuse Working Group*. This group consists of a number of Internet service providers (ISPs), network operators, telecommunications companies in Asia, Europe, North and South America, which are active against the misuse of IT networks.

6.2.2 Germany

By principle, Germany wants only a small extension of the ITU mandate, by including the technical experience of ITU in this environment. This is so that worldwide access to information and communications technology will be further developed and

---

[10]Note: From the US perspective, this statement is comprehensible because the IETF is clearly dominated by the USA.

therefore being beneficial for all from the perspective of the *Millennium Development Goals* MDG on UN level.

Firstly, the extension of the *International Telecommunication Regulations* (ITR) from 1998 should be mentioned in this context, which was presented during the *World Conference on International Telecommunications* (WCIT) in 2012 in Dubai covering Internet governance issues. Germany's view was that the ITRs were not adequate at the time for governing Internet issues. Although the legal texts of the ITRs are not directly related to the Internet and Germany is concerned that these laws can be exploited by states to legitimize the restriction of Internet freedom. Similar to the position of the US, Germany has not signed the new ITRs. This position is also supported by many non-governmental institutions in Germany, which are well-organized across the country.

From Germany's perspective the advantages of the multi-stakeholder model outweigh the advantages of Internet governance via the ITU, i.e. giving the responsibilities to governments. Nevertheless, Germany considers this model not without controversy. Here the German government shows a form of global empathy (see Chap. 7) by putting themselves in the position of other countries, especially a number of the emerging and developing states currently under represented in governance issues. Germany recognizes that these countries do not have sufficient technical and human resources to be adequately represented in the current processes of Internet governance. These countries view Western states, especially the US, as the dominant players in this process characterized by asymmetry and inequality. This German understanding is positive from the perspective of the present book for a direction towards a cultural convergence in the future, although basic competency questions must be answered if international organizations should influence the Internet Governance.

Compared to the US, which often prefer informal or bilateral agreements rather than international institutions, Germany supports multilateral mechanisms as provided by the UN system, for example, when it comes to the setting of standards. Germany intends to maintain the multi-stakeholder approach primarily for another reason. The responsibility of an organization like the ITU for the regulation of the Internet plays into the hands of authoritarian governments that are very interested in increasing control over cyberspace in limiting Internet freedom. Such a change would be counterproductive from the perspective of democratic states, because authoritarian governments then would be in a better position to effectively carry out control measures if they had more to say on the establishment of Internet rules.

The federal government of Germany takes the position that the Internet is a *public good* and a *public space*, which should be *free*, *open* and *transparent*. Of particular interest are the issues on *freedom of expression*, *information*, *assembly*, *religion* and *right to privacy*. In order to promote these rights, Germany has joined the Freedom Online Coalition [111], which is an association of 24 states committed to the promotion of Internet freedom.

Germany has repeatedly emphasized that governments must not use their concerns, for example on terrorism, as a pretext for interference in the rights of individuals. The revelations in this context by Edward Snowden about the *National Security Agency* (NSA) has motivated Berlin to take a leading position on strengthening privacy rights of individuals. Together with Brazil, Germany promoted a UN General Assembly Resolution on the right to privacy in the digital age. The phone-hacking scandal regarding the German Chancellor Angela Merkel have influenced the discussion so that some countries have started to exert additional pressure on the US to recognize individual users' rights to privacy more than before.

Organizationally, the Federal Ministry for Economic Affairs and Energy has the representative role of the German position with respect to the ITU. This ministry represents Germany in the ITU Council. The federal government has distributed a lot of questions regarding Internet governance in various ministries [51]. Questions on cyber security are dealt with by the Federal Ministry of the Interior, access and infrastructure by the Federal Ministry of Transport and Digital Infrastructure, and data protection by the Federal Ministry of Justice and Consumer Protection. The Federal Foreign Office represents the German standpoint at the United Nations, the Federal Ministry for Economic Affairs and Energy is negotiating with ICANN regarding IP address assignment. The distribution of these tasks on several ministries is not an issue, but a central coordination site at the Federal Chancellery is necessary. In the future, with respect to the way of becoming an information and knowledge society, this subject will gain more importance.

6.2.3 China

Considering the number of Internet users in 2013, China has facilitated Internet access for almost half of its population with an IPR of about 42% (see Chap. 4). By the end of 2017, the number of Internet users was about 1 billion people. In absolute terms, China has the world's largest online market. This figure alone highlights the importance of China's clear vision on the issues surrounding Internet governance. The trend of growing Internet users in China will continue and increase the weight of China in the Internet Governance Ecosystem. Additionally, the Chinese language and writing system play an increasingly special role in this market.

In this context it is also worth mentioning that Chinese companies, which are directly or indirectly connected with the Chinese government, are one of the leading players in the Internet world now. The Internet is used in China as a source of economic and social development, but also for ensuring the political stability of the party. The Internet service providers, both the Backbone-ISPs and the Last-Mile-ISPs, play an important role within China's Internet regime [170]. In regards to Internet use, in its *"White Paper on Internet in China"* the Chinese issued the following statement [59]: *"The Internet sovereignty of China should be respected and protected. Citizens of the People's Republic of China and foreign citizens, legal persons and other organizations within the Chinese territory have the right and the freedom to use the Internet, at the same time they must obey the Chinese laws and regulations."*

The visions of the Chinese government are basically incompatible with the multistakeholder approach of the Internet Governance Forum, because currently most decisions (technical and organizational), with respect to the development of Internet, are not only taken by government employees but also by engineers, members of non-governmental organizations, lobbyists or other interested individuals, for example, those engaged in the IETF, ICANN or the IGF. In general, China looks to such organizations rather critically since they consider them a tool of the Western world, specifically the US. For example, China has rejected the expansion of the Internet Governance Forum mandate under the status quo within the UN in 2009, for the following reasons [185]. The IGF would not have the substance of solving the problems of the management of critical Internet resources. Furthermore, the developing countries are underrepresented in IGF meetings due to lack of human and financial resources for participation, which has led to the downplaying of the development agenda according to WSIS. In addition, the IGF has focused on many topics which have been previously worked out in other international organizations.

After the IGF mandate was extended for another five years in 2010, the efforts of China run towards that the IGF is becoming more and more a government affair. At a high-level meeting of the UN General Assembly (GA) in December 2015, the GA presented a document on the overall review of the implementation of the results of the World Summit on Information Society (WSIS). In a specific paragraph of the document, the *GA extended the IGF mandate until 2025*. It remains to be seen what China's approach will be in this context.

From China's point of view, there are partly political reasons for a critical position against ICANN having (technical) supervision functions over the Internet in the past. For example, China was dissatisfied that ICANN did not take into account the Chinese position in the China-Taiwan conflict and therefore refused to send governmental representatives to ICANN's *Governmental Advisory Board* between 2001 and 2009. As of 2009, China has sent people from its own circle to ICANN after a compromise was reached where Taiwan was renamed Chinese Taipei [11].

As a consequence, China would like to see Internet administration in the hands of the United Nations. The white paper [59] states: *"China wants the United Nations' role to be fully extended with regard to international Internet administration. China supports the establishment of a major and international organization for Internet administration under the UN system with democratic processes on a worldwide scale."* The ITU should take the lead role. The official big picture in China is that security is more important than freedom, which means that a *secure* Internet is preferred over an *open* Internet. There are diametrically opposed ideas between China and the US or the EU. China believes that the Internet should be subject to the sovereignty of national states with each country paying attention to their own sovereignties. This also applies to topics such as censorship, with regard to China, a model of *self-discipline*, in which every company or provider or person is responsible for the disseminated and/or web-based content themselves and should therefore perform self-censorship [77]. The rules governing what is to be censored are determined primarily by state power. With the revelations regarding US surveillance and intelligence services such as NSA, China has further strengthened its position and wants to play an even stronger role in the debate on Internet governance.

Until the digital age today, it has been rather unusual for Chinese officials and ministers to address a global audience, but it seems that the Chinese government has changed their strategies and rhetoric on the subject of Internet governance. The focus here is primarily on the *de-Americanization of the Internet*, which is ultimately a reflection of the general international policy of China aimed at global competition with the US.

6.2.4 India

From the Indian government's point of view, the increase in Internet users in India to about one billion in the next few years is another crucial point. In addition to China, India will become the country with the largest number of Internet users. The increase in Internet users, coupled with the rapidly increasing volume of data increases the pressure on telecommunication infrastructure and raises questions about security aspects and legal systems in India.

The ITU should play a crucial role in the *Internet Governance Ecosystem*. One reason for this is that worldwide communication via the Internet cannot be treated separately from conventional communication, e.g. over the fixed telephone or via mobile phone. This reason makes sense from the perspective of the technological convergence of the Internet and mobile phone to the mobile Internet (see Chap. 4). On the other hand, India does not completely oppose a multi-stakeholder model. India recommends that the different players in the multi-stakeholder approach assume different roles and tasks. These players include the ITU.

In general, India is striving to clarify the relationships and interactions between the Internet and telecommunications networks. According to this argument, the Internet, with its functions and services, is regarded as a network that has been built on other networks (network of networks) and is still running there. So initially, the existence of the Internet is a result of using existing networks, in particular telecommunications networks. Looking at results of the empirical studies on the relation between the Internet and the fixed telephone (see Chap. 4 Sect. 4.4.2), one can well understand this point of view. They show that the Internet and the fixed telephone were closely linked in the year 2000, although this link has decoupled somewhat by the year 2013, but still exists. This empirical link is manly attributed to the fact, that a close physical infrastructure relation exists between the Internet and fixed network.

For the Indian government, the question is not whether the ITU should work on issues such as *routing, address resolution, network architecture* and *standardization*. India is familiar with ITU's many years of international experience in the field of telecommunications. India is proposing that the rights for IP numbering should be passed on to national authorities. The *Internet Assigned Numbers Authority* (IANA) assigned the IP addresses to five *Regional Internet Registries*. These, in turn, distribute the addresses to local or *national Internet registries* (see Sect. 6.1.2). The Indian government argues that IP address management should be easier at the national level, for example, when it comes to identifying IP addresses by country. National Internet traffic should be managed in the respective country or flow within national borders. This would remove the concerns regarding international jurisdiction of data flows only affecting Indians. The proposal leaves open the complex issue of competence for cross-border data flows [400].

6.2.5   Russia

Russia has developed into a major player in the international Internet governance debate, with the Russian government pushing for a centralist, hierarchical, and government-imposed model based on the inviolability of state sovereignty. Russia's position on Internet governance is determined by the country's *security concerns* and reservations regarding *security policy*. In their opinion, independent, uncontrolled sources of information may offend and threaten the state and society. Furthermore, Russia has a clear aversion to the status quo of Internet governance because the status quo, from a Russian perspective, involves a clear dominance of the US government, in particular over ICANN [312]. But this argument is no longer valid since the IANA-transition is completed in October 2016.

Russia favors the UN and especially the ITU as an organization that is most likely to address issues related to Internet governance. As a matter of principle Russia is similar to China. This idea does not fit into the conceptions of the Western world, nor to the non-hierarchical nature of architecture of the Internet. The multi-stakeholder model is also incompatible with the ideas of Moscow, however this does not seem to bother them. On the contrary, Russian foreign policy holds the view that the *world moves from a unipolar set-up*, which is strongly dominated by the USA, *towards a multipolar state*. Russia's intentions are to play a key role in this multipolar world, similar to the power of the former Soviet Union. This is why Russia has argued against the privileged position of ICANN which has acted as a form of US authority. Such management of the global Internet is no longer compatible with adequate *Global Governance* from a Russian perspective.

While the ideas on this issue differ between Russia and Western countries, Russia is building an *alliance* of like-minded states, mostly ex-USSR, such as *Kazakhstan*, *Uzbekistan* and *Tajikistan*. In this context, Russia launched the so-called *Russian Internet Governance Forum* on 7 April 2017 for the eighth time. Russia, together with these countries, aims to increase the pressure on the Western world in the direction of a more centralized global Internet administration.

6.2.6 Arab World

Since the establishment of the IGF in 2006, a number of actors from the Arab world have actively participated in the multi-stakeholder process. These include *The Economic and Social Commission for West Asia* (ESCWA) and the *League of Arab States* (LAS). With high media attention, the ESCWA launched the *Arab Dialogue on Internet Governance* (ArabDIG) initiative together with a publication of an Arab regional roadmap for Internet Governance in 2010 [98]. This timetable was then adjusted in 2012 with the emphasis on the need for an *Arab IGF* [10].

The Arab countries generally *advocate a strong authentication of all Internet users* meaning that they are primarily *against the current anonymity of the net*. Within the framework of the NSA intelligence agencies and the recent affairs surrounding the NSA, the Arab countries complain that national barriers should be built up with the active role of the ITU. Edward Snowden's revelations have demonstrated the root causes of the Internet blackout in Syria in its entire dimension of spy attacks that affect not only private users, but entire countries [1]. On this issue, the Arab states are very close with Russia. The Arab states demand that the work of the ITU should include possible legal norms against intelligence services and given a greater role in Internet governance.

The Arab states have complaints regarding the status of Internet governance. They view the multi-stakeholder model primarily as American dominated. Therefore, since December 2012, Arab states have been pushing for greater Internet responsibility for the UN and its sub-organization of the ITU. Within the ITU, they could make use of their rights and further restrict Internet freedom in their own countries due to the decision-making power of member states within the ITU. The Arab states want to have the opportunity to inspect communications (e.g. *Deep Packet Inspection*). For this reason, they propose to regulate the *flow of Internet traffic within national borders*. This is parallel to the position of India.

6.2.7 Latin America

The tenth *Internet Governance Forum in Latin America and the Caribbean* (LAC IGF) was held in August 2017 [245]. This provided the opportunity to expand deliberations and contributions regarding regional Internet governance as well as developing a common position of Latin American states at a global level. Latin America supports the development of participatory, inclusive and democratic processes within a multi-stakeholder approach. This is because broad consensus-building for a complex issue such as Internet governance can only be achieved through this model in the future. This is also a result of the *NETmundial conference* held in Brazil in April 2014 [308]. The major principles recognized at NETmundial include: *freedom of expression, privacy* and *protection of personal data, access to information technology, freedom of association, human rights, open standards, multi-stakeholder participation, respect for cultural and linguistic diversity, transparency* and *accountability* [450].

At the *World Telecommunication / ICT Policy Forum* (WTPF-13), held in Geneva in May 2013, Brazil, as one of the main Latin American states, took the lead in implementing a precise operationalization of the role of governments in the multi-stakeholder model for Internet governance. The ITU should provide means of government intervention in daily operations and the long-term policy of the Internet. In Latin America in general there is a wish to achieve a harmony of both concepts (multi-stakeholder and multilateral). On the one hand, they are convinced that the multi-stakeholder process is necessary. On the other hand, there are some aspects of the Internet that need to be negotiated more multilaterally between governments. These aspects include important questions such as *tax*, *security* and *jurisdiction*. Such international issues in the Internet world would have to be dealt within a multilateral context. However, this does not mean that a multi-stakeholder process can occur. In the end, decision-making in multilateral organizations must be approved.

6.2.8 Africa

In Africa, especially in southern regions, the coordination of Internet governance nationally and internationally is almost non-existent. The biggest problem is the relative low Internet penetration rate in Africa compared to other regions (see Chap. 4). In addition to limited Internet access, lack of financial resources prevent governmental and non-governmental organizations from getting actively involved into Internet governance issues. In addition, disinterest in this field is a major reason, partly promoted by the view that subjects are dominated by primarily Western powers and cannot be changed in the near future [53].

A number of questions related to the effectiveness of Internet governance structures and processes in Africa concern the activity of non-governmental organizations for human rights, Internet access, the development of Internet standards and technical protocols for (cheap) mobile Internet access, as well as Internet security and data protection. From the point of view of African states, organizations such as ICANN, IEEE or ISOC are considered (technically) competent in addressing issues related to those mentioned above. Apart from this, their own attitude is primarily characterized by lack of competence (in Africa) and sufficient financial resources in pursuing these issues in the long term [388]. The Internet Governance Forum and the multi-stakeholder approach is seen as an effective platform to discuss upcoming issues. Topics such as Internet security and data protection should be better addressed at the governmental level, which are then, for example, under the auspices of the International Telecommunications Union (ITU). From a general African perspective, *governments should be also better placed to tackle cybercrime*. Although the ITU is seen as the most appropriate international organization for these issues, most African countries also believe that both the global IGF and the African IGF [458] are the leading organizations for Internet administration. Moreover, Africa is of the opinion that the ITU is the most effective organization for research and development as well as the broad implementation of Internet access in the context of development programs and projects. An example of this are the expired *Millennium Development Goals* (MDGs). The African states have also emphasized decision-making rather than debate in the context of Internet governance issues. Through this perspective, the ICANN and ITU are perceived as important committees. The multi-stakeholder approach is widely seen as an innovative governance model that allows greater transparency, accessibility for decision-making and awareness of responsibility in the African public. It is also interesting that the multi-stakeholder model is seen as an opportunity to minimize US dominance of Internet governance.[11] Additionally, the multi-stakeholder approach is seen as a model for strengthening human rights in Africa since NGOs can articulate themselves at these international meetings. Most African countries are otherwise of the opinion that effective, democratic and participatory governance structures in the digital world are unlikely without first being realized in the non-digital world.

The African perspective on Internet governance can be summarized as follows: Many questions should be addressed by a coordinated multi-stakeholder approach. One of the most important issues remaining is the spread of Internet access, in particular broadband access via mobile phones. African affairs should be better coordinated between African states. The ISOC, which has many sections in different African countries, is considered an appropriate organization to distribute information and data on Internet governance issues while promoting Internet rights. Questions regarding Internet security should be better represented by multilateral organizations such as the ITU with its decision-making powers. The technical management of the Internet should continue to be governed by the IETF, but with the focus on tackling technical problems of the Internet (also) from an African perspective.

6.3 Change of Power Structures on the Way to a Global Information Society

Humanity has repeatedly created new power structures throughout history. Thereby the *government* side has played a dominant role as the *central authority* in these power structures. The *Internet is structurally a decentralized system* which knows no explicit central authority at the pinnacle of its power. This is particularly associated with the history of the Internet (see Sect. 6.1.1). The Internet is a networked system, a *network of networks*, in which every single server (e.g. router) has a specific control or power function. However, a central authority is missing, except for the root servers. This is why, from a political or economic point of view, it is so difficult to understand who or what has the decisive power in the Internet.

---

[11] Other regions of the world fear the multi-stakeholder model could increase the influence of the United States on the issue of Internet governance due to the dominance in manpower at these meetings.

In the Internet Governance Ecosystem this question is not easy to clear up. It is a system in which no one alone has control over the whole system, not even the US. The US control single aspects through the NSA or other Secret Services. American companies control individual areas, e.g. the search engine market through Google or social networks through Facebook (see Sect. 6.1.3). The IANA functions, which were connected with the supervision of the root servers by *ICANN* until end of 2016 did not mean that it does have total control over the Internet, although the *trojans function* resulting from the control over ICANN the US Department of Commerce had, assigns the United States a special role. This (only) means one restricted technical function (*notary function*) of the authorization of so-called *Zone-Files* for the *top-level domains* (see Sect. 6.1.2).

The United States dropped this responsibility [316], because it does not constitute real power. Therefore, it is easy for them to surrender the control on that matter. This system will therefore be decentralized further. At the Internet Governance forum in 2014 in Istanbul there were suggestions to dissolve the authoritative routes in millions of route servers into a decentralized system in which no one could control a single element. As we have in October 2016 the US did not extend the contract on IANA functions and the stewardship transition to a multi-stakeholder group is completed.

Thus, in perspective, total control of the Internet when it comes to the technical side, is rather illusory. Everyone does their own thing which is necessary, if one wishes to obtain the advantage that anyone can communicate with everybody anytime. This requires a collaborative approach. Stakeholders from various segments must come together in developing a new power structure for Internet governance. First, new ways must be found in the technical field in further developing global policies. This concerns the IP addresses, the Domain Name System, country codes or the generical domain names and further protocols. If one looks at the technical infrastructure and the *regulations* on or off the Internet, one discovers political questions like *freedom of expression* and *opinion, data protection, e-commerce, cyber safety*, etc. These are issues that fall within the scope of Internet application and usage.

Of course these influence *puplic policy issues*. There is a sort of a *(technical-political) clash of cultures*, because puplic policy issues have procedures that have grown for many years, either in a democracy or dictatorship, in a parliamentary or only in an executive area. It is about how laws are made, interests are put through and contracts between governments are treated. This political culture clashes with the other technical culture which has grown through the Internet. Here in particular, differences between decision-making in the political sphere and within the technical community are relevant. The *bottom-up, transparent open multistakeholder* processes of the technical area illustrate a strong participatory and democratic approach because it relies on the decentralization of decision-making and inclusion of experts and participants [314]. Regarding the politics of representative democracy, decisions are delegated to the parliament. Ideally a *habitation*[12] is created, where representative and participatory elements complement each other and result in a participatory decentralized system of decision making. There is not the one ideal way for all these issues. It might be, that in the end one come to governance models and regulations for privacy, which differ from those that address safety issues or intellectual property today.

This *co-habitation* can be better understood, if one devotes oneself with how technical aspects are regulated on the Internet. An understanding of these issues is often too little developed and deplored as a deficiency at the IGF 2014 in Istanbul. If one had a deeper understanding of how the IETF, the *Regional Internet Registries*, the ICANN, the IAB, and the IEEE, ergo the so–called I-Star organizations work (see Sect. 6.1.2), better decisions regarding the future of the Internet could be made.

What is clear is that more dominant and technical forms of decision-making processes can not be regulated in future Internet governance. It certainly works from the bottom to top, in the manner of technology. This is not universally accepted, especially in countries governed by authoritarian regimes. The effects of the *multi-stakeholder approach* are interesting at this point where *government employees are forced to better understand the technical aspects* of this field. The *technical community is increasingly confronted with political issues*, so the *cultures politics and technology converge*. An alternative would be the multilateral policy approach: governments make laws in their countries and negotiate treaties with other countries. However, in today's Internet world it would not be possible to find a consent.

Today's transformation is a long process that can be compared with the first industrial revolution. Up to this time, there was for example the royal family as a governance system of authority. The King consulted with his ministers and then made his decision. The industrial revolution brought up further stakeholders. The persons who had made large investments in certain production processes, wanted to have a say in certain subjects, e.g. in military questions. They demanded that the king can not make decisions before consulting them. Ultimately, the goal of these investors was a parliament, which should decide about politics. This resulted in a power struggle between *palace* and *parliament*.

In the following, there has been a slow *powershift* over many decades, namely the shift away *from monolithic king systems* to *parliamentary governance* systems. This change in policy was the result of changes in the economic base of a society. A similar development is now to be observed in the digital age. We are experiencing changes due to the digital revolution on the economic basis of our society. The change in the (digital) technical infrastructure with changes to the entire economic process leads to further change in the political superstructure in the medium term. The king has not disappeared to this day and this is even more true for the parliaments in the future. They are, however, supplemented or superimposed by new multi-stakeholder models of governance. Does the Internet and the development of multi-stakeholder models and changes in power structures mean a movement towards a certain kind of global

---

[12]Based on the French word for the term for a joint exertion of government responsibility by two contrary political camps [410].

culture? At the very least, the Internet has the potential to bring the diversity of cultures world-wide into closer relationships and stimulate convergence processes (see Chap. 8). In this context, it is important to move from the current *multi-stakeholder model* to a *multi-cultural-stakeholder model*.

6.4   Conclusion

The US has traditionally exerted more influence on the Internet than any other culture as we look back at history. Many countries in the world are calling for a model based on the interplay of different interest groups in a society as a whole, based on a multi-stakeholder approach. In this model, the different actors of the state, economy, science and civil society have different (cultural) perspectives with regard to the future of Internet governance and its ecosystem. On the one hand, this is the position of the US, EU and Japan, which to a certain extent stand for self-regulation according to the multi-stakeholder approach. On the other hand, the BRIC states and the Arab world want to see an increasing role of international organizations within the UN system regarding Internet governance. Here a similar perspective or convergence of positions can be seen across cultural boundaries. Conflict exists in the area of power and control questions regarding the dominant position of the US.

On October 1 2016, the contract between ICANN and NTIA to perform the IANA functions officially expired. This point in time marks the transition of the coordination and management of the Internet's crucial resources to the private-sector, a process that started 1998. The final chapter of the privatization process began in 2014, when NTIA asked ICANN to convene the global multistakeholder community, which is made up of private-sector representatives, technical experts, academics, civil society, governments and individual Internet end users, to come together and formulate proposals to both replace NTIA's historic stewardship role and enhance ICANN's accountability mechanisms. The package of proposals developed by the global community met the strict criteria established by NTIA in its March 2014 announcement. The proposals reinforce ICANN's existing multistakeholder model and aim at enhancing ICANN's accountability. The improvements include empowering the global Internet community to have direct recourse if they disagree with decisions made by ICANN the organization or the Board.

The Internet Assigned Numbers Authority (IANA) stewardship transition is completed now. In the night of September 30, 2016 to October 1, 2016, the telecommunication authority of the US Department of Commerce (NTIA) terminated the contract on the IANA functions with ICANN. For the first time since its founding in 1998, the multi-stakeholder-community controls and manages the so-called zone files of the top-level domains.

Chapter 7
The Global Superorganism as an Intelligent Human-Technology System

In history humanity has always used tools and technologies to guarantee its survival and to make life easier. Compared to the strong influence of earlier forms of technology on human culture, today's technologies, especially ICT, have greater effects in their promotion of the collective human intelligence and behavior. Communication processes caused by mobile telephony and the Internet (of Things) leads system-theoretically to the effect of the development of humanity to a kind of *global superorganism* which is a form of an intelligent *human technology system* [446], equipped with a so-called *digital nervous system*. The question of a cultural convergence to a global culture will be decided in the process of arrangement of this global superorganism (see Chap. 8).

For a better understanding of the developments taking place (see Sect. 7.1), this chapter describes a thematic approach to higher-order systems in the form of so-called superorganisms. This is one possible interpretation of the events. Such a superorganism originates from the *coordination* and *cooperation* of many (vital) individuals and (intelligent) technical components and emerges into something bigger. This paradigm is represented by different groups such as mammals, whose existence assumes, as in humans, a coordinated interaction of billions of cells; also ants and colonies of bees [231]. These are similar to groups of people, organizations, companies and [162, 285, 358] nature in its entirety [269]. To get a better understanding and interpretation of the phenomena of current global networking processes in the area of information and communications technology, the future human-machine system is described in context of current examples such as the *Internet of Things*, *artificial intelligence* (AI) or *big data* (see Sect. 7.2).

In addition, the question of the nature of a global superorganism humanity as a whole opens up new starting points to explore upcoming *cultural challenges* for human beings. This theoretical framework is helpful and considers cultural influences (see Sect. 7.3). Furthermore it is about the main questions such as: which power structures and its distribution are sustainable, how is the inclusion of global

© Springer International Publishing AG 2018
H. Ünver, *Global Networking, Communication and Culture: Conflict or Convergence?*, Studies in Systems, Decision and Control 151,
https://doi.org/10.1007/978-3-319-76448-1_7

communication processes significant for sustainability and whether worldwide *cultural empathy* becomes possible for humanity. Is there a global consciousness of cultures [247], which is beyond the sum of single consciousness of the individuals? [42] A better understanding of the questions regarding the intelligence of superorganisms enables a chance for a better understanding of cross-cultural empathy.

7.1   The Development of Humanity to a Global Superorganism

The idea that humanity with its different civilizations and cultures can be understood as a complex organism, a superorganism, goes back to ancient times.[1] There are many similarities between the roles of different actors in a *social body* and the roles or functions of cells, organs and structures in a *biological body* [162, 358, 366, 369]. In this context and metaphorically, for example the military and police fulfill the function of an *immune system* to protect society against threats. Protective functions in times of hunter-gatherer cultures had other characteristics when compared to modern times, because there was e.g. no (longterm) warehousing of goods and monetary system [431]. The immune system of humanity was able to develop to today's level throughout millions of years. Thanks to the advances in the area of health technology and pharmacy, the potential of this system increased in recent times. Even so, we sometimes experience illnesses. Humanity, as a global superorganism is not safe from menaces, because our social *immune systems* are not always able to recognize every threat or danger in order to counteract them successfully.[2] (International) Terrorism can be seen metaphorically as a disease in the body of *superorganism humanity*. Secret services and military are instances of the immune system that sometimes fails and sometimes overdo things. The latter metaphorically can be seen as an *autoimmune reaction*.

This book concentrates on global networking processes in the area of information and communication. The communication infrastructure, similar to power lines and other infrastructures in this context, can be seen as the (digital) nervous system of humanity. The *digital nervous system* of *superorganism humanity* in particular is composed of the *Internet* or the *Internet of Things*. In the literature it talks about the *World Brain* [497], *Global Brain* [163] or *Global Mind* [133]. This is a metaphor for a (central) nervous system of mankind since the invention of the telegraph. It should describe adequately, which quality a (global) electronic communication

---

[1] Aristotle was fascinated by the bee and ant colonies as a reference point for understanding societies. *"The whole is more than the sum of its parts"* indicates, in a certain sense, the dimension of a complex superorganism. The interaction of two components can generate independent new qualities. In systems theory one speaks of *emergence* or emergent properties [267, 268].

[2] It is interesting in this context, that the immune system protects the people even if they know nothing about its existence or partially understand its function.

infrastructure shows. One of the strengths of the Internet is its (permanent) availability as a whole system. Flexibility of access and high efficiency (inexpensive) are reasons why the Internet quickly became the standard for communication [30]. Other communication methods such as radio, telephone and television, amplified by *technological convergence*, are in this sense no longer the main artery of the (digital) nervous system of the *superorganism humanity* [95].

The concept of the superorganism contains a certain view of the basic functioning of this system. The evolution of humanity is in this context rather an *evolution of cooperation* [12–14] than an *evolution of competition* [71], however competition (*survival of the fittest*) is an important mechanism in supporting efficiency. With this in mind, the *emergency* of civilizations can be understood as a kind of formation of humanity to a system, basically as an evolution of cooperation between humans, supplemented by a mechanism of cultural conformity [162]. Cultural structures so far probably are just an intermediate step on the way to the next evolution step towards a global *superorganism humanity* [446].

7.1.1 Superorganism Humanity as Autopoietic System

The general systems theory distinguishes living (social or cultural) systems from non-living (physical or technical) systems by the feature of *autopoiesis* [265, 267]. The term autopoiesis is a composition of two ancient Greek words (*autos = self*, poiein = create, make) and stands for *selfproduction* or *selfpreservation* [278, 279]. The term originally appeared in the context of investigations on biological systems. Machines and computers (up to now) are not able to produce or reproduce themselves. Humans as organisms (systems of 2nd order) and humanity as a superorganism (system of 3rd order), however, have to organize and reproduce themselves in their states frequently. By reorganization they can transform states.[3] This *selforganization* also means a creation of infrastructure to avoid disorder where structures of information and communication play an increasing role. Organization extends order and decreases the entropy of a system [513]. In this sense autopoiesis means the processes that achieve a development of the system through self-organization and self-production. In human history such a development can be noticed in the transition from a society of hunters and gatherers to an agrarian society as well as in the transition from an agrarian to an industrial society [431]. At the moment we experience the transition from an industrial society to an information and knowledge society [369, 371].

Autopoietic systems are constituted of a network of processes, which generate their own components and competences (recursively) and detach themselves from their environment in some degree [267, 279]. Thus an autopoietic system is an autonomous (closed) unity, which ensures its own maintenance, growth and extension. Besides, environment can be seen as means of required resources as well

---

[3]With cells or protozoa in this connection one speaks about first-order systems [280].

as a source of setpoints and disturbances. In this context the internal stability and functionality of the autopoietic system matters. Therefore, the environment does not tell the system how to organize itself. Rather the system disposes of its own knowledge about how to organize itself as network of production processes.

In regards to using this theory on social or cultural systems, the demand for seclusion of autopoietic systems is controversial, including the distinction of what is inside and outside the system [159, 162]. Here, seclusion or closure property means that every component of the system is produced by one or more other components of the overall system [279]. This is because the definition of autopoietic systems by *Maturana* and *Varela* demands a production of own spatial and topological limits by the system itself [278]. In contrast to biological organisms, however, most social systems do not have clear spatial and topological limits. Although, currently national states constitute a form of (political) definition, these limits are not clearly defined socially [267]. This becomes clear, if one looks at the situation of families with migration background in such countries. These groups maintain their social relationships to other countries, i.e. their countries of origin, over national borders using money, communication, technologies and travel. Furthermore, a country is able to produce its essential products and services internally, but will also import and export many products. Germany should be mentioned here in its role as a long-standing champion in international export. A spatial and topological distinction is even more difficult if one considers cultures or cultural systems. Cultural spheres tend to extend over several states and non-contiguous areas. Boundary questions arise if one considers multicultural national states. In the age of mobile telephony and Internet, communication across cultural borders is unproblematic. This is also the case since people from different cultures use mobile phones and half of the worlds population uses the Internet (see Chaps. 4 and 5).

In the end, any attempt to draw borders between social systems and cultures proves to be porous and blurred. Even if one tries to use religion as a dividing line between one culture (e.g. Christianity) and another (e.g. Islam), one will identify similar patterns within the clearly defined dividing lines. This ultimately indicates a basically equal transcendent entity or equal universal values (see Chap. 5). Although the fathers of the theory of autopoietic systems refer to the consideration of biological systems [278, 279], other scientists like *Luhmann* [265], *Robb* [390] and *Hufford* [172] describe the application of these theories to social systems. Also interesting are posts by *Mersch* to this subject in the context of so-called systemic evolutionary theory [285]. In general, this form of theory of evolution maintains, that (autopoietic) reproduction of components of social systems induces its competences (communication). In the technological world not the technical products, but the companies as organizations, which produce these devices, are the autopoietic systems. The machines are only the competence of such companies. Of particular interest is the possible change of actor's objectives in the interaction of various levels. For example, possibly stronger interest of a person in influencing the thinking as well as the competence of many other people in comparison to the own biological reproduction and concentration on (usually relatively less) own descendants.

It appears that in the area of human social systems the character of an autopoietic system applies especially to humanity as a super-system, i.e. for the so-called *global superorganism* [162]. If one follows the application of the theory of autopoietic systems on social systems according to the works of Luhmann [265, 267], in particular communication constitutes and sustains societies. In other words: society is the sum of the communication processes (in a very general sense) between the elements of the system. It is precisely communication that changes rapidly in the wake of latest innovations in the field of information technology. Consequently our society changes, however, the question is where does this change take place?

7.1.2 Communication as a Driving Force in Superorganisms

Considering the organization of different cultures and civilizations, there are different points which seem to have an universal character [371]. On the one hand, in every culture *time* has a particular importance. It is a determining point for the description of all growth processes.[4] In addition, the *size of population* is a determining factor. Among the rest, it is a coarse measure of the performance of single cultural systems or even humanity as a whole at a specific time. With regard to communication, the mobile phone as a communicative device is used across all cultures by nearly everyone for the organization of professional and everyday life as well as for maintaining social contacts [55] (see Chaps. 4 and 5). The Internet is becoming considerably more important for the organization of life across cultures and life is even inconceivable without it in more developed countries (see Chaps. 4 and 5). Without using these forms of communication one would probably not be able to maintain current prosperity.

The *driving force* for the protection of social coexistence and the development of societies lies in *communication* and *interaction* [267]. Mathematically considered, the number of (potential) communication relations grows proportionally with the square of the number of participants in a system.[5] *The quadratic impact explains many network effects* [20]. Innovations for the utilization of the potential of squaring interactions generate much quicker effect than the growth of people. The Russian physicist *Kapitza* describes this type of a square growth in the context of world population development over the last thousands of years [210, 211]. In essence, his observation says that the numeral growth of people is far over-proportional surpassed by the growth of the interactions between people and in particular that these interactions promote social progress. Furthermore, the exchange of knowledge and experience between people increases where earlier it was based on *books* and now it

---

[4]Time is a common denominator for cultures, even if there are differences, for example the physical measurement or also the perception of time.

[5]The number of possible connections between N nodes in a graph is $\dfrac{N \cdot (N-1)}{2}$. This term grows proportional to N^2 for N large.

relies heavily on the *Internet*. This boosts the transition of the entire system humanity towards an intelligent superorganism massively [369].

Looking to the survival capability of cultures, people are born and die, but humanity as a superorganism continues, even if the superorganism transforms itself into other states over and over again.[6] When people are dying, much of their knowledge and skill is retained as long as other people preserve and develop the knowledge in the form of a *collective (cultural) memory*. This is done through communication for future generations [371]. Here, the primary collective memory of humanity from time to time can move from one culture to another, as it was apparently the case several times in the past (e.g., the role of Islam for the rediscovery of the antique Greek heritage in 800–1000 A.D.). Communication channels can be temporarily defined and controlled by a leading culture. The different cultures, however, adapt relatively quickly to the new forms of communication if they want to assert themselves (see Chaps. 4 and 5). Otherwise, they may die as a distinct culture and are integrated into other cultures or can even be exterminated. In the context of this book, globalization increases the risks of a cultures survival if individuals do not participate in the international interaction processes, i.e. the process of communication. This is especially true in the view of the future of communication and interaction. The *Internet* or the *Internet of Things* in this case form an important basis and a new quality for the expression of a *collective (cultural) memory* or of the knowledge of humanity.[7]

7.1.3   Evolution Based on Cooperation

The number of people on earth has grown in the past million years. The population increase over the last few hundred years reminds of exponential growth. Currently there are about 7.5 billions people on earth. It is estimated that the number of all people ever lived on earth is about 100 billion. A self-similar process toward such a number of people was described for a period of 4 million years [210, 211]. The key for growth and for overcoming resource limits is communication, copying and generation of knowledge within humanity, which is understood as a global superorganism [358, 359, 362]. In a certain sense, all people have transported ideas, experiences and innovations that together constitute a *knowledge-generating* and *knowledge-transferring* system with a *collective (cultural) memory* [371]. Besides individual contributions, these ideas and innovations are to be understood (only) in the context of *(cultural) cooperation* [12, 13]. Only at this level a handing down of the generated knowledge and skills is possible. Even if (technical) achievements and the discovery of countries and planets can be attributed to single individuals, these people were formed by a personal sphere that made the later innovations possible.

---

[6]Here is again an analogy to the biological body of a human.

[7]The capacity of information of a person may in future even go far beyond what people naturally are equipped with. One thinks of possible chip implants with connection to the central nervous system [512].

It often happens that individuals have the strength and propulsion to drive innovation and cooperation on a mass scale.

Currently, the pioneering innovations in the area of information technology and transport systems cause an immense burst of growth towards globalization, as it never happened in human's history before [133]. We experience a duplication of performance-price ratio in the speed of (digital) processors every two years – a factor 1000 every 20 years (*Moore's Law*) [297]. We find ourselves in a process of digital miniaturization (*downscaling*), which causes a transition phase in the development of the *superorganism humanity*. This is because more and more technical components support or speed up the (autopoietic) organization in this superorganism. In this transition phase we experience a high (cultural) dynamic in conjunction with a further growing world population to about *10 billion people in 2050* and a *rising consumption of water, food, energy* and *oil*. This results in a higher demand for human *learning abilities, efficiency* and *adaptability* [369, 371]. In the end, one feels the need for a *global cultural cooperation* to generally master these dynamics.

This historical transition phase is closely associated with limits. In this context the considerations of the *Club of Rome* in relation to scarcity of resources [284, 359, 380], information and knowledge [57] and the rapid pace of social and cultural changes are of high importance [210]. This begs the question of why we currently experience many cultural tensions and conflicts as we move forward in shaping the *superorganism humanity* even further. Maybe humanity has to be exposed to such *stress* in order to facilitate a transformation of the entire system to a new level of performance in terms of emergence. One of the problems behind the form of present globalization processes is *"that dominant cultures spread very quickly and force profound changes to people with different backgrounds"* [371]. This topic is explored in Chap. 8 in the context of three different fundamental future scenarios (balance, inequality or collapse) for humanity.

The theory discussing a *superorganism humanity* in this book is oriented to the goal of an efficient (cultural) cooperation in ensuring survival and coexistence. It has to deal with the fact that we still live in societies and cultures, which in addition to a collaboration or division of labor and cooperation, are permanently in competition and conflict situations. Problems sometimes arise from simple misunderstandings (i.e. unintentionally) or different interests (i.e. intentionally) between involved participants. At the same time, the insight gains space that a global (cultural) cooperation and division of labor using a tightly linked network of technical and social systems across borders between countries and cultures is necessary, if a sustainable development is to be ensured. This is a chance for a cultural convergence. Possibly the *superorganism humanity* in this context as an intelligent human–technology system creates a *global culture* which does not seek the dominance of one culture over another but the cooperation and interaction of different cultures as a manifestation of a global culture – a *global ethic* (see Chap. 8).

7.2   Always More Intelligent Human-Machine-Systems

In the ongoing processes towards a *superorganism humanity* we observe the constant increasing of the efficiency of the entire system in creating technical or mechanical forms of intelligence. What does this mean? To answer this question, one first has to discuss some thoughts about *intelligence* and *consciousness* in a system-theoretical view referring to the works [353, 356, 366]. In the context of machine intelligence, *Alan Turing* formulated the so-called *Turing–test* in 1950, which should answer the question, whether *machines* are able to *think* or are *intelligent* by having the ability to exhibit intelligent behavior, equivalent to or differentiable from that of humans [457]. Turing formulated his test in such a way that a person puts questions, that are answered by a person and a machine (in writing). If it is not possible to determine whether the respondent is human or a machine, then the machine would have to be regarded as intelligent. At the time, Turing certainly already had an idea of the enormous potential that would unfold by using computing machines in our social systems. However, the machines were not yet linked up together. What happens if intelligent machines are now interlinked among themselves and with humans through the Internet or the Internet of Things?

7.2.1   Intelligence of Social and Technical Systems

Here we first choose a system-theoretical approach for a specific understanding of intelligence in the context of evolution and thus an understanding of today's forms of *machine* or *artificial intelligence*. Intelligence generally constitutes a typical quality of mankind that according to the opinion of many people only occurs within humans. There is also the question of whether intelligence anyway can be understood by humans themselves?

The concept of *artificial intelligence* (AI) tries to transfer intelligent behavior of humans to machines. However, in the area of machines, intelligence and cognition are different than within humans [352]. Looking to machines, much has been achieved in the processing of symbols with regards to mathematical questions, or the game of chess as it concerns many aspects of *human intelligence* (*imagination, intuition, logic, ability to make decisions, learning, creativity, combination, empathy,* etc.).[8] Humans and machines play chess very differently from one another. Humans are obviously not able to bank in this extent on a massive parallel evaluation of halfmoves like a machine that entails a corresponding computing power [384]. Instead, people rely on other strengths, for example, on forms of intuition based on the brain as a neural network, which we still do not fully understand to this day. In chess computers appears that in certain types of analysis combined with the ability to check a large

---

[8]Already Konrad Zuse said that one day there would be calculating machines (*chess brain*), which would be able to defeat the chess world champion. However, it was for him not only about the calculating machine, but about the corresponding chess program [527].

number of alternatives, a exceptional performance can be achieved. Today machines are superior to humans in chess [300]. Especially laymen and ordinary chess players have no chance against a machine. Thereby, the machine power does not only result from *brute-force*, but from the evaluation function which plays a central role for the particular constellation on the chessboard. This function has become very powerful. Most chess players do not have such a mighty personal evaluation function in their neural network. Most recent AI systems can win against human players at Go and Poker. Chess and Go have in common that they are *perfect information games* which means both players (human and machine) know exactly what the other is working with. Poker has a more random nature and is a *imperfect information game*.

The highest forms of intelligence originate in a coordinated and appropriate interplay between the above-mentioned various intelligence phenomena [356]. Today the real demarcation of man against machine occurs with regard to the dimensions of creativity, emotionality, intuition, instinct and location in the world [352]. In this sense, intelligence of machines is not the same as the intelligence that manifests in a human brain in the form of neural networks [433]. Especially with the simple skills of intelligence services, such as the recognition of faces. The human being compared to the otherwise distinct and faster machine is still far more superior [326]. On the other hand, the machine is superior to human brains in many other ways [418]. The computer has become an irreplaceable tool for the use of certain mathematical methods and often for the derivation of evidence. The empirical analysis in Part II of this book would not have been possible if there were no numerical and statistical programs to implement corresponding algorithms by using a computer.

Ordinarily we see that the technical solution differs from the biological one [305]. One has needed and still needs to this day the ingenuity of individuals and groups to find suitable technical solutions; one can sometimes find suggestions in nature (bionics) [304]. In this case people discover what there is already and then the solution found is the scheme, or in other words, an algorithm that does the job. This is a way of how to make use of the incredible processing power of modern and powerful computers. Thus, if the algorithmic scheme is combined with the now economically available computing potential the results will more frequently surpass the achievements of humans [418], even the ones of the inventor. But this also means that we, as a highly developed civilization, can find a technical solution, which can be superior to mankind. On the other hand, surely humans can combine their skills with the skills of a machine. This is evident nowadays by world champions in chess while analyzing their moves and playing against others with the support of machines. Certainly it is also true that machines allow people who do not play chess to play brilliantly. Thus quality differences between people can potentially be overcome, which in some ways will be helpful. This pushes the imagination of a great helper, but also of a hard concurrent for jobs if one combines this with the fact that machines are able to work permanently and do not require a salary. They do need energy, maintenance but do not go on holidays, become ill and can be updated and improved constantly and do not resist changing (except potential problems with technical migration) [69, 418]. In the modern world, strongly influenced by information and communications technology, machines remove many cumbersome activities from everyday life. *Search engines, social networks, on-line traders, commercial systems, automotive cars, airplanes, ships* and *linguistics translations*, in the future, will unite in the Internet (of Things). These machines and their processes develop a huge intelligence regarding their evaluation and combination capacity for storing large amounts of data (*Big Data*).

7.2.2   The Internet (of Things) as a Digital Nervous Network

Linked up sensors and actuators, so-called *embedded systems*, will be prevalent almost everywhere in the future. Today, they are present in the engines and bodies of vehicles while monitoring combustion and performance as well as driving ability. They will be in our clothes and accessories. They are already in our smart phones, buildings and houses, hospitals, offices, companies, airplanes, ships and trains. *Intelligent* refrigerators can independently note that the milk or butter has passed its expiration date. We will have washing machines, which precisely start a wash cycle when electricity prices are low. On the *Internet of Things* objects become *intelligent* and can exchange data between each other. They produce huge volumes of data streams which will be used by industries and individuals for energy consumption, agriculture, economic processes, logistics and our health. *Algorithms* process these streams where insights and reports are created in real-time analyses, controls and by operating in the physical world.

Concerning this *Karl Steinbuch*[9] has written the following already in 1966 [434]:

> *"There will be in a few decades hardly any industrial products in which the*
> *computers are not interwoven."*

Behind the idea of the *Internet of Things* (IoT) is the linking of digital *information* in a network with real *physical products* (intelligent objects) and machinery [277]. The aim of the Internet of Things is to combine the virtual with the real world. A basis for this is for example the development of RIFD technology, through which goods and devices not only receive their own identity in the form of a code, but are also able to capture states and perform actions [110].

As described at the beginning of this chapter, the metaphor of a (global) nervous system of humanity has been used already in connection with a (global) electronic communicational infrastructures since the invention of the telegraph. In this regard, for example, *Hawthorne* in 1851 wrote [150]:

> *"Is it a fact – or have I dreamed it – that by means of electricity, the world of*
> *matter has become a great nerve, vibrating thousands of miles in a breathless point*
> *of time? Rather, the round globe is a vast head, a brain, instinct with intelligence:*

---

[9]Karl W. Steinbuch (1917–2005) was a German computer scientist, cyberneticist, and electrical engineer. He was an early pioneer of German computer science, in particular in the field of artificial neural networks. He also wrote about the interaction of society and modern media.

> *or shall we say it is itself a thought, nothing but thought, and no longer the*
> *substance which we dreamed it."*

A little later, *Wells* wrote the following with respect to a *World Brain* [497]:

> *"a sort of mental clearing house for the mind, a depot where knowledge and ideas*
> *are received, sorted, summarized, digested, clarified and compared."*

Even then *Wells* expressed that *"the idea of a permanent world encyclopedia"* could soon become a reality. He primarily criticized that encyclopaedias in the past were mainly written *by gentlemen for gentlemen* a world in which universal education for all was unthinkable. Today the former image of Wells has in a certain way come true with Wikipedia. Also search engines like *Google, Yahoo* and *Baidu* produce appropriate achievements.

With regard to the comparison between machines and the nervous system *Alan Turing* in 1950 critically noted: [457]:

> *"The nervous system is certainly not a discrete–state machine. A small error in the*
> *information about the size of a nervous impulse impinging on a neuron, may make*
> *a large difference to the size of the outgoing impulse. It may be argued that, this*
> *being so, one cannot expect to be able to mimic the behavior of the nervous system*
> *with a discrete-state system."*

A relativization of Turing's observation is to be found in *Palm* [326]. It has been worked out here that the *neural system brain* has developed in itself a small computer, or has accurately emulated it on the neural network. This interacts sufficiently with the rest of the neural network and at the same time points to the existence of a (fairly robust) *discrete–state systems* in our analogue neural network brain.

If we consider nowadays humanity with its narrow communication relationships and the link of billions of components via the Internet, massive changes are about to come. One could say that humanity, *technological components*, *artificial intelligence* and the *algorithms* in a *digital nervous system* are connected in a complex and intelligent superorganism. In this way billions of people and even more technical devices are linked to a unique large information network. In 2020 we can expect nearly 8 billion people and about 25–30 billion active technical components that together will generate huge amounts of data [422]. The rapid increase of data originating from the Internet and the Internet of Things is currently discussed under the theme *Big Data* [67–69, 283, 329, 377]. With *Big Data* we refer to the use of extremely large *volumes* of data with huge *velocity* and *variety*. We will experience Big Data as *Smart Data* and can profit from this development in society as a whole and in economics. Already today Internet communication takes place mainly between machines, to a considerable part also between people and machines. A huge, mighty and intelligent digital system is established based on never ending streams of communication where the communication between people is only a small part of the system. Today we can ask whether this system as a whole is already the most important carrier of intelligence in this world.

7.2.3 Technical Intelligence: Example Automotive Cars and Algorithmic Trading Systems

What is the reason for most of the technical improvements. Through information and communications technology, we have the highest rate of innovation and the greatest penetration rates of new technologies. At the heart of this development is the extreme speed of price reduction of a basic arithmetic operation (*Moore's Law*) [297]. Since decades we have experienced the increasing efficiency of processors. This results in the previously mentioned improvement by a factor of 1 billion over the last 60 years. These are incredible successes, such a fast technical improvement has never been there in human history. Ultimately, this is a consequence of the possibility of a miniaturization of the coding of information, which means that the coding of a single unit of information (one bit) requires less physical memory or space. This is because the link of *information and their physical manifestation is mostly decoupled*. We can make the coding of information (e.g. numbers) decrease without changing the results of the following algorithmic calculations, no matter if they arithmetic or boolean operations. This implies that the size of the physical representation of the number is not as important for the processing of numbers, however compared with the construction of a car the corresponding limiting size is that people have to sit in there afterwards. So, the size of the car is generally (at least in the downward direction) no variable.

In the digital world, the progress of hardware is also connected with a huge advance of application-oriented software systems as well as with input and output devices, networks and networking technologies with standards, protocols, platforms, etc. All this is coupled with permanent accruing data and provides access to a large amount of diverse data. This development is in a certain sense unavoidable and allows impressive applications. The OECD Synthesis Report on *"Data–Driven Innovation for Growth and Well–being"* [320] addresses two powerful examples: Algorithmic trading systems and driverless cars.

The fact that cars at some point in the future will be able to drive without a human driver is a big step forward. It will have an immense impact on peoples lives but could also lead to the loss of jobs in the traffic sector [320]. It is doubtful, whether new fulfilling jobs will follow. This is the subject of occupational survival in a *race between education and technology* [131]. One of the main reasons for this sudden breakthrough of these technologies is also due to the *Internet of Things*. It is characteristic that infrastructure and other cars tell the car what it needs to know to drive in a correct and appropriate way [8]. It is not necessary to equip a car with a technical image processing system, which is just as powerful as the image processing system of a human in order to drive in a safe manner. The power of the human image processing system is so high that it will still take a long time to develop an alternative technical solution with comparable performance [23]. This is not necessary for a car to drive itself therefore the car will get a large amount of information from the environment. The car will know more about the world around us than humans, although the image processing system of the car will be weaker.

Subsequently the autonomous driving will be more sure than a driving of people. It is conceivable that driving of people will be eventually completely forbidden except on special courses for safety reasons. The following is also to be expected: The autonomous driving of vehicles then has again appropriate effects on the number of employment of the public transport or of taxi drivers [320].[10]

The permanent receipt of information from the network is also the reason why a lot of robot applications will be possible soon, many of which were unattainable before. It is not in such a way that the sensor systems of a robot would be extremely good rather all devices in the room will deliver information for it to finish its actions. Consequently the robot will know more about its environment as a human who uses its biological sensor systems in the same situation. Surely we will also see progress if we make such information accessible for humans. In this way one improves the human skills, as for example with *Google glasses* [2].

The second example of technical forms of intelligence concerns the financial sector. Today algorithmic trading systems are already very effective. They are instruments of the financial market and almost irreplaceable in the area of *high frequency trading*, (HFT) [61]. In this area machines steer other machines and during milliseconds huge amounts of purchase and selling orders get issued on the stock exchanges. On the European stock exchanges high-frequency trade steered already about 40% of the whole commercial volume in 2013 and 70% on the US stock exchanges [81]. The *European Securities and Markets Authority* (ESMA) numbers the portion of the HFT at European stock exchanges depending on the measuring method to about 24–43% of the whole traded volume. In this type of trading algorithms are particularly programmed to so-called *arbitrage*. These algorithms look for minimal differences between prices of finance titles on worldwide stock exchanges to use them for gainful buying and selling [3]. The whole profit results from the sum of the profits of every single transaction which probably lie in the amount of a few cents. Finally, the mass of transactions are crucial for the profit. The effects of the HFT on transaction costs, liquidity and market quality remain in spite of many scientific efforts unpredictable. Because of the fact that in this business speed is of significant high importance, owners of algorithmic trading systems try to maximize their profit with so-called *collocation*. In this case high-capacity servers of the traders are positioned in the computer centers of stock exchanges for high charges.

But who are the losers in this game? These are the conventional institutional investors, i.e. fund societies, insurances, pension funds, but also private small investors because the yields of their retirement arrangement are reduced. European regulation activities such as the *Market in Financial Instruments Directive* (MiFD) as well as indirect regulation activities at the national level including transaction costs in *Italy* and *France* has provided the possibility to get over the effects of HFT. Additional regulation activities also include *Germany's* high-frequency trade laws. *Germany* has a pioneering role with the introduction of the high-frequency trade law

---

[10]On this questions originate with regard to the juridical arrangement. Who has the responsibility in case of an accident? Currently the situation is in such a way that the (human) driver must be able to take over the control of the car any time and he also has the (sole) responsibility [257].

in May 2013. This obliges every high-frequency trader with the so-called permit duty to apply for an authorization from the *Bundesanstalt für Finanzdienstleistungsaufsicht (BaFin)*. The acquisition of a *German* banking license with the associated minimum capital requirements is another provision, which entered into force with this law. This should serve to improve transparency and market integrity. The results from the first analysis on the impact of such a law show that trading activities significantly decreased, because, e.g., the trade of German DAX equities became more expensive in terms of the implicit transaction costs. In particular the number of orders on the order book tip has disproportionately decreased. Among other things, this is due to high-frequency traders with their superior precision and speed and are known to act mainly on the order book tip. The impact of the German high-frequency trade law provided a first indication, which consequences direct regulation might have in the area of HFT. From January 2018, new notification requirements will apply to investment enterprises as a result of the transposition of the Markets in Financial Instruments Directive II (MiFID II).

Here one could describe some more examples in the area of *Internet of Things* and *Big Data* that would result in the overall image that such developments transform humanity into an intelligent human-technology system as a *superorganism*. In this way, there are central *cultural challenges* at the macro level which will be discussed further.

7.3   Cultural Challenges

The cultural development within the scope of *superorganism humanity* is to be looked at from an institutional background compromised of a wide variety of national and cultural societies. Culture manifests itself, as shown above, by the cooperation of different components of a complicated entity (see Chap. 2). These structures are held together by interactive processes of a communicative type. Culture structures social and material processes on networks of social and technical interactions. Part of these networks are dynamic braids of power structures, which are (often) transferred or inherited intra- and intergenerationally. In the history of humanity these power structures have transformed from small groups of power holders (e.g., royal families, governments, business elites, etc.) into today's global governance structure and the economic exercise of power. Apart from the power holders there were and are always larger groups of people as a functioning system which fulfill tasks in keeping the power mesh and related material processes moving. These type of power mechanisms are to be expected for the future.

Turning now to the power structures in the information and knowledge society in the 21st century, there are certain power processes on the national and supranational level such as the EU, NATO and WTO and the economic sphere including large private and national Wealth Funds administrating billions of dollars. Additionally there are *multi-stakeholder attempts* for the configuration in, for example, the *Internet Governance* that are essential for the future *backbones* of our communication systems (see Chap. 6). The complexities of the Internet need complex governance structures in order to produce appropriate solutions. Applying this idea of the distribution of power on the political process, the question arises of which political power structures globalization in connection with information and communications technology will be created in the long run. Perhaps the most important question is whether humanity is able to construct an *Inclusive Global Governance with a balance between central and distributed control and power*. This also takes into consideration the cultural sensitivities and protects nature. Do we want a world that most people, in future possibly 10 billions, will conceive as balanced, fair and sustainable? Or are social and economic conflicts a threat? This is possibly exacerbated by environmental degradation and/or an ecological collapse, as in the wake of a worsening climate catastrophe. In this context it is also important to clarify how much power and control we want to give or will have to give to machines and algorithms.

7.3.1 Inclusion Instead of Exclusion

Considering the *"problematic consequences of functional differentiation"* described by *Luhmann* in connection with the *"rapid proliferation of human bodies"* and the *"own differentiation, specialization and high-performance orientation"* [267] in the context of current globalization processes, the concepts of *inclusion* and *exclusion* can be understood as a description of either systemic integration or disembodiment of individuals and states in the context of global communication structures. This is in the sense of a *superorganism humanity* also about the fair inclusion or exclusion of parts of the system regarding the participation in successes of the entire system [162].

It was also *Luhmann* who extended the concept of social integration, also used by *Parsons* [328], by the dimension of the inclusion or exclusion in communication issues. Here, however one has to ask in advance whether individuals or nations want to be part of a global system at all. One accepts in a certain extent, if groups like the Amish in the United States or also people who retreat to the monastery elude mostly from usual social processes under the condition that the *outsiders* do not compromise or threaten the *insiders*.

Nevertheless, the more humanity develop towards a superorganism. There is more of an affiliation of individuals and a nation to the overall social system that brings about advantages. These advantages raise the readiness to fit (so-called *enslavement principle*) [144]. The benefits include collectively achievable levels of security, freedom, prosperity, education, religion, etc. If we do it right, modern forms of slavery could be prevented and poverty could be overcome by ensuring education for all [520, 523]. In all these areas, in particular transport and information systems, there is a high potential to integrate individuals and states into the system.

However, a higher degree of freedom is usually accompanied with restrictions. For example, to ensure the safety or the balance of freedom of different actors, synergies between themselves differentiating components, of the one part, and interacting components, of the other part, have to be achieved [162]. The loss of freedom as a result of the inclusion into a global system with *constraints* compensates for the costs that would result from exclusion. However, the respective *constraints* are to be picked out as a central theme. Here, positions vary from the perspectives of different cultures.

If one considers inclusion at the individual level, one can imagine, for example, a woman wearing a veil in public despite liberal dress in that country. This veiled woman may come originally from a society with other constraints. In contrast, how does the situation looks like for a woman that comes from a clothing permissive country, when she (temporarily) lives in a country where women are more likely to be veiled? How far will, or can, or do individuals have to adapt? The pressure to adapt and willingness to adapt in a mosque is different than in a restaurant.

At the level of nations the global inclusion requires a harmonization of laws and constraints to allow a regulated exchange of human potential, money, goods and services (e.g., global trade within the policy of the WTO). This is also used to maintain a stable supranational political structures such as the EU [368]. During this inclusion process some groups of states want to disconnect from others or do not want to exist as part of the larger whole (e.g. BREXIT with Great Britain leaving the EU). Currently, there are dividing lines between Eastern and Western countries such as the US and Russia but also between Islamic and Christian countries. It is interesting to speculate how all this will evolve in the upcoming decades as well as the three future scenarios (*balance, inequality* or *collapse*) discussed by the *Club of Rome* [284, 359] (see Chap. 8).

7.3.2   Balance Between Central and Distributed Control

One of the most common objections to the mental model of a *superorganism humanity* is the totalitarian and collectivist system architecture [162]. In particular the freedom of individuals within the system is considered to be endangered. This, often spontaneous, negative connotation is comprehensible, but nevertheless inadequate. Under certain circumstances, these are only advanced arguments (if strong actors seek to avoid certain constraints by selecting between competing states) or at least a series of misunderstandings. *To live in freedom* is of great importance for our individual perception as well as for performing religious duties, the legal system and our general self-conception. It belongs to the determination of self-awareness and freedom. From this view of the world, there results a big motivational strength [356, 366]. Ultimately the ability to make responsible decisions is on the basis of our legal system. Many forms of freedom are linked to this such as the freedom of belief, expression, information, assembly, association and contract. These are all closely related to the issue of human dignity. The constitution of democratic states protects rights and freedoms without a direct reference to human dignity, such as the *Bill of Rights* from 1776, that postulates the *"unalienable right of life and freedom and also the possibility to acquire and retain property and to strive and attain happiness and security"* [6].

There is, however, the view that there is no freedom in a general sense. The processes and life had to take place afterwards as they took place, even if they are determined by individual decisions and (random) factors. In the end it results in personal decisions, which do not have to be free in the scientific sense [42]. Freedom is a good description for what happens, if the interaction between natural laws and cultural contexts together with the processing procedures in the brains (as *discrete–state system* like a analogue neuronal network) in the end yield a decision in the sense of a causality about what has to be done now. In other words: The possible concretely existing form of non-freedom in this case is what most people intuitively imagine under freedom [42, 366].

In a *global superorganism* the interdependence between international organizations and nations will increase in the coming years. In this situation it will become difficult for individual players to dominate the world. This means a movement towards a multipolar world as a unit, according to an understanding of the *unity in diversity* [33].

The *superorganism humanity* requires a balance between centralized and decentralized control mechanisms based on cooperation and communication. A certain degree of centralization is necessary in allowing for a homogeneous system [159]. The most important advantage of a central control is the speed at which decisions (depending on demand) can be made. The more centralized apparatus of power in *China* seems to promote the economic recovery better than the democratic, multicultural system of *India*. It remains to be seen how far this applies for the long-term prospects of such states.

It is certain that the *global superorganism* today needs more central mechanisms of coordination if sustainability is the goal. There is a lot to do, from avoiding a *climate catastrophe* [376] to preventing *financial crises* [426]. Here the *invisible hand* by Adam Smith[11] is no solution, because it only leads in the right direction if the governance conditions are right, which is not the case nowadays. This particularly applies to the subjects of taxation and *containment of tax havens* [526]. The attempt by the international community to deal with such global themes at the level of the G20 member states is a step in the right direction [371]. G20 countries represent 2/3 of the world's population where almost 90% of world economic output is generated.

---

[11]One has to note that *Adam Smith* has been both an *economist* and a *moral philosopher*. In his work *"An Inquiry into the Nature and Causes of the Wealth of Nations"* [424] he does not claim that (rational) egoism (always) leads to the prosperity of a nation, but that this is the case only if the self-interested behavior is limited and long-term (meaning sustainable in the sense of this book) focused.

7.3.3   Global and Cultural Empathy

More specifically empathy denotes the ability of a human brain to understand what another is thinking. When it comes to empathy it is assumed that the other is able to think, more specifically, think in a different way. This is sometimes discussed also within the scope of a *"Theory of Mind"* [42], which is more than just the attempt to conclude from the observed behaviour of other their actions in the future. One can also call this *behavioral reader*. Compared with this much more sophisticated is *mind reading*. Probably only the (empathic) person is able to this [42]. In particular, empathy involves an understanding that others might think differently than oneself. This topic is particularly important for today's globalisation processes, for instance for the overcoming of conflicts between the West and the Islamic world. Especially due to recent events such as September 11, 2001, the Israel-Palestine conflict [419] or the issue of Islamic cartoons. Building up empathy on the individual level is not easy. How is a construction of empathy then for whole cultures together conceivable? From the point of view of the author of this book the development of an imagination is helpful for a sort of global culture to promote more empathy between cultures. This is the main subject of the next Chap. 8.

7.4   Conclusion

The information and communications infrastructure can be considered in the context of global networking processes as a *digital nervous system* of mankind. Humanity as a whole is in the process of digitization on the way becoming a *global superorganism* as an *autopoietic system*. In this system, communication is the driving force for humanity in reaching the next stage of evolution. This development is based on *cultural cooperation*, whereby humanity with the *Internet* and the *Internet of Things* in connection with *Big Data* is increasingly becoming an intelligent human–technology–system. A central cultural challenge humanity has to deal with is the theme of *inclusion* as a systemic and fair integration of people and countries in global communication processes. In addition, the possibility of *Inclusive Global Governance* with a *balance between centralized and distributed control consistently protecting cultural sensitivities and nature*. This is possible in the context of global sustainable development and facilitates global empathy. The emergence of a global culture includes a variety of worldviews that coexist and are tolerated under an appropriate worldwide economic, educational and religious culture (see Chap. 8).

Chapter 8
Convergence Towards a Global Culture?

Fast and reliable Internet and its cross-cultural use, social networks and digital contents build an information technology basis for the development of all cultures in becoming one *global culture*. Conceptually this does not mean a global culture implies equal global behavior or worldviews everywhere. For example people will still have their religious beliefs or not even adhere to a specific religion in the future. The engraving role of religion can change as a result of cohabitation, way of life, and the question of how and where one works and whom to marry. This is already the case in countries such as Germany and the US. The diversity of philosophies of the people living in these countries does not imply a lack of a common culture. What is allowed for people from different cultures, what they do and how they will live is less important for cohabitation, if it even comes to this.

In the same way the letterpress stimulated the (nearly) worldwide distribution of ideas, for example such as the enlightment principles and literacy, the Internet supports and even demands the faster distribution of knowledge as well as ability to use this access. Typically, this way enables a single person to acquire extensive knowledge over a wide variety of subject areas in an inexpensive and comfortable manner. Moreover, some information systems accumulate knowledge about persons without their contribution therefore they reinforce the effect of IT systems. The latter one happens today (with advantage and disadvantage for the concerned people) for example under the term *Big Data*.

Currently it is still uncertain if an assimilation of cultures can be seen as a consequence of today's processes concerning the World Wide Web, which is often and successfully used as propaganda tool against *cultural convergence*. Therefore the Internet can be a catalyst for convergence, conversely it could be responsible for increased conflicts as well. In all probability general tendencies regarding *cooperation* and *conflicts* between people respectively are crucial factors here. This means convergence is not (only) determined by the World Wide Web as a communication

© Springer International Publishing AG 2018
H. Ünver, *Global Networking, Communication and Culture: Conflict or Convergence?*, Studies in Systems, Decision and Control 151, https://doi.org/10.1007/978-3-319-76448-1_8

vehicle, but primarily by broader settings. Such factors and thereby resulting developments will be analyzed in this chapter. Most impacts of the digital networking process referring to the future cannot be specified now. They will rather ensue in the forthcoming decades and centuries – in a path-determined process.

The following sections describe three potential, broadly defined scenarios for the future, which are influenced by the work, discussions and ideas of the Club of Rome [57, 210, 284, 359, 371]. Thereby the implications of digitization as a process of globalization will be studied. Subsequently the author dedicates himself to the terms of: *world polity/global culture* and *world ethos*, which are describing the issue of a consensus on values between different cultures in connection with the organization of the *economic*, *education* and *religious culture* [239]. Finally this topic takes up the discussion on the *Clash of Civilizations*, proclaimed by Huntington. The following will discuss some reasons why this concept of Huntington is often misunderstood by people. This will be clarified by leading to the point of view declared by Huntington himself [174] and also Amartya Sen [416].

8.1   The Term Global Culture

What is understood about the term *global culture* or the civilization of the whole human race? Cultures are social systems that arrange and regulate the co-habitation of humans (see Chaps. 2 and 7). The essentials of life are to be found in cultures: The *cooperation of genders*, the interaction between *young and old people*, how one falls in *love* with someone, *marriage*, *divorce/breakup*, *heritage* and exposure to *death*.

Today, big culture systems all around the world already covers, to various extent, together realized elements like the world commerce and accordingly the *global economy*, the dealing with and transfer of *knowledge*, *educational systems*, proportions of *power* and *hierarchy* levels. The development of different *religions* and their togetherness and conflicts and ideologies are also very essential. If this is the understanding of culture, then questions related to *wealth*, *poverty*, *education* and *religion arise*. These are distinctions, which have an influence on the behavior of humans – one can find these differences in every cultural context [416]. *Language, religion, skin color, eating habits, social manners*, how one presents himself in public, dealing with children, what is allowed and not allowed, are assigned to cultures.

Nowadays, people still prefer to marry someone from their own cultural background. Admittedly also because one can most likely meet a partner within his own cultural group. Relationships within a culture generally facilitate the love attachment and the situation at home in everyday life, for example the organization of meals, celebrations, the interaction with family members, relatives, and guests/friends. The affinity of people coming from a humble background towards the things one is used to, is also determined by the fact that there are barely people from other background in the direct social environment [135, 256, 261].

The question about convergence of cultures is especially interesting, if programs such as the Erasmus exchange program in Europe or even travel for leisure or work lets people study and work in other countries, where they meet with people from other cultural background [423].[1] With the modern possibilities of information and communications technology, with which the people are becoming increasingly mobile and can be permanently online, such phenomena are amplified. One may wonder what triggers this. Are there effects on tolerance, e.g. related to behavior towards others, a higher appreciation and more empathy? Are for granted taken opinions possibly scrutinized? How high is the probability that it leads to transcultural friendships, partnerships, economic cooperation or love relationships and how long do these relationships uphold and how stable are they?

Any direct and tight worldwide communication has to accept the essential values of the different cultures in the sense of a global culture [83]. Up to now, we have not reached this point and we might not even will in the future. However if we move in this direction, the author of this book says, there has to be a economical minimum standard (*minimum daily allowance*) for every human, so that nobody needs to suffer of hunger any more [342]. This means that overcoming hunger needs to be top priority. According to Nobel Peace Prize winner *Muhammad Yunus*, poverty is something that belongs into a museum [520]. This problem could be solved with something like HARTZ IV for the whole world (universal basic income, UBI) after the idea of an unconditional basic income for everybody in Germany [134]. Thereby all states must participate in the financing. In order for a global culture to take hold *general education for all* is needed. This provides the chance for employment and enables or even enforces (e.g. via traveling) a gathering of knowledge and information about one's own culture as well as other cultures and techniques [235, 496]. The minimum level of education may be based on a common commercial language.[2] At the same time, we need to ensure that every single language available today is protected as a cultural good and value. In this context AI systems can be very helpful in future, in particular when it comes to mutual understanding between different languages. AI and speech recognition can provide massive improvements. So, in the near future, it is highly probable that machines and APPs can (simultaneously) translate a specific language into any other language. Eventually, widespread access to the digital world and the resulting communication will be a major part of a global culture, provided that they once emerge. After this, we can expect norms and values in accordance with a global ethics [234, 236, 239].

---

[1]In context of the *Young Leaders Group* of the *European Institutes of Innovation and Technology Foundation* the author of this book in cooperation with other people introduced at the *Annual Innovation Forum 2013* in Brussels a concept about the future mobility of students between European universities with the usage of the possibilities in the area big data [477].

[2]Language barriers for a (simple) international exchange and transcultural understanding could also be overcome with the increasing use of help of machine translation. But even if future machines are able to translate spoken words relatively well, it will not be possible in the foreseeable future to let a machine emulate the cultural different ways of communication like gestures or countenance, if it even will be possible.

8.2 Three Fundamental Future Scenarios

The situation today shows that global social processes is going in a direction, which diverges from a global balance, global democratic governance structures and a global convergence [371]. This corresponds with the fact that current developments oppose sustainable development in many aspects [359]. This includes aspects from the social division between North and South in the form of a global apartheid [3] and a (neo-) feudal future perspective (inequality) that leads to overworking an a subsequent ecological collapse i.e. effects of the growing climate crisis. These are possible future scenarios for humanity, discussed by the *Club of Rome* (CoR) in their works. These scenarios are explained below [284, 359, 380].

The CoR has long been working on future considerations that deal with the question of how to proceed the global efforts of reaching a balanced world and sustainable development. At the UN level (Rio +20, current UN processes of phrasing of the so called *Sustainable Development Goals 2016–2030* (SDGs) [460] as following processes of the so called *Millenium Development Goals 2000–2015* (MDGs) [462]) or on a OECD level (*Better Life Index*) [259, 322], the consensus aims for a *green and inclusive economy* as well as *green and inclusive growth* [507]. In Germanophone regions one speaks of ecologically and socially regulated markets (*Ökosoziale Marktwirtschaft – Eco-social market economy*) [359, 361, 372, 389]. This conception of the future has technical requirements and requires massive technological innovation and investment, for example, in new energy systems (widely available, inexpensive, environmentally friendly and climate neutral), but then also in political innovation. The technological innovations are probably easier to achieve than the political ones. It is about a suitable *Global Governance* [361, 368] that *encloses* the world's economic processes like the financial structure [426] within the meaning of *green and inclusive* [507]. This could be for example the reduction of CO_2 emissions [376] or setting up minimum social standards for each [134].

If one addresses the question, whether the global networking and international communication leads to the creation of a global culture, one concludes a world, where religion, ethnics, language, skin color and also ideology have decreasing influence on, if humans build a tight relationship with each other, or form a cooperation. This is called the *cultural convergence* fall. In this case, less extreme differences are observed relating to lifestyles, eating and behavior habits in former cultures than it is today. This is an effect of the increasing interactions and communications across cultural borders of today's state. This happens with a high cultural diversity and the increasing interdependence of different cultural systems, whereby the cultural diversity is destined through the complexity of the different cultural systems, by communication technologies, media and the increasing worldwide net product networks [146]. Expressed differently: For a prediction of the behavior of human beings, religion,

---

[3] According to Pogge [342], Radermacher [359], Ziegler [523] and many other observers the current world is in a state of glabal apartheid: High relative income differentials allow material feedthrough in many social, cultural and ecological contexts and attack human rights directly, which will not lead to a peaceful and sustainable state in the long term.

ethnicity, language and skin color could be significantly less important indicators in the future, in comparison to today. This statement is aimed at the above-described scenario for the future balance.

The (international) communication may also be an instrument for conflict or amplify in a case of conflict. With increasing social inequality and the scenario of global collapse, the Internet will more likely become a mechanism of action.

8.2.1 Collapse – Overshoot and Decline Enforced by Nature

Throughout human history, between great successes there were also occurrences of collapse from time to time, as described in *Jared Diamond's* book *"Collapse: Why societies survive or go down"* [88]. Diamond's book refers to the history of human civilizations, which repeatedly lead to states of total degradation, e.g. of the ecological systems or mass deaths. This can be understood by the example of Easter Island. This collapse coheres with the fact, that there were no governance structures with which the inhabitants could have sustained themselves with the limited resources they had. Instead there were concurring partial systems (clans) with very different potentials ready to fight for their needs at the expense of others. Therefore they primarily worked against one another instead of cooperating in dealing with the challenges they faced.

A similar situation today can be seen with sovereign states with different positions primarily following their own interests. Ultimately, they compete with one another rather than work together in addressing common problems [368]. The game theoretic situation is often also not favorable [138]. We are in a so called *Prisoner's Dilemma* situation, in which states make false decisions and do not cooperate, even if it appears that it is in their best interests to do so. [13, 14, 344, 382]. In history there were attempts of individuals to secure resources for themselves, that would have potentially been able to provide a base for everybody. Tragically the resource base was strained to the point of collapse [88, 367]. What can be expected, if global cooperation on topics such as in the context of Internet governance and the threat of the mass surveillance fails? The Club of Rome discusses *overshoot and managed decline* and alternatively *overshoot and decline enforced by nature* as realistic possibilities [380]. In a broad analogy this means the *neo–feudalization* of the world or an *ecological collapse* [359]. These alternatives are also to be found in contributions of advocates for a worldwide *Ecosocial Market Economy* as well as a *Global Marshall Plan* [132, 361, 389]. In this context, the position of the former sustainability council in the state of Baden-Württemberg on the subject of future energy and critical infrastructures, is interesting [303].

Going back to nature is not possible any more anyway, regarding the soon to be 10 billion of humans that want to live in a civilized and humane way, which is for example stated by Reichholf [383]. Since man has burst all natural limits due to its large number and the rapid development of technology – particularly in the field of information technology as an engine behind most innovations of recent decades – it now has to develop global government structures as an *artificial biotope* with it's mind. This structure must steer mankind's behavior toward sustainability and balance [309, 371]. The goal hereby is a harmonious hybrid human-technology system integrated into the biosphere as a superorganism [397] (see Chap. 7). Because regardless of whether humanity as a whole does or does not cooperate with each other, whatever it decides, there is just a single ecological base, only one planet [88].

The preservation of this livelihood must happen with today's existing people in accordance with the political structures. Changes have the nature of an *operation on a living body*. With every crossing of boundaries, more and more people will be confronted with a variety of problems – mainly pollution and exploitation [451, 506].[4] This is also an effect of the so called *rebound effect*, that has been discussed insightfully, among others by Jaques Neirynck [309]. This states that humans are consuming a higher amount of resources with continuous technological improvements and breakthroughs. This effect is based on the fact that we are becoming more efficient per unit of net product as a result of technology, we consume and need more and more units absolutely. This means, that the profit in relative efficiency is compensated or overcompensated by increasing absolute consumption. This effect is also demonstrated in the chip industry. Hereto, *Moore's law* was previously introduced, describing the duplication of the chip performance every 18–24 months [297]. This exact *downscaling* causes that todays computer chips are to be found in many devices, from appliances and cars to industry processes. In the future, all the devices will be connected with each other, known as the Internet of Things [422]. The impact of the *rebound effect*, triggered by the progress in the area of information technology can lead to a situation of collapse, while it seems impossible at the same time to secure a sustainable future for 10 billion people without using these forms of technical progress on a mass scale [369].

The question to be asked is if the political structures in the respective societies, especially the coordination between each other, is able to react to upcoming problems together and appropriately. Worldwide one must adapt oneself to a limitation of available levels of resources that are to be used. This is especially true for the utilization of fossil fuels. The best way would be a new energy system or energy carrier, similar to 300 years ago with the switch from wood to coal which rescued a majority of forests [373].[5] If this is not possible, there is the danger of a collapse, which *Radermacher* estimates at 15% [359]. From the perspective of the *Club of Rome*, *Randers* estimates the probability for a collapse to be even higher in his scenario of *overshoot and decline enforced by nature* [380]. Therefore one of the biggest challenges will be how to collectively cope with shortage, e.g. through an *honest* price mechanism on the markets, which on one hand is linked with measures of social balance and on the other hand, yields required innovations timely, because

---

[4]From the perspective of the former UN Special Envoy for the Right to Food, Jean Ziegler, the following is to be mentioned: Let's regard external debts of the circa 100 so called developing countries. The revenues these countries make with exports, go right back to creditors abroad. There is nothing left for the promotion of agriculture and the fight against hunger (...) there is already a war conducted, the third global war – against the third world [523, 524].

[5]Today the forest is not endangered because of the usage of wood for energy generation, but because forest areas (e.g. rain forest areas) are needed for purposes of agriculture or settlements.

prices tell the (economical and ecological) *truth* [393]. In this context, one speaks in economical theory about an *internalization of external costs* [274]. This must also happen against the interests of those who do not want this, preferably in a *fair deal* between the involved states and its citizens. This is to be done especially in correlation with countries with a high level of wealth with a *carbon–based economy* [157, 438], specifically the OECD states. In the context of this book the following thought experiment is worth considering: *What consequences would the shutdown or the collapse of the Internet cause worldwide?* Certainly this scenario will not only affect those with Internet connection, but everyone in the world. This scenario gets more dramatic if we consider the future Internet and the Internet of Things.

8.2.2 Increasing Social Inequality – Overshoot and Managed Decline

The pattern of increasing social inequality and the relative impoverishment of many people, even in rich countries, is increasingly visible in recent years and has led to a new dimension of the discussion on poverty and wealth. As examples one could cite the situation in Southern Europe as a consequence of the financial crisis [427], the rising gap between the rich and the poor in Germany [50] and the situation in the US with declining income in the last 20 years despite high economic growth rates [341, 437]. However, in no country is social inequality as high as it is on a global scale. The reason for this is due to the greater differences in prosperity between countries than within each country [155, 201, 206]. This applies to notoriously unequal countries like Brazil and South Africa. Surprisingly the situation in South Africa barely improved after the end of the apartheid system. The extreme *two-class society* has remained. However, not a part of the small prosperous group are people of color. Ultimately the *global apartheid* is a consequence of the historical development, especially the colonialization in combination with the present globalization under the neoliberal paradigm of broadly free markets [348]. The recently defunct sociologist *Ulrich Beck* from Munich describes the tendencies to a transition of the post industrial Western societies in the direction of a *two-class society* as a process of assimilation to the developing countries. This is linked with a *loss of the middle* and a disintegration of the bourgeois society [29, 371]. This is a *downwards assimilation* instead of an *upwards assimilation* which is anticipated in less developed countries. Beck's sketch of the future shows a society in which, like in South American states like *Brazil*, a ruling class is facing a serving class. This is characterized through scenarios of poverty that are yet only known in third world countries. He describes this process also as a *brazilianization of the world*. However, considering the recent improvement of the balance in Brazil as a consequence of the former president Lula da Silva's politics [252], maybe today the term *South Africanization* would be more appropriate than *Brazilianization*. However, this would lead to misunderstandings caused by the importance of skin color for many people in South Africa.

How did Ulrich Beck characterize this situation? He says that *"it would be disastrous, if the fatalism of the postmodernism and the neo-liberal globalism develops towards a self–fulfilling prophecy. Then the worst visions could become reality, which already occupy the public imagination. One of them would be the Brazilianization of Europe"* [26].

In context and connection with the digitalization the following hint is important: Something similar can happen when intelligent systems replace well paid *intellectual jobs* in a great number [45, 118, 374, 377]. This raises the following questions: To what extent will the digitalization affect the supply and demand of jobs? Will the (small) property side of the machines use the technical progress for further increase of property concentration [341]? If this happens, can one hinder this politically? On one hand, today we witness increasing wages for professionals with *digital knowledge*, while on the other hand, *digital companies* employ cheap employees, e.g. in commerce and shipping [283]. Brazilianization would mean that the scarce resources are handled without the necessity of yield technical innovations. This results in a more and more exclusive access of few at the cost of many to solve the problems of shortage and to secure the environment.

Just as formerly, deer hunting was prohibited for normal citizens (delict: poaching) and became a privilege of a feudal class in order to reduce the pressure on the resource deer. Prospectively something similar to this could happen to energy and other resources in the future. Ultimately a new world structure could constitute, reflecting a globalization of markets, but without a democratic and balancing governance. This would be an increasingly historically developed supranational social system, with a powerful and rich global elite that controls on the one hand, the world financial system and on the other hand, the physical side of the economy. With the help of the media and money even politics is manipulated. In this context, former US vice president Al Gore [133] and the economy Nobel laureate Joseph Stiglitz [445] denote the US as a *plutocracy*. Today, this process is manifested in the accumulation of huge property components in the hands of a small number of families, the so called *super-rich*, that have great political power which they wield covertly [128, 341, 437].

This distribution structure includes a specific interaction with power hierarchies of strong states. In particular, the *United States* provide the *backbone power structure* that enforces ownership on the globe and declares these structures legitimate. If the situation does not change, especially regarding the taxation, a *world two class society* is very likely. It's core is basically the active *hemorrhage* of the *middle* of the OECD states. The resource requirement is thereby decreased massively. At the same time *global justice* is surprisingly achieved by a convergence downwards instead of upwards. This reduces resource requirements in a massive way, while a convergence upwards would induce a multiple of the current requirements, unless new technologies solve this dilemma with a massive increase of resource efficiency. However, significant price signals in the market are required for such technology, which require appropriate cooperative *global governance*. Here we've come full circle: If a global governance compatible with sustainability does not succeed, the world is forced to become a two class society in order to prevent resource collapse.

In a *case of conflict*, a majority of the world population would not participate in the technological development of the modern world with it's various possibilities [369]. The required innovations that are needed in order to secure a comfortable level of wealth for all people are not developed due to an inadequate global governance [365]. In this case, large parts of the population also are not educated according to the necessary level of knowledge, and therefore cannot participate in appropriate education and exchange programs [133]. Basically, they are assigned to limited knowledge and narrower local contexts where they are practically cooped up economically. This doesn't have to be a slum, but is not a very attractive place. The opportunities of education are limited. The jobs are more of a serving, subprime sort. This happens in a prosperous environment oriented around the elites, where they are organized worldwide and feel at home anywhere on the world. Until then, they have regulated all relevant economical questions with supranational treaties so their position cannot be endangered by any democratic process. The political processes will be steered from this side *(global plutocraty)* [133, 445]. If anything should run out, the elite simply have the luxury to leave the country and shift their assets elsewhere.

Since the overarching (cross-cultural) exchange will not happen for most people in the situation of increasing social inequality or collapse, it comes to relapse into closer cultural patterns and radical reforms of the respective dominant religions or worldviews. Rigid structures give people the false sense of protection and identity. Strangers are made into scapegoats, especially economically successful strangers who (apparently) take away the attractive jobs. It comes to a radicalization – against foreigners, against people of other (or no) religions. Aggression, conflict and violence, fear and intolerance are gaining ground. In this situation, people will try to organize a minimum of protection and care in their narrow cultural groupings. The case of collapse however, would be even worse than *Brazilianization*, because a collapse with it's resulting shortage of vital resources (like water, food, energy) and especially the *Brazilianization* in aggravated form, would lead to hundreds of millions of people who will not have enough food.

What does this mean in the context of global networking and Internet? In such a crisis-ridden environment politics may preferably seek to avoid damage with limited means. Politics is no longer an active formative, but rather only responsive actor in permanent crisis management, which binds all the spare capacity, yet never achieves to calm the situation sufficiently. Today, policies which oppose a form of digital and social division are mainly intended to expand the (mobile) access or broadband access to the Internet for the general population, which is addressed with keywords like Internet for All or Broadband for All [193]. Naturally, the world community is struggling more to achieve this than a country like Germany [34]. The situation worldwide is far from providing every human a (broadband capable) Internet access (see Chaps. 4 and 5). Even if all members of societies would have Internet connection, which may be possible in perspective, this does not state anything about the quality of this access, e.g. net neutrality or open internet.

Social stratification or outright discrimination, for example, in the field of education, will lead to different usage level corresponding different usage patterns [525]. Taking the differentiation of users based on their Internet use into account, the way of Internet usage is highly dependent on the status of a person in a society [158]. The handling of information varies greatly between educated and uneducated people, but also between those with power and influence and those without. In the future, a key question will ask to what extent observed differences in usage are a momentous turn out for the distribution of socially relevant resources, such as the distribution of money, social capital or access to relevant information.

Similar to today, people who are socially better off will profit more from the connectivity to the Internet than people who are not well off, due to social inequality [493]. At the same time, access to the Internet and an increase of Internet competence could lead to a decrease of *social inequality* through a higher level of education. However, in a global information society coined by educational and wealth differences, a social and therefore digital inequality is present in the predominant majority of the world population [525]. In other words: *Increasing social inequality leads to the growth of digital inequality and also to an increase in cases of conflict.*

In this case, cultural differentiation patterns lead to a strengthening of national cultures without taking the intercultural and transcultural relations into account and promoting cross-cultural communication. Thereby, the potential for cultural convergence is overridden. The level of prosperity includes access to digital technologies and in particular the Internet with differentiated ways of usage, while countries with lower level of prosperity are restricted in their usage. At the same time, the educational level of a country reflects the digital competence of the people living in them. Ultimately, this is expressed, in combination with the higher or lower usage possibilities, given by the level of prosperity in a respective country, as the differentiated Internet usage. The *cultural habitus*, which takes effect through wealth and educational differences, plays a key role in determining the Internet usage [158]. To some extent, the culture-specific conditions for Internet use are crucial if all people have access to the Internet.

If the world develops in the direction of increasing inequality or collapse, many people will be restricted in mobility, exclusion processes will be strengthened and they will relish less freedom and variety of experiences. These cultural limitations additionally reinforce the tendency of social inequality. It comes to a negative feedback effect, whereby cultures of lower status would be in a worse position compared to the cultures of higher status despite the availability of the Internet [493]. Overall, the (relative) position of cultures with lower status would deteriorate in the world community. Tendentially, the global spread of the Internet would preclude cultures of lower status out of global communication processes, although they participate in the Internet and use it as a tool for economic, political, education and religious purposes. Increasing social inequality – and even the situation of collapse – leads therefore to an inferior qualitative usage of the Internet and the amplification of conflicts between the cultures.

8.2.3 *Balance* – Sustainable Future

Overlooking the negative scenarios described above, there is a possibility of a positive future with a sustainable and peaceful perspective with states and cultures making fair contracts with one another. This is necessary to reasonably regulate the use of scarce resources and strains on the environment for example, through a smart climate contract that caps and sometime decreases the CO_2 emissions and ensures fair participation in arising adaption costs (climate justice) [370]. Emission rights, therefore, can be a significant instrument, as it is in the case of the European Union [46]. These rights could be configured tradable [375, 376]. Thus, adequate (scarcity-) prices for greenhouse gas emissions will be incorporated into the world economy. This affects people's behavior, lifestyle as well as the nature of the economy, technological developments and innovations which are all key for a global balance. However, this can only be brought out under adequate governance conditions. At best, all people have adequate income and thus an respectively specific and sufficient access to scarce resources. This route would also ensure a balance between cultures. The EU enlargement processes shows how one could take action globally [365]. The appointment of common standards at the supranational level in democratic processes (partly on supranational level), linked with elements of cross-financing seems currently to be a good. This could serve as a model for the world, because the development of the *EU* has the character of a *globalization in small* [359]. However, even the Europeans struggle with this, particularly illustrated through the world financial crisis [427] or the Brexit. How can this succeed worldwide if the disparities are even bigger than in Europe? Considering what will come in the next years and decades, it still remains uncertain whether balance will be attained. Will future more likely bring balance or destruction [359]?

System theoretically, this question is equivalent to whether the implementation of the market model of a worldwide eco-social market economy, a *green and inclusive economy* linked with a *green and inclusive growth* [507], can be implemented internationally or not. To this end, standards for environmental protection and social and cultural arrangements have to be agreed on and cross-financed internationally. These should be compatible with sustainability and represent an effective framework of the global economy. A consequent regulation of the world financial system [426], the containment of tax havens [526] and adequate transboundary taxation as economic processes have to be added [368]. An important element of balance and thus social equality is a prerequisite for a convergence of cultures and provides a good degree of education for all. However, it is very expensive to facilitate broad prosperity. The democratization of knowledge and education through the digitization of knowledge content, for example, on the Internet, can help [258]. From the perspective of one individual, education is mainly the basis for the defense of one's own position in relation to access attempts by others in positions of privilege. At this point, the Internet is instrumental as a tool for global education and training. Additionally, it can also contribute to greater transparency and disclosure [472], as well as promote supranational coordination [4].

The empirical analyses from part II of this book regarding the relation between educational level and Internet penetration rate of countries show that between 2000 and 2012, the for the access to Internet required education sunk, as measured with the Education Index. The ability to read and write is still a crucial requirement for Internet usage. In addition, in complicated issues, the understanding will be the bottleneck in the future. No matter how user-friendly Internet access will become, factual requirements of subjects demand a certain level of education [525].

For a global sustainable development incorporating central media functions, especially the Internet, on a basis of democratic principles, is important [108]. In essence, it is necessary that the Internet serves to provide relevant information and knowledge to the people who can then develop both, the appropriate specific knowledge and the intercultural skills on the basis of this information. The Internet offers never before existing possibilities to quickly and inexpensively acquire knowledge [313]. It is however questionable, whether the constant availability of information and knowledge actually leads to the correct knowledge for the world population. In the digital age, information asymmetry is a key issue when it comes to cultural understanding [140]. *The participatory effect of the Internet is not (only) a question of technical access or the constant availability of information. This effect is ultimately only a necessary condition to allow cultural convergence.*

An eco-social model for balance requires consensus between the states and cultures. The economically weak states must participate too, when considering the associated costs for environmental protection [216]. The decisive condition for this are the powerful socially supported education systems, which require international cross-financing through state contributions. All this does not mean egalitarianism, a fortiori neither socialism nor communism. It's about cultural balance, cooperation, development of potentials and the respect of human dignity. This balance is based on a mechanism of cooperation, which intelligently uses the principle of competition for the promotion of net outcome [139, 367]. Ultimately, cooperation was always the basis for the success of humanity [12, 162]. The only question is: who belongs to the circle of those with whom we collaborate? In other words: who is a part of the system and who is not?

If one tries to assess what has been said, one can conclude that there is a chance for an *eco-social and sustainable future*. Associated with this would be a convergence to a form of *global culture* [412] in the sense of a *world ethos* [239]. However, this approach requires political patterns that differ from what is the habit now. But there is still reason for hope: The shift of discussions on specific domestic policy issues after the *world financial crisis* from the level of the G7/G8 states to the level of G20 states, which make up 65% of the world's population and generate about 90% of the worlds GDP, is a step into the right direction [371]. This also applies to the current propositions of the OECD/G20 level for an enclosure of *tax havens* [526], the new regulation of transboundary economical processes [368] as well as the *automatic data exchange between banks* and financial services authorities, currently agreed upon by 50 states [321].

A global sustainable development contains economical, ecological and socio-cultural aspects. It is about how to effectively implement the economical side of life such as the wishes and expectations of people in terms of goods and services. This can be done in an ethical manner by taking into account social and cultural concerns. Sustainability is therefore crucial to a development towards a *global culture* [236, 371]. If *sustainability* succeeds, it is itself a strong unifying common element of a *global culture*, as a functioning Internet for all and a viable global transportation infrastructure that everybody benefits from. All such structural elements will need the *functioning cooperative interaction between the governments of the world*, the (multinational) *corporations* and *global civil society* as a basis. In Chap. 6, which disputes global networking processes in the context of power and controlling issues, the so called multi-stakeholder approach for future Internet governance is discussed as an exemplary model of a worldwide cooperation in governance issues. From the perspective of the author, the multi-stakeholder approach is not only a crucial element in the field of Internet governance, but also for the establishment of a functioning global governance in the area of politics. Just like our democracies cannot succeed without a strong civil society and dedicated companies [213, 486].

The potential emergence of global culture will make the preservation of the diversity of cultural historical experience and manifestations become a core theme [428, 469]. *Global culture would mean a largely common material culture* (airplanes and automobiles as well as the Internet) *with great diversity and mutual tolerance between worldviews, therefore no dominance of one culture over all the others* [413]. Rather, it is about developing a common basis of values (like human rights and tolerance for diversities) with tolerance and cooperation on the basis of all cultures and worldviews [459]. The developments in the European Union indicate what work has to be done for this. Therefore the EU pursues the preservation of cultural diversity as an explicit policy aim [104], as well as UNESCO at the UN level [469]. Obviously, this has its price. Efficiency may be greater in cultural homogeneity, but the diversity of life and the creativity of the entire system is rather poor. From an economic perspective, cultural diversity requires a certain extent of social compensation mechanisms for balancing for economic differences, which are the result of the different affinity of different cultural forms of organization of value creation processes. As an example, one may mention that in Israel devout Jews are largely funded by the state [310], because the extensive exertion of religious practices leaves little time for economic activities. The situation of financing monks in Buddhism is not much different [230]. Regarding the whole of society, it is about cross-financing between these two aspects.

Looking into the future, characterized by balance, the megatrends of the *Global New Economy* are to be worked out. Everything that is said below only applies under the condition that a global peace is kept and intact ecosystems and human rights are respected everywhere. One megatrend is the current rapid growth of the world population [210], which will be about 10 billions in the year 2050 [461]. In the context of global networking processes, the empirical analysis in part II of this book shows that the population explosion is heavily influenced by the factors of education and level of prosperity. Conditions of poverty generate large families with many children, who mostly remain poor and have access to limited education [166] – a vicious circle. However, in case of balance, there are good chances of overcoming such patterns while inequality and collapse would lead to rather more problems. The population would continue to grow, the danger of a two-class society would threaten and conflicts regarding resources would be foreseeable. However, the analysis also comes to the conclusion that the steady growth of the population is no conflict for the continued use of digital resources worldwide (see Chap. 4). Respective analyses of the relationship between the major religious groups and the use of modern communication technologies also show no conflict situation (see Chap. 5).

8.3   Global Culture, World Ethos and World Consensus

The term *global culture* refers to a the system of humanity as a whole [289, 487]. As described above, an important prerequisite for the constitution of a global culture is digital networking in the context of global sustainable development. Culture in form of worldview is only limitedly controllable through social power structures. This especially applies under the conditions of the modern digital age [411]. It is not surprising that even the power mechanisms in the former socialist state structures, who wanted to suppress religions, were only partially successful [275]. In a sense, this was the equivalent of the *missionary of faith* through church-doctrinaire facilities [307]. The methods of influencing people resemble each other. Who stood up against the teachings of the church in it's sphere of control, was accused of heresy and forced to a revision of his attitude by the *holy inquisition* through mental and physical restrictions. Who turned to a belief in *God* in a real-socialist state, reduced his possibilities of social advancement. Both forms of discipline have indeed partly controlled the social behavior of people, but the faith itself could not be eliminated.

Amartya Sen states that we belong to many cultural contexts, not just one culture or one worldview [416]. *Dahrendorf* spoke of *ligatures* in which we are integrated individually [70]. In a world of cultural convergence, these many contexts and ligatures, so our professional– and vocational orientations, our individual eating habits, our sport and travel preferences and relationships determine way more, than what we associate with culture in a strict sense, like language, religion, ethnicity, skin color and also worldview. Against this background, it seems that in addition to Internet access, social conditions have also been brought in the right direction due to adequate global governance. It can be said that the capacity of the Internet to promote the development of humanity towards world culture is limited. It is important to determine the nature of the global governance system and therefore the Internet governance. *A global governance that is promoting cultural convergence, includes the worldwide consensus to establish a economical, educational and religious culture.*

8.3.1   Economical Culture as Part of a Global Culture

A possible future for the world, as already mentioned, includes a global eco-social market economy as an economic system compatible with sustainability [359, 361,

371, 389]. In the Anglo-Saxon language it is about *green and inclusive economy* [507] in the context of a world culture in the sense of a world ethos [236, 238, 239]. A central thesis of this book is that in the time of globalization, we are moving towards a common global culture in the sense of a tolerant cooperation of many cultures and different worldviews with approximately similar material culture levels. This happens under the conditions of open information systems and an open international policy that is aiming for balance, which is only the case if the politics are green and inclusive, aiming for global economic balance.

Statistically the existing simultaneous occurrence of certain cultural patterns, composed of language, ethnicity, religion, eating habits and so on will keep on equalizing. This means that today's cultures will partially lose their characteristics, which at the same time means winning other characteristics, especially on a material level. The according specific cultural context will therefore play a much smaller role for the material reality of the people [145]. The different worldviews will stay [18] and will influence everyday life specifically. However, through tolerance, conflict free cooperation will be possible [237]. The different cultures themselves will be less homogeneous than today, and will be more colored and mixed, due to traveling, greater interaction and communication between people (3T's: talk, travel, trade). This also means that the questions of interaction, cooperation, relationships between people will be then determined by other factors than by factors that are close to a specific culture today.

A basic cooperation between different cultural systems can only succeed on a long term, if the economical interests of different parties are integrated into a world civic order, perceived as fair and just. This presupposes a corresponding global framework. These conditions have to specifically organize markets and competition on the basis of standards and principles that meet world ethical values [238]. This includes ethical standards of economic activity and transactions under specific guidelines which ensure the basic needs for a dignified life for all people and induce cultures [134]. Furthermore, these shall actively promote the development of skills for everybody, even over the Internet [97]. Development of skills is a key issue for a good working environment and economical cooperation. Especially the abolition of unworthy business and working conditions, the exploitation and utilization of people in form of (economically based) slave labor and child labor must succeed [238, 239, 371, 520]. All this goes hand in hand with the enforcement of international standards for labor, for example, based on the internationally adopted ILO rules at the UN–level [5, 73]. In this matter, the EU can serve as an example, however the enforcement of social standards is only simpler at the European versus the global level [72, 315]. In the information and economic society, occupational safety has to be ensured with state of the art technology [499]. The product and process safety can be further promoted with the help of the Internet of Things, in order to protect the health of people and in parallel enable a sustainable approach to the environment and the preservation of the resource base [21]. This will additionally lead to a debate on the political level. In terms of environment, tendentially every form of unnecessary waste of resources and pollution of natural resources should be avoided. Bribery, cartels, corruption, tax evasion, industrial espionage, patent infringement generally damage public interest, because on one hand they may systematically lead to misallocations of goods and services and on the other hand, set the wrong incentives [276, 485]. Therefore they must be eliminated within the scope of a global economic culture. This can happen among other options such as setting international sanctions.

All of these goals can be promoted through an intelligent use of the enormous potential of the large amounts of data in the digital age. This means, big data analytics should not only or primarily be used for the optimization of economical success, but also for all other mentioned issues, such as making circumstances transparent (*ground truth* in the language of AI) and supervision of relevant parameters such as ecological in the area of agreed restrictions. At the same time it is important to prevent accumulating vast amounts of data of a private nature for use by companies, investors or within the scope of politics [69, 329]. As the Internet and its applications exceeds national boundaries, compliance and joint enforcement of the applicable national and international law must be implemented as a central principle. In this, self-commitment and self-control are desirable instruments in the establishment of an economic culture as part of a world culture [239].

8.3.2 Educational Culture as Part of a World Ethos

In parallel of the establishment of a economical culture based on world ethics, the establishment of effective institutions that aim for a (high) education and training of all citizens worldwide is important [239]. The importance of a sufficiently high level of education and knowledge as part of *cultural capital* [40] of all people should not be underestimated. It is not just about work and value creation and indirectly the production of the necessary technical and social innovations for sustainable development [354, 496, 503], but also to reach a successful coexistence of people, nations and cultures. This also has educational requirements, like a historical and enlightening world orientation, the formation of a self-reflective and humble attitude towards the relevance of one's access to education and tolerance to other world orientations and worldviews [311]. At best, all these conditions should be formulated in a common discourse on the world in the process of globalization and digitization. Whether we like it or not, all cultures in the era of globalization and information society are faced with the challenge of coping with the coexistence of people of different cultural backgrounds and beliefs or worldviews in a process of self-organization, if the aim is a good future for all. Even more so, because in many countries already today people with different cultural backgrounds (must) live together.

Coping with this challenge successfully requires global concepts in the field of education that do not polarize, but offer alternatives for a multipolar and multicultural world in the future. The project *World Ethos* [239], but also pioneering school projects like the *Club of Rome schools* [64, 378] or the UNESCO project schools [229, 401] are thinking worlds and examples that lead into this direction. Within these thinking worlds, children are not only prepared for an increasingly complex world with a broad interdisciplinary and intercultural training. In addition, they are supported in their physical and mental development and health in the context of sporting and musical subjects. Simultaneously, the development of intercultural understanding and corresponding empathy is an essential objective. This aim is primarily achieved through arousing curiosity for other cultures, while at the same time teaching knowledge and values about one's own and other cultures. Therefore conditions for a peaceful coexistence are created on the basis of knowledge of each other. The earlier, for example in kindergarden, the respective educational systems in this direction can take effect, the better [235]. In schools, colleges and universities as well as in vocational training, intercultural training must be given a high priority. The same applies to an interdisciplinary education, which is indispensable for sustainable development.

The practical side of mediation of a educational culture in the context of a world culture can be strengthened distinctly with the possibilities of information and communication technology [97, 496]. Hereby one should not only focus on the younger generations like pupils and students, but also older generations including parents, teachers, trainers, professional and retirees [44]. Thereby the networking of generations over mobile telephony and Internet and also via forums and learning platforms, is helpful.

The human being is a creature of culture, a *Leonardo creature* [296], and therefore a creature that (potentially) constantly develops through perpetual learning and capturing experiences. Learning processes and own experiences are more important than ever in the fast paced world of the Internet and mobile telephony, both in school, professional and private life. More "intelligent" technical systems that constantly replace labor demand new kinds of humans in the future and therefore change our educational system in the described *race between education and technology* [131]. The human increasingly finds its place in areas that require the concurrency of *brain, body and heart*, like creativity or intuition, because logic and computing capacity is increasingly dominated by machines. Therefore it is obvious that besides the ability to deal with people of different cultures the skill to intelligently cooperate with machines adequately has to be promoted (technical culture). This is one more partnership.

This means the following: In the future, children will have to deal with elements of higher mathematics and computer science, languages, history and sport from early on, but also be specifically trained towards a *homo economicus cooperativus* as social beings [392]. This development gives the child the necessary potential to make decisions regarding long-term objectives and while dealing with uncertainty. At the same time, cooperation competence and willingness to take responsibility for others is promoted [391]. The current situation in the educational area is more defined by a high performance pressure and high competitiveness. This leads to the situation that the standards people set to themselves and others are constantly strained, which partly leads into over working, selfish and anti social behavior. Against this background, schools like the Club of Rome schools [64, 378] as well as the UNESCO project schools [229, 401] teach that actions of individuals have a significant weight, but are primarily orientated on (intelligent) cooperation. The positioning of the individual, his (digital) networking, communications, even across cultures, are crucial for the chance of a sustainable development and world culture.

8.3.3 Religious Culture and World Ethos as a Basis for a Global Culture

The constitution of a *global culture* and, consequently, the creation of world peace cannot exist without peace between religions [232, 237]. This is the central position of the world ethos movement [234, 239]. There cannot be a religious peace without mutual (digital) communication between the religions. In this mutual (digital) communication process the focus should be on the practical behavior, the ethics, instead of focusing on issues of faith in each religion. Because in matters of ethics the great religions are closer than in matters of faith. What is the common ethos of the major global religions from which the world ethos movement has developed [232, 239]?

On the way to a worldwide information and knowledge society a civilization has to be created which has to unite all cultures and religions in a increasing mesh of interactions and communication [369, 371]. There are similarities in worldviews that are all in a sense *messages of salvation* and answer fundamental questions and factors relating to the human entity. These fundamental factors of human existence are love and avoiding hatred, after joy and sorrow, happiness and destiny, after justice and self-indulgence, and finally life and death. Religions are, metaphorically speaking, important moral agencies [239] that disembogue in a form of evaluation of the rich and poor, educated and uneducated, powerful and weak while promoting and enabling cooperation. They serve as stabilizers between different forms of people and systems, but also the many coincidences of existence with its various consequences. Peace and justice are at the center of all religious communication, however in the competition of belief systems, people often interfere with one another. Religions often served only as a stage for expressing interests. Predominantly, the main challenge is to alleviate conflicts, for example, in unequal distribution of wealth and resources on a personal, national and international level.

With a perspective of attaining world peace with the support of a global culture one must consider that the Western world often postulates Christianity as an obvious carrier of human rights, justice and constitutionality [239]. However, in the *Global Framing* of the West, the Islamic world is often seen as a source of terrorism and fundamentalism [381, 521]. To be fair, one also refers to the Crusades, which were conducted in the past on the part of the Christian world, but sees this as *cultural confusion* of the past [349]. However, this two-class view is not appropriate. *Correctly interpreted and lived, Christianity and Islam not only share prophets and values, but also both provide best conditions for the realization of humanity and altruism* [387], also in cooperation with each other. This has been described in *Lessing's "Nathan der Weise"* [251] involving the Jewish culture, that demonstrates anything but high ethnical standards through the oppression of the Palestinians [359, 419].

In the correct interpretation and under the right environmental conditions, *religions have powerful potential for peace*. It is necessary to implement these intelligently if one wants to achieve a sustainable development [232, 237]. This has to become a part of the collective awareness and memory. This requires the major religions to agree upon a common religious culture or a consensus of values and commit to a basic document or contract basis for compliance [239]. The prime principle being a culture of nonviolence, prohibition to kill, having reverence of life and tolerance and respect towards others [234]. At the same time, the basic document must include solidarity and a fair and inclusive economic order. *A cross-cultural consensus on values is not only to be found in the three Abrahamic religions, Judaism, Christianity and Islam* [243, 244], but also in *Buddhism and Hinduism* [404]. This minimum consensus and its principles can also be shared by non-believers and atheists. Like many world ethos works [233–236, 239] show, a last common and sufficient basis is the *"Golden rule: One should not treat others in ways that one would not like to be treated"* [24] or the *"Categorical imperative: Act only according to that maxim whereby you can, at the same time, will that it should become a universal law"* [209]. With perspective on the dignity of the individual, this rule results in the requirement of inclusion. In view of future results regarding the requirement of ecological constraints for a world economic system, this means, a eco-social orientation of the markets (*green and inclusive economy*).

8.4   No Clash of Cultures

Currently, many global developments portend more at a rising inequality and collapse than at a balance [359, 371, 380]. These different scenarios were described earlier. Not surprisingly, conflicts increase. Especially actors from different cultures collide in this world, particularly if it is regarding the situation between the Islamic and Western world, for example, the assassinations in Paris or the September 11 attacks in New York City. However it has to be mentioned that from the perspective of the author of this book, these assaults should not be interpreted as a part of a fight between the cultures. They are rather a *fight between cultures of the world* on their way to a convergence, *against the enemies of a peaceful coexistence of all cultures* on the basis of tolerance and common principles like in the sense of a world ethos (like described above with reference to Küng [237, 239]).

Samuel Huntington contributed many important shares about the possible *Clash of Civilizations* [173–175]. However, he was often misunderstood. In many ways he was right but also wrong. He thought it was unlikely to achieve global harmony under the conditions of Western democratic philosophies. He criticized the idea of a universal world culture based on the Western model and sees an overestimation of cogency of Western principles in this wishfulness. His theses primarily say, that there will still be conflicts on this world, however not of an ideological, but of a cultural nature. Hereby he divides mankind into great religious coined world cultures, which he calls civilizations that have several noticeable differences in the way people think. However, not in the field of material culture, as was clearly shown in this study with reference to telephone, mobile phone and Internet usage (see Chap. 5). Concentrating on religion, there is a big difference between the in this book used term of culture (see Chap. 2) and the cultural division by Huntington. Culture does not only or primarily consist of religion, but of parts of business, science, politics, media and also different worldviews within a society. Religion is therefore only part of a culture and historically occupies a substantial role.

Huntington indicates that the current conflicts, that exist between the cultures, are in a certain sense triggered by the West, because people in the West are *convinced of the universality of their culture and believe that their [...] power imposes on them the obligation to extend that culture throughout the world* [174]. This observation shows that the intended model of a world culture looks completely different in the context of this book. In this worldview, there is no domination of one culture over all others [290]. It is also not about a culture worked out beforehand or agreed upon by everyone. Rather, it is about a variety of worldviews in a tolerant environment of a common material culture. The ability to communicate of representatives of different worldviews is the basis for peaceful coexistence of religions. Here, a common *world ethical basis* in the sense of *Kant's imperative* or the *categorical imperative*. This imperative, formulated as *"Act only according to that maxim whereby you can, at the same time, will that it should become a universal law."* Reference [209] is anchored in all cultures worldwide.

In this context, Hans Küng, as described above, has worked out the existence of the *"golden rule"* in every religion of the world [24, 239]. It is a common component in the minds of people all over the world. Overlooking the living generations, it leads to the requirement to respect the dignity of all human beings, which must lead to a social or inclusive rulemaking for the markets. Looking at future generations, it is all about protecting the environment and avoiding a climate catastrophe which implies an ecological or green regulation of markets. This again leads to a eco-social market economy or to a *green and inclusive economy* [507]. Also Helmut Schmidt points out the special importance of this imperative in the context of his book regarding human rights [406].

According to this logic, it is important to send a message to the Western world to not hurt the religious beliefs and cultural values of others, as long as they are tolerant towards other worldviews. This logic may be in a conflict with freedom of expression, as it is extremely pronounced in the Western world, at least more than in other cultures. This form of Western freedom of expression must be spelled out in a consensus with the other cultures about the level of desired tolerance. The attack on the satirical magazine Charlie Hebdo in Paris can be primarily attributed to a justification context where Islamic fundamentalists but also ordinary believers can or cannot bare the insults made towards their prophet. One could also mention that many Christians and Jews have a similar problem if something similar happens to their prophets or god himself. Here, one should take into account that no pictorial representations of the prophets (including Adam, Abraham, Moses, Jesus, etc.) is allowed in Islam.[6] However, how would the *prophet Mohammed* react to these pictures in person if he were alive today? With indulgence, not with acts of violence. Here, solutions can be derived and possible potentials for conflict transformation can be found.

---

[6]In this context, the so called *ban on images* or *prohibition of images* exists, which is a provision in the context of the monotheistic religions and is there to counteract a polytheistic idolatry.

Regarding Huntingtons contributions one must also consider, that the original title of his book is *"Clash of Civilizations"*.[7] To avoid conflict or war, states of specific cultural backgrounds should stay out of the affairs of states with cultures that are significantly different. Thereto Huntington wrote *"For preventing wars in the emerging era, it is necessary that core states do not intervene in the affairs of other civilizations"* [174]. The theory behind the *Clash of Civilizations* could, but should by no means be understood as a plea for a fight between cultures. If one tries to understand and analyze the conflict potentials that are described in Huntingtons work, possibilities for conflict transformation and prevention emerge.

The major issue is as follows. The cultures of the world should recognize other cultures in a tolerant manner and join efforts to peacefully live together in coexistence. Thereby one should grant self-determination in accordance with the human rights to every human being so that everybody can decide which forms of culture(s) he wants to choose for himself. For many, the materialistic side will be more and more similar in the case of convergence and balance. The cultures in a worldwide information and knowledge society or civilization will learn from each other with the help of vast potentials that arise through communication and the Internet, therefore striving for every human to live a good life. In a sense of *Global Emphaty*, cultures can hardly avoid to understand each other better and better because of the consequences of more knowledge and personal experience. This strengthens the chances for a convergence, reduces the risk of conflict and is a benefit for everyone.

From the author's perspective regarding Samuel Huntington [173–175] and Amartya Sen [416], two essential aspects are to be added. Firstly, in the history of humanity it has come to conflicts between nations of almost similar religiously coined culture (e.g. France and Germany). Even in one single country, conflicting forms of religion almost led to the collapse of a country. One has to remind, for example, of the conflict between the Protestant and the Catholic Church in Germany in the *Thirty Years War* between 1618–1648 [52]. At the end of the war, the country was destroyed and one-third of the original population was dead. From time to time, there have also been *cross-cultural coalitions*, for example, in *World War* I between *Germany* and the *Ottoman Empire* [218].

Furthermore another important point: The mutability of cultures has often been underestimated, one just has to think about the abdication of the Catholic Church from the monopoly on truth against the background of the rapid changing processes of the modern age driven by the progress of science. In a similar way, the modern information and communications technology may enable changes in the assimilation of worldviews and a cooperation between people of different cultural areas.

---

[7]The author of this originally in German written book criticizes that the translation *"Kampf der Kulturen"* is more sensational than adequate. Huntington himself was not pleased with this title, however the German publisher won through regarding the intended success at the markets. Huntington critizised, that *Clash* does not mean *Kampf*, but *Aufeinanderprallen*, *Aufeinandertreffen* or *in Konflikt sein* in German [272].

8.5 Conclusion

Balance includes inherent mechanisms to *bridge cultural differences* and to enable *convergence* which will greatly affect and shape the material side of life. The Internet and the way in a global networked society will support this balance. Rising *inequality* and *collapse* are possible but not unlikely alternatives. Tendentially, they promote less the assimilation across cultural borders and more the cultural separation and withdrawal into the corral of losers of the globalization process.

In case of balance, the transition to a global culture happens with three mechanisms:

1. It comes to a massive convergence of living standards throughout the world in the context of comparable material culture.
2. Within each culture there will be a higher flexibility and less stringent conditions, a greater tolerance towards differences, and the inclusion of elements from other cultures into their own. This applies to all aspects of the practical life from food to sports to the practising of worldviews i.e. declining church attendance of believers. Some differences remain in the ideologies and religions however, marked by diversity and a decreasing discipline in the practice of certain expressions.
3. In such development, people switch between cultures throughout their lives. Therefore they themselves, but also in exchange with their love and parter relationships, embody more intercultural bridges and live in various, partly even foreign cultural contexts, which are on the way in losing the sense of foreignness.

In summary, on the level of individual humans, it comes to a more diverse cultural experience in combination with the easing of one's own cultural imprint. At the same time the interactions with other cultures increase, which in turn reacts on the relaxation of their own culture and promotes and simplifies the coexistence with other cultures. All this happens in accordance with wealth and therefore comparable material experiences and a adapting high level of education. This contributes to tolerance and balance. Obviously, the case of balance and a sustainable development are helpful conditions as well as a positively fertilized path of development. This all looks very different in the case of a global two class society and/or an ecological collapse.

Chapter 9
Closing Remarks

The conclusion begins with a summary of the preceding chapters and emphasis of the essential results of this book with regard to global networking, communication and culture. Subsequently, an outlook in future works and especially in more open questions in the context of the subject is given. The book completes with final closing remarks.

9.1 Summary

In Part I the initial situation and the conceptual foundations of globalization, culture and communication as well as the concepts of *digital divide* and *digital inequality*, *Internet Governance*, *superorganism humanity* and *global culture* are put within a suitable framework. Part II follows a comprehensive statistical–empirical analysis to connect between general key figures from the areas of *economy, education, religion* and the state of digitalization or access rates to information and communications technologies (ICT) of countries in a global perspective. Part III discusses questions of *cultural cooperation* and *power issues* with regard to the *Internet Governance ecosystem*. Furthermore, considerations on possible future developments are reviewed in the context of the humanity as a *global superorganism*. Against this background, a possible convergence to a kind of global culture is discussed.

In Part II the empirical analyses of (functional) relations between economic performance of states and the access rates for the fixed telephone infrastructure, mobile phone, Internet and mobile Internet are carried out, starting in 2000 over more than a decade until 2013. The results of the investigations demonstrate that the *convergence with regard to ICT usage worldwide is (much) faster than the convergence with regard to prosperity in general.* The rapid spread of (modern) ICT across the world, and thus across all cultural systems, is becoming clear. Despite global differences

© Springer International Publishing AG 2018
H. Ünver, *Global Networking, Communication and Culture: Conflict or Convergence?*, Studies in Systems, Decision and Control 151,
https://doi.org/10.1007/978-3-319-76448-1_9

in the economic realm, the convergence speed is high in ICT usage, i.e. there is a rapid catch–up in ICT use across the world population. Almost the entire world population is using the mobile phone and about 50% of the world population is online. Nowadays, differentiation between states and cultures concerns access quality and usage patterns (digital inequality) rather than access itself (digital divide). Thus, communication media initially have a convergence–promoting effect in corresponding sub–areas of material culture.

Using the example of mobile phone use, the thesis stating that the catch–up in ICT use is faster than prosperity in general has been convincingly confirmed. In regards to Internet use, it has been demonstrated worldwide for all cultures that the Internet, especially in connection with the technological convergence of the Internet and mobile telephony to the mobile Internet, spreads faster than prosperity or wealth. In 2000, a few rich countries were represented on the Internet in a disproportionate manner in relation to the distribution of the world's population to countries. In 2013, the situation was better balanced. In this context, studies on country's social inequality level show that the Internet user rate increases the more balanced a society becomes in terms of income and wealth. The worldwide fixed telephone user rate is declining over the period between 2000 and 2013 under investigation, although the absolute number of fixed telephone connections worldwide is growing slightly, as the world's population grows faster than the number of fixed–line connections (especially in developing countries). In rich countries the reason for the drop in fixed telephone connections is the technical possibility to bundle fixed lines. Additionally, a primary reason for the decrease in the fixed telephone rate worldwide is the rapid spread of the mobile phone. This spread, also called a "mobile miracle", was also promoted by prepaid cards as a business model.

A high level of education is generally a necessary condition for a high penetration of technology in a country, but by no means sufficient. It has been shown that the adult literacy rate has a weak correlation with ICT use, while a high literacy rate remains a necessary condition for high ICT user rates from 2000 to 2012. The Education Index developed by the UN measures the level of education of countries in a different manner (in greater detail) than the literacy rate and is also much more closely related to the use of ICT. From the functional relationship between the education level and the Internet user rate of countries, one can deduce that between 2000 and 2012 the required level of education for access to the Internet, as measured by the Education Index, has fallen. Thus, the educational requirements for use are now less than, for example, ten years ago. The ability to read and write, as expressed by the literacy rate, is still a necessary condition for Internet use. This relationship is unlikely to change, particularly in the next ten years, as the use of the Internet and the access to knowledge are still largely text–based. It is also likely that the level of education will continue to be a key issue when it comes to a high quality use of the Internet, no matter how user–friendly Internet access will become.

A system–theoretical approach for determining the interaction of influencing factors and key figures was presented based on the so-called multi-step method from numerical mathematics. The implicit assumption made in the empirical analysis, that the economic performance of a country is closely linked to the education level, has been confirmed. The multivariate analysis show that ICT use is most closely related to economic performance, followed by the Education Index, which contributes only slightly less to ICT use than the economic performance. Compared to the these two variables, the adult literacy rate hardly has any influence on ICT use.

Furthermore, the interaction between the different ICT technologies was investigated. It became apparent that a fixed telephone connection in 2000 had a close relation with the Internet usage, also due to the physical coupling of the two infrastructures. This relationship has declined somewhat over more than a decade, but still exists. The reason for the decline over time is access to the Internet via the mobile phone, in particular via smartphones. Mobile telephony has hardly any connection with fixed telephony in 2013, since both infrastructures are physically decoupled from each other and mobile telephony in addition has spread worldwide.

The empirical analysis of the relation between the different worldviews of *Christianity, Islam, Judaism, Buddhism, Hinduism,* and *other group* (which consists of those without a confession of religion, e.g. atheists, and people with a folk faith) and the different access rates to *ICT show no significant relation.* This is a great result in terms of global networking, communication and cultures aiming to convergence. It can be concluded that *the participatory effect of ICT usage is accepted by all worldviews and cultures,* where this effect is ultimately (only) a necessary condition to allow cultural convergence. It has been revealed that the diversity of religious groups in a country has a positive impact on the level of ICT use. This is made plausible by the economic performance of the respective countries. Overall, the thesis has been confirmed that religions have no or very little relation to the different access rates to ICTs in a global perspective. Convergence in the use of ICT among the major religions is not yet complete in 2010, but has already advanced considerably.

In Part III, the study of the bargaining positioning of a number of major states or cultures on the global *Internet Governance* debate has shown that, in the area of power politics, similar positions can cross cultural boundaries and thus convergence patterns can be observed. On the one hand, the *US, EU* and *Japan* have a similar position, namely the favoring of a low level of government control and no, or little, responsibility to ITU in the *Internet Governance Ecosystem.* On the other hand, *China, Russia* and the *Arab world* as well as *Turkey* and *Iran* argue for more state control and more responsibility for the ITU. A central condition for convergence of cultures in the Internet governance issue was the transfer of the administration of critical Internet resources, so-called IANA transition, by the US to a broader international structure of responsibility. A possible surrender was officially announced in March 2014. In October 2016, the National Telecommunications and Information Administration (NTIA) terminated the contract on the IANA functions with ICANN. The Internet Assigned Numbers Authority (IANA) stewardship transition is completed now. For the first time since its founding in 1998, the multi-stakeholder-community controls and manages the so-called zone files of the top-level domains. This fact is particularly pleasing from the point view of this book, because it had already required exactly this fact when the dissertation was published in 2015. In the context of this book, it is argued that the information society in connection with communication technologies like the Internet and mobile phones could produce stronger participatory governance than is the case by the current distribution of power structures at state levels and cooperation across borders. A *social transformation* in the direction favored by the author could be similar to the slow *power shift* from palace to parliament at the time of industrialization. But parliament will not die out by having a power shift from parliament to people.

In Part III, humanity as a whole in the process of digitization was investigated on the way to a global superorganism as an autopoietic system. The term *technological singularity* (or simply singularity) postulates an acceleration of the speed of technological change with extreme growth. The invention of artificial super-intelligence will abruptly trigger uncontrollable technological growth, resulting in incredible changes to human civilization, and more generally, all evolutionary processes. In this context, communication, especially through ICT, has been regarded as the driving force of the further development of mankind to the next evolutionary stage. The superorganism humanity could develop into an intelligent human–technology system on the basis of cultural cooperation. *Inclusion* as a systemic and fair integration of people and states into global communication processes is a *central cultural challenge* on the way becoming a *balanced global superorganism*. This book argues for an *inclusive global governance with a balance between central and distributed control mechanisms promoted by ICT.*

In this framework, a possible convergence to a specific type of world culture was presented in the context of three widely conceived future scenarios (*balance, increasing social inequality, collapse*) following the work of the Club of Rome and examined in connection with the digitization process. The two possibilities of increasing social inequality and ecological collapse, which are undesirable for humanity, tend not to promote balance over cultural boundaries for the vast majority of people worldwide, but rather cultural separation. The balanced future scenario is different. It incorporates inherent mechanisms to bridge cultural differences and enable convergence between cultures, which will then determine and shape the material side of life in particular. The Internet and the direction towards a global networked society can promote such a balanced development.*Realistically, it has been described that current global developments tend to point to increasing social inequality or ecological collapse, but not to balance.* Overall, the capacity of the Internet to promote the development of humanity towards a global culture is limited by the nature of the global governance system, including Internet governance and telecommunications regulation. At the core of a *global culture*, in the sense of common values and broad consensus between different cultures, is a corresponding formation of an *economic, education* and *religious culture* on the basis of a global ethic. Finally, the investigations came to the conclusion that there is *"not a Clash of Civilizations"*. In considering the influence of potential religions, there is a great difference between the concept of culture used in this book and the concept of culture in the well–known book *"Clash of Civilizations"* by *Huntington*. At this point, however, it should be pointed out that Huntington himself did not want to see his book as an appeal to the struggle of cultures. The attempt to understand and analyze the conflict potentials between cultures described in his work should lead, rather, to hints for conflict transformation, such as to avoid conflict and promote cultural convergence, such as in the area of ICT usage.

9.2 Outlook

Due to the broad and interdisciplinary orientation of this book as well as in part to the empirical data situation, some important questions could not be addressed in depth. The primary aim was to provide a comprehensive (empirical–functional) understanding of the global networking processes through ICT and to classify this understanding into a cultural framework while identifying possible trends in future developments. This results in a number of interesting questions, which can be investigated in further work.

In this book, four different types (motivation, material, skills, usage) of digital divide or digital inequality were distinguished. The worldwide empirical investigations essentially address the types of material access (based on economic performance) and skills access (based on education). In further studies, a comprehensive global empirical analysis on motivational access and usage access could be undertaken. In principle, one would then have to pursue the following questions: what primarily motivates people in different cultures to use the Internet or mobile phones? How do people use the Internet in different cultural systems? So far, there are isolated papers that examine these questions in selected countries. According to the author's knowledge, there are currently no investigations of this kind in a worldwide perspective. It is to be expected that similar motivation grounds will also be present across cultural boundaries, but due to the variety of offers on the Internet and the quality of access, very different usage patterns can result. The question might be whether the usage patterns of people around the world differ more than the patterns within a single country, which also depend on the degree of religious diversity.

Another point that should be investigated is the extent to which the interdependence or coupling of different ICTs will, in the future, promote *technological convergence* in both technical and social terms. The question then is whether the spread of the (mobile) Internet will be accelerated this way, or whether, for example, the fixed network infrastructure will gain importance again in connection with fiber optic cables. Certainly it will to be used at least as a backup infrastructure in future.

Another future work is the investigation of the technical conditions and consequences associated with various developments in the *Internet Governance Ecosystem*. The Internet multi-stakeholder world is faced with the challenge of providing the right *secure infrastructure* for the additional consumption as the numbers of Internet users and *data* applications are growing. In this situation, for example, centralized control by a few backbone ISPs is no longer viable. A research question could be: What technical consequences would the political approaches of different countries have for the Internet's (backbone) infrastructure?

Important questions for the future concern the consideration of *humanity* as a *superorganism*, i.e. as an *autopoietic system* in the context of information and communication: how far will the Internet and the Internet of Things influence the way we think and work? What types of jobs will potentially be replaced in the future by machines and algorithms? How and where will it be possible to combine people's abilities with that of (intelligent) machines?

In the area of a possible *global culture*, most questions are probably open to the future, because the nature of culture requires culture to develop over a long period of time. The following questions are open to the fields of *economy*, *education* and *religion*, based on a world culture that has been considered within the scope of this book: Is a global eco–social market economy, which is also mainly driven by innovation and digital markets, as the sustainable economic system also the system that most likely promotes cultural convergence? Through which elements does the eco–social market economy concretely combine the different cultural systems of the world, especially in the areas of economy, education and religion? There are certainly many other questions and points that should be examined and discussed in future work. A clarification of the above mentioned questions would make sense with the present book as a starting point.

9.3 Conclusion

The invention of languages many thousand years ago was most probably the invention behind all inventions in the field of communication. Certainly the fixed telephone as a place-bound communication has contributed to a decisive step in the process of constructing an information society. It is not without reason that the worldwide spread of the mobile phone is described as a miracle. This miracle was possible due to the fact that it is a location–independent and person–bound communication technique. It has changed the way people are connected to each other in a unique form. A similar or more important change also applies to the Internet. Its invention and global spread as well as its diverse fields of application have created a new era in state, economy, science, religion, media, art and sport, thus a new cultural age. This cultural era of permanent communication and exchange (anyone with anyone) regardless of time (anytime) and place (anywhere), is further extended in its dimension by the technological convergence of mobile telephony and the Internet to the mobile Internet. It is expected that in the near future the overwhelming majority of people will use the Internet and its (mobile) services. Through the use of the Internet, the people and their data can increasingly be monitored, in order to predict or even control human behavior. This development involves opportunities as well as risks and calls for a strong organization through a global multi-stakeholder community, which must nevertheless develop the necessary consensus and corresponding global governance structures through cultural cooperation. It is not clear whether we as humanity will succeed in the foreseeable future to develop a global culture that is more than the sum of all cultures. It is also open to the question of whether a largely (digitally) networked world society will form a worldwide community, constituting a global culture or not. From the author's point of view, however, one thing is true: *"There will be no global culture without a (digitally) networked world society"*. At this point, the various thematic strands written in this book are closely linked.

Appendix A
Categorization of Countries

All's well that ends well

The single countries investigated in this book are assigned to different categories. Categories include the economic performance, the Education Index or the adult literacy rate. Following section will describe in detail, how the categories are defined exactly.

A.1 Economic Performance (GDPpC)

The assignment criteria is the GDP per capita (GDPpC) in relation to the average World-GDPpC.

$$\text{World-GDPpC} = \frac{\sum_i \text{GDP}_i}{\text{World population}}, i \in \forall countries$$

A	Highest	$\text{GDPpC} > 2\times \text{World-GDPpC}$
B	High	$\text{World-GDPpC} < \text{GDPpC} < 2\times \text{World-GDPpC}$
C	Middle	$\frac{1}{2} \text{World-GDPpC} < \text{GDPpC} < \text{World-GDPpC}$
D	Low	$\frac{1}{4} \text{World-GDPpC} < \text{GDPpC} < \frac{1}{2} \text{World-GDPpC}$
E	Lowest	$\text{GDPpC} < \frac{1}{4} \text{World-GDPpC}$

The assignment of each country i to a corresponding GDPpC–category (A – E) can be found in Table A.1.

© Springer International Publishing AG 2018
H. Ünver, *Global Networking, Communication and Culture: Conflict or Convergence?*, Studies in Systems, Decision and Control 151,
https://doi.org/10.1007/978-3-319-76448-1

Table A.1 Categorization by GDPpC: List of countries

A	B	C	D	E
Australia	Antigua Barbuda	Albania	Angola	Afghanistan
Austria	Bahamas	Algeria	Armenia	Bangladesh
Bahrain	Chile	Argentina	Belize	Benin
Belgium	Croatia	Azerbaijan	Bhutan	Burkina Faso
Bermuda	Cuba	Barbados	Bolivia	Burundi
Brunei Darussalam	Cyprus	Belarus	Cabo Verde	Cambodia
Canada	Czech Republic	Bosnia	Congo, Rep.	Cameroon
Denmark	Eq. Guinea	Botswana	El Salvador	Cent. African Rep.
Finland	Estonia	Brazil	Fiji	Chad
France	Gabon	Bulgaria	Georgia	Comoros
Germany	Greece	China	Guatemala	Congo, Dem. Rep.
Hong Kong	Hungary	Colombia	Guyana	Cote d'Ivoire
Iceland	Israel	Costa Rica	India	Eritrea
Ireland	Italy	Dominica	Jamaica	Ethiopia
Kuwait	Japan	Dominican Rep.	Kosovo	Gambia, The
Luxembourg	Kazakhstan	Ecuador	Lao PDR	Ghana
Macao	Korea, Rep.	Egypt	Moldova	Guinea
Netherlands	Latvia	Grenada	Morocco	Guinea-Bissau
Norway	Lithuania	Indonesia	Nigeria	Haiti
Oman	Malaysia	Iran	Pakistan	Honduras
Qatar	Malta	Iraq	Paraguay	Kenya
Saudi Arabia	New Zealand	Jordan	Philippines	Kiribati
Singapore	Panama	Lebanon	Samoa	Kyrgyz Rep.
Sweden	Poland	Libya	Swaziland	Lesotho
Switzerland	Portugal	Macedonia	Tonga	Liberia
United Arab Em.	Puerto Rico	Maldives	Ukraine	Madagascar
United States	Romania	Mauritius	Uzbekistan	Malawi
	Russia	Mexico	Vietnam	Mali
	Seychelles	Mongolia		Marshall Isl.
	Slovak Republic	Montenegro		Mauritania
	Slovenia	Namibia		Micronesia
	Spain	Palau		Mozambique
	Trinidad Tobago	Peru		Nepal
	Turkey	Serbia		Nicaragua
	United Kingdom	South Africa		Niger

(continued)

Table A.1 (continued)

A	B	C	D	E
	Uruguay	Sri Lanka		Papua New Guinea
		Suriname		Rwanda
		Thailand		Sao Tome Principe
		Tunisia		Senegal
		Turkmenistan		Sierra Leone
		Venezuela		Solomon Isl.
				Sudan
				Tajikistan
				Tanzania
				Timor-Leste
				Togo
				Tuvalu
				Uganda
				Vanuatu
				Yemen
				Zambia
				Zimbabwe

A.2 Education Index (EI)

For creation of the categories, the EI of a country i is compared with the maximum EI (Max), the minimal EI (Min) and the average (MW). Further boundaries between the categories are:

$G_1 = ((MV+Max)/2 + Max)/2 = (MV + 3 Max)/4$

$G_2 = (MV+Max)/2$

$G_4 = (MV+Min)/2$

The boundaries for 2000 and 2012 are:

	2000	2012
Max	0.974	1
G_1	0.874	0.914
G_2	0.775	0.828
MV	0.575	0.655
G_4	0.353	0.416
Min	0.131	0.177

Categories are therefore defined as follows:

Category	Condition
A	$G_1 \leq EI \leq Max$
B	$G_2 < EI \leq G_1$
C	$MV < EI \leq G_2$
D	$G_4 < EI \leq MV$
E	$Min \leq EI \leq G_4$

The countries and their corresponding assignment into a specific EI–category can be found in Table A.2.

Table A.2 Categorization by EI: List of countries

A	B	C	D	E
Korea, Rep.	Hong Kong	Mauritius	Kiribati	Yemen, Rep.
Slovenia	Ukraine	Ecuador	El Salvador	Nepal
Germany	France	Mexico	Morocco	Comoros
Ireland	Japan	Peru	Cameroon	Pakistan
Israel	Belarus	Brazil	Honduras	Bhutan
Estonia	Georgia	Romania	Iraq	Benin
New Zealand	Kazakhstan	Albania	Cambodia	Gambia, The
Australia	Russia	Malta	Namibia	Papua New Guinea
Canada	Cuba	Turkmenistan	Nicaragua	Mauritania
Iceland	Spain	Bosnia Herzegovina	Lesotho	Ethiopia
Czech Republic	Greece	Uruguay	Dominican Rep.	Niger
Netherlands	United Kingdom	Algeria	Indonesia	Mali
United States	Hungary	Bolivia	Guyana	Bangladesh
Norway	Latvia	Armenia	Lao PDR	Burkina Faso
Sweden	Belgium	Argentina	China	Senegal
Denmark	Austria	Saudi Arabia	Vietnam	Eritrea
	Lithuania	Tajikistan	Guatemala	Rwanda
	Italy	Mongolia	Cabo Verde	Afghanistan
	Slovak Republic	Panama	Equatorial Guinea	Chad
	Switzerland	Libya	Madagascar	Congo, Dem. Rep.
	Finland	Iran	Swaziland	Sudan
	Montenegro	Moldova	Paraguay	Burundi

(continued)

Table A.2 (continued)

A	B	C	D	E
	Palau	Tonga	Angola	Mozambique
		Venezuela	Uganda	Sierra Leone
		Azerbaijan	Zimbabwe	Cent. African Rep.
		Sri Lanka	Thailand	Guinea
		Barbados	Togo	Cote d'Ivoire
		Botswana	Egypt	Haiti
		Portugal	Zambia	Guinea-Bissau
		Kyrgyz Rep.	Ghana	
		Micronesia	Tanzania	
		Uzbekistan	Malawi	
		Lebanon	Liberia	
		Philippines	Nigeria	
		Croatia	Congo, Rep.	
		South Africa	Suriname	
		Gabon	Oman	
		Fiji	Kenya	
		Bahamas	Solomon Islands	
		Seychelles	Tunisia	
		Costa Rica	India	
		Chile	Vanuatu	
		Colombia	Maldives	
		Trinidad Tobago	Qatar	
		Bahrain	Kuwait	
		Grenada	Turkey	
		Malaysia	Sao Tome Principe	
		Luxembourg	Timor-Leste	
		Belize		
		Jordan		
		Dominica		
		Bulgaria		
		Jamaica		
		Brunei Darussalam		
		United Arab Em.		
		Antigua Barbuda		
		Singapore		
		Poland		
		Cyprus		
		Macedonia		
		Samoa		
		Serbia		

A.3 Adult Literacy Rate (ALR)

The categories are defined by comparing the adult literacy rate ALR of country with
the maximum ALR (Max), the minimum ALR (MIN) and the average ALR (MV).

Category	Condition
A	$ALR \geq 99\%$
B	$(Max + MV)/2 \leq ALR \leq 99\%$
C	$MV \leq ALR < (Max + MV)/2$
D	$(Min + MV)/2 \leq ALR < MV$
E	$Min \leq ALR < (Min + MV)/2$

The ALR–categorie assignment for each country can be found in Table A.3.

Table A.3 Categorization by adult literacy rate: List of countries

A	B	C	D	E
Armenia	Albania	Bahrain	Algeria	Bangladesh
Australia	Argentina	Bolivia	Angola	Benin
Azerbaijan	Austria	Brazil	Bhutan	Burkina Faso
Belgium	Bosnia Herzegovina	Dominican Rep.	Botswana	Cent. African Rep.
Bulgaria	Brunei Darussalam	Ecuador	Burundi	Chad
Canada	China	Gabon	Cambodia	Cote d'Ivoire
Chile	Colombia	Kenya	Cameroon	Ethiopia
Cuba	Costa Rica	Lebanon	Cabo Verde	Gambia
Czech Rep.	Croatia	Lesotho	Comoros	Guinea
Denmark	Cyprus	Mauritius	Congo, Dem. Rep.	Guinea-Bissau
Estonia	Equatorial Guinea	Namibia	Egypt	Haiti
Finland	Greece	Oman	El Salvador	Mozambique
France	Indonesia	Peru	Eritrea	Nepal
Georgia	Israel	Puerto Rico	Ghana	Niger
Germany	Jordan	Saudi Arabia	Guatemala	Senegal
Hungary	Kuwait	South Africa	Honduras	Sierra Leone
Iceland	Macao	Sri Lanka	India	
Ireland	Macedonia	Swaziland	Iran	
Italy	Malaysia	Zimbabwe	Lao PDR	
Japan	Mexico		Liberia	
Kazakhstan	Mongolia		Madagascar	
Korea, Rep.	Panama		Mauritania	
Kyrgyz Rep.	Paraguay		Morocco	

(continued)

Table A.3 (continued)

A	B	C	D	E
Latvia	Philippines		Nicaragua	
Lithuania	Portugal		Nigeria	
Luxembourg	Qatar		Pakistan	
Moldova	Romania		Rwanda	
Netherlands	Singapore		Sudan	
New Zealand	Spain		Tanzania	
Norway	Suriname		Togo	
Poland	Thailand		Tunisia	
Russia	Trinidad Tobago		Uganda	
Samoa	Turkey		Vanuatu	
Slovak Rep.	United Arab Em.		Yemen	
Slovenia	Uruguay		Zambia	
Sweden	Venezuela			
Switzerland	Vietnam			
Tajikistan				
Tonga				
Turkmenistan				
Ukraine				
United Kingdom				
United States				
Uzbekistan				

A.4 Worldview

See Table A.4.

Table A.4 Categorization by Worldviews

Country	Pop.[Mill.]	RDI	Christian [%]	Muslim [%]	Jew [%]	Hindu [%]	Buddist [%]	Other [%]
Afghanistan	28.4	0.1	0.1	99.7	0.09	0.09	0.09	0.27
Albania	3.2	3.7	18	80.3	0.09	0.09	0.09	1.69
Algeria	37.1	0.5	0.2	97.9	0.09	0.09	0.09	1.98
Angola	19.5	2	90.5	0.2	0.09	0.09	0.09	9.39
Antigua/Barbuda	0.1	1.5	93	0.6	0.09	0.2	0.09	6.3
Argentina	40.4	3	85.2	1	0.5	0.09	0.09	13.3
Armenia	3.0	0.3	98.5	0.09	0.09	0.09	0.09	1.49
Australia	22.4	5.6	67.3	2.4	0.5	1.4	2.7	25.7
Austria	8.4	3.8	80.4	5.4	0.2	0.09	0.2	13.69
Azerbaijan	9.1	0.7	3	96.9	0.09	0.09	0.09	0.27
Bahamas	0.4	0.9	96	0.1	0.09	0.09	0.09	3.7
Bahrain	1.3	5.4	14.5	70.3	0.6	9.8	2.5	2.19
Bangladesh	151.1	2.1	0.2	89.8	0.09	9.1	0.5	0.58

(continued)

Table A.4 (continued)

Country	Pop.[Mill.]	RDI	Christian [%]	Muslim [%]	Jew [%]	Hindu [%]	Buddist [%]	Other [%]
Barbados	0.3	1.1	95.2	1	0.09	0.4	0.09	3.39
Belarus	9.5	4.7	71.2	0.2	0.09	0.09	0.09	28.78
Belgium	10.9	5.7	64.2	5.9	0.3	0.09	0.2	29.29
Belize	0.3	2.6	87.6	0.1	1	0.2	0.5	10.5
Benin	9.5	7.2	53	23.8	0.09	0.09	0.09	23.19
Bermuda	0.1	4.6	75	1.1	0.3	0.09	0.5	23.2
Bhutan	0.7	4.5	0.5	0.2	0.09	22.6	74.7	2.08
Bolivia	10.2	1.3	93.9	0.09	0.09	0.09	0.09	6
Bosnia/ Herzegovina	3.8	6	52.3	45.2	0.09	0.09	0.09	2.68
Botswana	2.0	5	72.1	0.4	0.09	0.3	0.09	27.2
Brazil	195.2	2.3	88.9	0.09	0.09	0.09	0.1	10.9
Brunei Darussalam	0.4	4.8	9.4	75.1	0.09	0.3	8.6	6.7
Bulgaria	7.4	3.5	82.1	13.7	0.09	0.09	0.09	4.38
Burkina Faso	15.5	6.2	22.5	61.6	0.09	0.09	0.09	15.89
Burundi	9.2	1.8	91.5	2.8	0.09	0.09	0.09	5.88
Cabo Verde	0.5	2.3	89.1	0.1	0.09	0.09	0.09	10.8
Cambodia	14.4	0.7	0.4	2	0.09	0.09	96.9	0.89
Cameroon	20.6	5.3	70.3	18.3	0.09	0.09	0.09	11.3
Canada	34.1	5.3	69	2.1	1	1.4	0.8	25.8
Centr. African Rep.	4.3	2.2	89.5	8.5	0.09	0.09	0.09	2.09
Chad	11.7	6	40.6	55.3	0.09	0.09	0.09	4
Chile	17.2	2.2	89.4	0.09	0.1	0.09	0.09	10.3
China	1,359.8	7.3	5.1	1.8	0.09	0.09	18.2	74.8
Colombia	46.4	1.6	92.5	0.09	0.09	0.09	0.09	7.49
Comoros	0.7	0.4	0.5	98.3	0.09	0.09	0.09	1.19
Congo, Dem. Rep.	62.2	0.9	95.8	1.5	0.09	0.09	0.09	2.6
Congo, Rep.	4.1	2.9	85.9	1.2	0.09	0.09	0.09	12.9
Costa Rica	4.7	1.9	90.9	0.09	0.09	0.09	0.09	9
Cote d'Ivoire	19.0	7.4	44.1	37.5	0.09	0.09	0.09	18.4
Croatia	4.3	1.4	93.4	1.4	0.09	0.09	0.09	5.28
Cuba	11.3	6.5	59.2	0.09	0.09	0.2	0.09	40.49
Cyprus	0.8	4.6	73.2	25.3	0.09	0.09	0.2	1.38
Czech Rep.	10.6	4.1	23.3	0.09	0.09	0.09	0.09	76.58
Denmark	5.6	3.3	83.5	4.1	0.09	0.4	0.2	11.98
Dominica	0.1	1.2	94.4	0.1	0.09	0.09	0.1	5.2
Dominican Rep.	10.0	2.4	88	0.09	0.09	0.09	0.09	11.9
Ecuador	15.0	1.3	94.1	0.09	0.09	0.09	0.09	5.89
Egypt	78.1	1.1	5.1	94.9	0.09	0.09	0.09	0.27
El Salvador	6.2	2.4	88.2	0.09	0.09	0.09	0.09	11.8
Equatorial Guinea	0.7	2.4	88.7	4	0.09	0.09	0.09	7.2
Eritrea	5.7	5.4	62.9	36.6	0.09	0.09	0.09	0.59

(continued)

Table A.4 (continued)

Country	Pop.[Mill.]	RDI	Christian [%]	Muslim [%]	Jew [%]	Hindu [%]	Buddist [%]	Other [%]
Estonia	1.3	5.5	39.9	0.2	0.1	0.09	0.09	59.78
Ethiopia	87.1	5.6	62.8	34.6	0.09	0.09	0.09	2.78
Fiji	0.9	5.8	64.4	6.3	0.09	27.9	0.09	1.39
Finland	5.4	3.5	81.6	0.8	0.09	0.09	0.09	17.78
France	65.4	5.9	63	7.5	0.5	0.09	0.5	28.5
Gabon	1.6	4.5	76.5	11.2	0.09	0.09	0.09	12.3
Gambia	1.7	1.1	4.5	95.1	0.09	0.09	0.09	0.28
Georgia	4.4	2.3	88.5	10.7	0.09	0.09	0.09	0.88
Germany	83.0	5.3	68.7	5.8	0.3	0.09	0.3	24.89
Ghana	24.3	4.7	74.9	15.8	0.09	0.09	0.09	9.3
Greece	11.1	2.5	88.1	5.3	0.09	0.1	0.09	6.29
Grenada	0.1	0.8	96.6	0.3	0.09	0.7	0.09	2.5
Guatemala	14.3	1.1	95.2	0.09	0.09	0.09	0.09	4.79
Guinea	10.9	3.1	10.9	84.4	0.09	0.09	0.09	4.59
Guinea-Bissau	1.6	7.5	19.7	45.1	0.09	0.09	0.09	35.29
Guyana	0.8	5.7	66	6.4	0.09	24.9	0.09	2.8
Haiti	9.9	2.7	86.9	0.09	0.09	0.09	0.09	13.1
Honduras	7.6	2.5	87.6	0.1	0.09	0.09	0.1	12.2
Hong Kong, China	7.0	7.2	14.3	1.8	0.09	0.4	13.2	70.4
Hungary	10.0	3.5	81	0.09	0.1	0.09	0.09	18.78
Iceland	0.3	1.1	95	0.2	0.09	0.3	0.4	4.2
India	1,205.6	4	2.5	14.4	0.09	79.5	0.8	2.89
Indonesia	240.7	2.6	9.9	87.2	0.09	1.7	0.7	0.49
Iran	74.5	0.1	0.2	99.5	0.09	0.09	0.09	0.39
Iraq	31.0	0.2	0.8	99	0.09	0.09	0.09	0.28
Ireland	4.5	1.7	92	1.1	0.09	0.2	0.2	6.49
Israel	7.4	4.5	2	18.6	75.6	0.09	0.3	3.4
Italy	60.5	3.3	83.3	3.7	0.09	0.1	0.2	12.59
Jamaica	2.7	4.3	77.2	0.09	0.09	0.09	0.09	22.7
Japan	127.4	6.2	1.6	0.2	0.09	0.09	36.2	62.1
Jordan	6.5	0.6	2.2	97.2	0.09	0.1	0.4	0.27
Kazakhstan	15.9	5	24.8	70.4	0.09	0.09	0.2	4.6
Kenya	40.9	3.1	84.8	9.7	0.09	0.1	0.09	5.4
Kiribati	0.1	0.7	97	0.09	0.09	0.09	0.09	3.09
Korea, Rep.	48.5	7.4	29.4	0.2	0.09	0.09	22.9	47.4
Kosovo	0.0	2.6	11.4	87	0.09	0.09	0.09	1.78
Kuwait	3.0	4.8	14.3	74.1	0.09	8.5	2.8	0.48
Kyrgyz Rep.	5.3	2.4	11.4	88	0.09	0.09	0.09	0.59
Lao PDR	6.4	5.4	1.5	0.09	0.09	0.09	66	32.3
Latvia	2.1	5.7	55.8	0.1	0.09	0.09	0.09	44.09
Lebanon	4.3	5.5	38.3	61.3	0.09	0.09	0.2	0.48
Lesotho	2.0	0.7	96.8	0.09	0.09	0.09	0.09	3.29
Liberia	4.0	2.8	85.9	12	0.09	0.09	0.09	2
Libya	6.0	0.7	2.7	96.6	0.09	0.09	0.3	0.38
Lithuania	3.1	2.1	89.8	0.09	0.09	0.09	0.09	10.18

(continued)

Table A.4 (continued)

Country	Pop.[Mill.]	RDI	Christian [%]	Muslim [%]	Jew [%]	Hindu [%]	Buddist [%]	Other [%]
Luxembourg	0.5	4.9	70.4	2.3	0.1	0.09	0.09	27.19
Macao, China	0.5	6.8	7.2	0.2	0.09	0.09	17.3	75.3
Macedonia, FYR	2.1	5.6	59.3	39.3	0.09	0.09	0.09	1.58
Madagascar	21.1	3	85.3	3	0.09	0.09	0.09	11.49
Malawi	15.0	3.4	82.7	13	0.09	0.09	0.09	4.29
Malaysia	28.3	6.3	9.4	63.7	0.09	6	17.7	3.2
Maldives	0.3	0.4	0.4	98.4	0.09	0.3	0.6	0.27
Mali	14.0	1.6	3.2	92.4	0.09	0.09	0.09	4.39
Malta	0.4	0.7	97	0.2	0.09	0.2	0.09	2.68
Marshall Islands	0.1	0.6	97.5	0.09	0.09	0.09	0.09	2.6
Mauritania	3.6	0.2	0.3	99.1	0.09	0.09	0.09	0.69
Mauritius	1.2	6.7	25.3	16.7	0.09	56.4	0.09	1.6
Mexico	117.9	1.1	95.1	0.09	0.09	0.09	0.09	4.88
Micronesia Fed.Sts.	0.1	1	95.3	0.09	0.09	0.09	0.4	4.3
Moldova	0.0	0.6	97.4	0.6	0.6	0.09	0.09	1.58
Mongolia	2.7	6.5	2.3	3.2	0.09	0.09	55.1	39.49
Montenegro	0.6	4	78.1	18.7	0.09	0.09	0.09	3.38
Morocco	31.6	0	0.09	99.9	0.09	0.09	0.09	0.27
Mozambique	24.0	7	56.7	18	0.09	0.09	0.09	25.39
Namibia	2.2	0.6	97.5	0.3	0.09	0.09	0.09	2.19
Nepal	26.8	3.8	0.5	4.6	0.09	80.7	10.3	4.09
Netherlands	16.6	6.4	50.6	6	0.2	0.5	0.2	42.5
New Zealand	4.4	6.2	57	1.2	0.2	2.1	1.6	37.8
Nicaragua	5.8	2.8	85.8	0.09	0.09	0.09	0.09	14
Niger	15.9	0.4	0.8	98.4	0.09	0.09	0.09	0.88
Nigeria	159.7	5.9	49.3	48.8	0.09	0.09	0.09	1.89
Norway	4.9	3.1	84.7	3.7	0.09	0.5	0.6	10.39
Oman	2.8	2.9	6.5	85.9	0.09	5.5	0.8	1.29
Pakistan	173.1	0.8	1.6	96.4	0.09	1.9	0.09	0.27
Palau	0.0	2.7	86.7	0.09	0.09	0.09	0.8	12.4
Panama	3.7	1.5	93	0.7	0.4	0.09	0.2	5.6
Papua New Guinea	6.9	0.2	99.2	0.09	0.09	0.09	0.09	0.69
Paraguay	6.5	0.7	96.9	0.09	0.09	0.09	0.09	3
Peru	29.3	1	95.5	0.09	0.09	0.09	0.2	4.3
Philippines	93.4	1.6	92.6	5.5	0.09	0.09	0.09	1.7
Poland	38.2	1.2	94.3	0.09	0.09	0.09	0.09	5.78
Portugal	10.6	1.4	93.8	0.6	0.09	0.1	0.6	4.99
Puerto Rico	0.0	0.7	96.7	0.09	0.09	0.09	0.3	2.8
Qatar	1.7	5.7	13.8	67.7	0.09	13.8	3.1	1.69
Romania	21.9	0.1	99.5	0.3	0.09	0.09	0.09	0.28
Russian Fed.	143.6	4.9	73.3	10	0.2	0.09	0.1	16.49
Rwanda	10.8	1.4	93.4	1.8	0.09	0.09	0.09	4.8
Samoa	0.2	0.7	96.8	0.09	0.09	0.09	0.09	2.99

(continued)

Table A.4 (continued)

Country	Pop.[Mill.]	RDI	Christian [%]	Muslim [%]	Jew [%]	Hindu [%]	Buddist [%]	Other [%]
Sao Tome/Principe	0.2	3.5	82.2	0.09	0.09	0.09	0.09	17.9
Saudi Arabia	27.3	1.5	4.4	93	0.09	1.1	0.3	1.3
Senegal	13.0	0.8	3.6	96.4	0.09	0.09	0.09	0.27
Serbia	9.6	1.6	92.5	4.2	0.09	0.09	0.09	3.48
Seychelles	0.1	1.3	94	1.1	0.09	2.1	0.09	2.79
Sierra Leone	5.8	4	20.9	78	0.09	0.09	0.09	0.99
Singapore	5.1	9	18.2	14.3	0.09	5.2	33.9	28.4
Slovak Rep.	5.4	2.9	85.3	0.2	0.09	0.09	0.09	14.48
Slovenia	2.1	4	78.4	3.6	0.09	0.09	0.09	18.18
Solomon Islands	0.5	0.6	97.4	0.09	0.09	0.09	0.3	2.2
South Africa	51.5	3.6	81.2	1.7	0.1	1.1	0.2	15.6
Spain	46.2	3.9	78.6	2.1	0.1	0.09	0.09	19.18
Sri Lanka	20.8	5.6	7.3	9.8	0.09	13.6	69.3	0.27
Sudan	45.6	2	5.4	90.7	0.09	0.09	0.09	3.89
Suriname	0.5	7.6	51.6	15.2	0.2	19.8	0.6	12.5
Swaziland	1.2	2.4	88.1	0.2	0.09	0.1	0.09	11.5
Sweden	9.4	5.4	67.2	4.6	0.1	0.2	0.4	27.4
Switzerland	7.9	3.7	81.3	5.5	0.3	0.4	0.4	12.09
Tajikistan	7.6	0.7	1.6	96.7	0.09	0.09	0.09	1.68
Tanzania	45.0	5.7	61.4	35.2	0.09	0.1	0.09	3.29
Thailand	66.4	1.5	0.9	5.5	0.09	0.1	93.2	0.48
Timor-Leste	1.1	0.1	99.6	0.1	0.09	0.09	0.09	0.28
Togo	6.3	7.5	43.7	14	0.09	0.09	0.09	42.4
Tonga	0.1	0.3	98.9	0.09	0.09	0.1	0.09	1.08
Trinidad/ Tobago	1.3	5.8	65.9	5.9	0.09	22.7	0.3	5.2
Tunisia	10.6	0.1	0.2	99.5	0.09	0.09	0.09	0.38
Turkey	72.1	0.4	0.4	98	0.09	0.09	0.09	1.49
Turkmenistan	5.0	1.5	6.4	93	0.09	0.09	0.09	0.68
Tuvalu	0.0	0.7	96.7	0.1	0.09	0.09	0.09	3.29
Uganda	34.0	2.7	86.7	11.5	0.09	0.3	0.09	1.5
Ukraine	46.1	3.1	83.8	1.2	0.1	0.09	0.09	14.88
United Arab Emirates	8.4	4.4	12.6	76.9	0.09	6.6	2	1.99
United Kingdom	62.3	5.1	71.1	4.4	0.5	1.3	0.4	22.4
United States	316.1	4.1	78.3	0.9	1.8	0.6	1.2	17.2
Uruguay	3.4	5.7	57.9	0.09	0.3	0.09	0.09	41.8
Uzbekistan	27.8	0.7	2.3	96.7	0.09	0.09	0.09	0.98
Vanuatu	0.2	1.5	93.3	0.09	0.09	0.09	0.09	6.7
Venezuela	29.0	2.2	89.3	0.3	0.09	0.09	0.09	10.29
Vietnam	89.0	7.7	8.2	0.2	0.09	0.09	16.4	75.3
Yemen, Rep.	22.8	0.2	0.2	99.1	0.09	0.6	0.09	0.28
Zambia	13.2	0.5	97.6	0.5	0.09	0.1	0.09	1.7
Zimbabwe	13.1	2.7	87	0.9	0.09	0.09	0.09	12

Appendix B
Statistical Tests

B.1 Correlation Between Economic Performance (GDPpC) and Internet Penetration Rate (IPR), Year 2000

The influence of the economic performance on the use of Internet is subject of investigation. A log–log–transformation shows a very good correlation (Fig. B.1).
Following model was applied for the regression analysis and statistical tests:

$$\log(\mathrm{IPR}_i) = \beta_0 + \beta_1 \log(\mathrm{GDPpC}_i) + \varepsilon_i \tag{B.1}$$

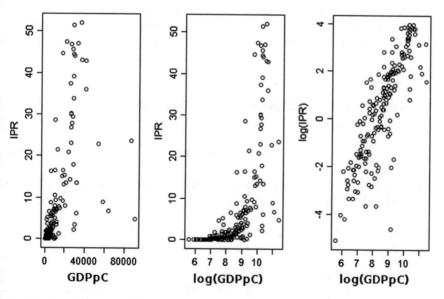

Fig. B.1 Level–Level, Level–Log and Log–Log–transformation of GDP and IPR

© Springer International Publishing AG 2018
H. Ünver, *Global Networking, Communication and Culture: Conflict or Convergence?*, Studies in Systems, Decision and Control 151,
https://doi.org/10.1007/978-3-319-76448-1

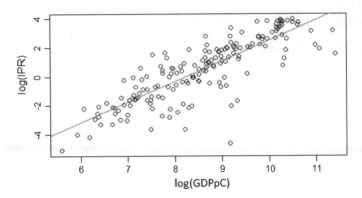

Fig. B.2 Regression line and Log–Log relationship

The estimation of the connection resulted in following values

$$\hat{\beta}_0 = -11.7 \tag{B.2}$$
$$\hat{\beta}_1 = 1.42. \tag{B.3}$$

In a next step the test is, whether there is a statistical relation between the variables. The null hypothesis

$$H_0 : \beta_1 = 0 \tag{B.4}$$

can be rejected with a 95% confidence interval, therefore a relationship between both variables is very likely (Fig. B.2). Following distribution results for the regression coefficients and the presumption of normally distributed, independent error terms:

$$\hat{\beta} \sim N[\beta, \sigma^2 (X^T X)^{-1}] \tag{B.5}$$

This can be used to determine following 95% confidence interval for the slope parameter β_1 of the model equation:

$$\beta_1 \in [0.559, 2.281] \tag{B.6}$$

Since the confidence interval does not include zero, the relationship between the logarithmic variables is significant.

Residual analysis:

In order to conduct the statistic tests, the error term have to be independent and distributed normally. Following, these two criteria are tested:

1. Gaussian distribution
   A QQ–Plot–analysis of the residues was considered. In following graph the theoretic quantiles of a Gaussian distribution are plotted against the quantiles of the

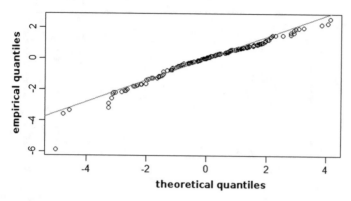

Fig. B.3 QQ–Plot

sample.

The QQ–plot in Fig. B.3 clearly implies Gaussian distributed residues.

2. Independency

The Durbin–Watson test checks the hypothesis of whether or whether not the residues are correlated with each other. Combined with the presumption that the residues are Gaussian distributed, a conclusion regarding the independency can be made, since Gaussian distributed random variables are independent, if they are not correlated. In the following, we will test the hypothesis

$$H_0 : \text{Residues are uncorrelated} \quad \text{vs.} \quad H_1 : \text{Residues are correlated.}$$

The Durbin–Watson test yields to a value of 0.4757. Therefore, the null hypothesis that the residues are uncorrelated cannot be rejected until an $\alpha \geq 48\%$ significance level. This value is to high for making a statistically exact statement. However, in our case it supports the hypothesis that the residues are uncorrelated. For this reason, the null hypothesis cannot be rejected.

3. Randomness

Beside independency and a Gaussian distribution, a good fit needs a randomness of its residues. The turning point test provides a corresponding test–statistic. Following hypothesis is tested:

$$H_0 : \text{Residues are random} \quad \text{vs.} \quad H_1 : \text{Residues are not random}$$

The evaluation of the results that we can not reject the null hypothesis on a 95% significance level. With a high likelihood, the residues are indeed random. Therefore, the test argue for independent, Gaussian distributed residues.

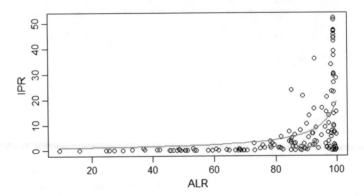

Fig. B.4 IPR as function of ALR

B.2 Relationship Between Adult Literacy Rate (ALR) and Internet Penetration Rate (IPR), Year 2000

As described in Sect. 4.2.5, a hyperbola was used to describe the relationship between ALR and IPR

$$\text{IPR} = \frac{1}{1 - 0.00095 \cdot \text{ALR}} \tag{B.7}$$

Figure B.4 illustrates the function (red) including the point cloud.

Based on following model, a linear regression analysis was performed:

$$\log(\text{IPR}_i) = \beta_0 - \beta_1 \log(1 - 0.0095\text{ALR}_i) + \varepsilon_i \tag{B.8}$$

The model above results in a hyperbolical relationship between IPR and ALR. Both sides of the model were logarithmized as preparation for the linear regression model. The manual regressions analysis yielded a positive influence of the ALR on the IPR and following estimation values:

$$\hat{\beta}_0 = -2.748 \tag{B.9}$$
$$\hat{\beta}_1 = 1.742. \tag{B.10}$$

Figure B.5 illustrates the transformed point cloud combined with the regression line.

The comparatively small coefficient of determination (0.5) implies that the IPR cannot exclusively be explained with the ALR.

On a 95%–level, the result is significant. On such a level, the null hypothesis that the ALR has no influence the IPR can be rejected. If this level is kept, following confidence interval is given for the regression coefficient:

$$\beta_1 \in [1.335, 2.149] \tag{B.11}$$

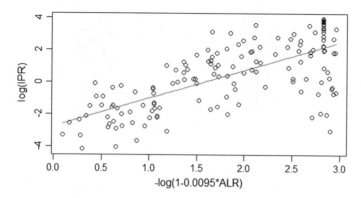

Fig. B.5 Regression line in the linearized model

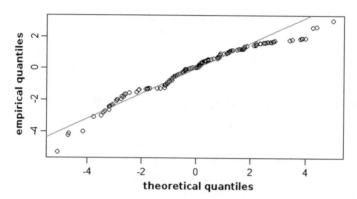

Fig. B.6 QQ–Plot

Following, a residual analysis is performed:

1. Gaussian Distribution
   In Fig. B.6 the empirical quantiles of the residues are plotted against the theoretic quantiles of a Gaussian distribution. In this case the QQ–Plot also implies a Gaussian distribution of the residues.
2. Independency
   The Durbin–Watson test yields to a p–value of 0.4751. Therefore, the null hypothesis that the residues are uncorrelated could be rejected not until an $\alpha \geq 48\%$ significance level. This value is to high for making a statistically exact statement. However, in our case it supports the hypothesis that the residues are uncorrelated. For this reason, the null hypothesis cannot be rejected.
3. Randomness
   The turning point test does not yield to the rejection of the null hypothesis. With a high likelihood, the residues in this case are random, too.

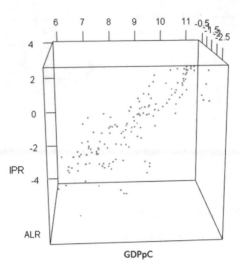

Fig. B.7 3D–Plot GDPpC, ALR, IPR

B.3 Multivariate Regression on the Basis of the Least–Square–Method

The results of the linear regression can be used for the multivariate regression. Here, we define GDPpC and ALR as independent variables and examine the IPR as response–variable. A suitable transformation function for the variables was found in previous section. Now it is statistically tested, to what extent a multivariate regressions suits to following model. The dataset is limited to $n = 151$ observations. The multivariate model is following:

$$\log(\text{IPR}_i) = \beta_0 + \beta_1 \log(\text{GDPpC}_i) + \beta_2 \log(1 - 0.0095\text{ALR}_i) + \varepsilon_i \quad \text{(B.12)}$$

The quality of this model is tested on the basis of statistical variables. Especially, a standardized residual analysis and the coefficient of determination are used. A 3D–plot allows a visual examination (Fig. B.7). Following estimation values are the result of a multivariate regression on based on the KQ estimation method:

$$\hat{\beta}_0 = -10.92439 \quad \text{(B.13)}$$
$$\hat{\beta}_1 = 1.19460 \quad \text{(B.14)}$$
$$\hat{\beta}_2 = -0.59151, \quad \text{(B.15)}$$

This results in the plane in figure B.8 as 3–dimensional regression function.

The estimated parameters are significant on a 1%–level, therefore, the null hypothesis $H_0 : \beta = (\beta_0, \beta_1, \beta_2) = (0, 0, 0)$ can be rejected on a 99%–significance level. Presuming, the estimated parameter vector corresponds a multivariate Gaussian dis-

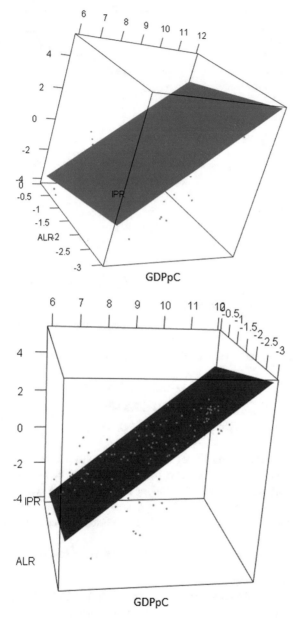

Fig. B.8 Fit–plane in 3-space GDPpC, ALR, IPR

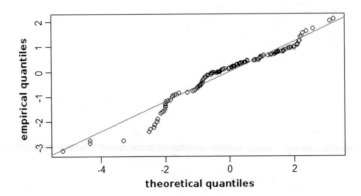

Fig. B.9 QQ–Plot

tribution, i.e.

$$\hat{\beta} \sim N(\beta, \sigma^2 (X^T X)^{-1}), \quad X = \begin{pmatrix} 1 & \log(\text{BIPpE}_1) & \log(1 - 0.0095 \cdot \text{ALR}_1) \\ \vdots & \vdots & \vdots \\ 1 & \log(\text{BIPpE}_{151}) & \log(1 - 0.0095 \cdot \text{ALR}_{151}) \end{pmatrix}$$
(B.16)

a 99%–confidence interval can be derived for all three parameters:

$$\beta_0 \in [-12.2988, 9.5499] \tag{B.17}$$

$$\beta_1 \in [1.0042, 1.3849] \tag{B.18}$$

$$\beta_2 \in [-0.873, -0.3091] \tag{B.19}$$

From a statistical standpoint it can be concluded with a 99% likelihood that the first and third parameters are negative and the second positive. The results can be interpreted as follows: Both the GDP and the literacy rate have a positive correlation with the Internet penetration rate. The correlation between the Internet penetration rate and the GDP is higher, compared to the correlation between Internet penetration rate and literacy rate.

The residual analysis is performed analog to previous section.

1. Gaussian Distribution
   In Fig. B.9 the empirical quantiles of the residues are plotted against the theoretic quantiles of a Gaussian distribution. The multivariate regression analysis suffices (like in the univariate case) the presumption of a Gaussian distribution.
2. Independency The independence assumption of the residues is a central characteristic that needs to be fulfilled. In the multivariate regression the Durbin–Watson test yields a very high value (0.9198), too. Therefore, the null hypothesis that the residues are uncorrelated cannot be rejected until an 92% level, which means, the

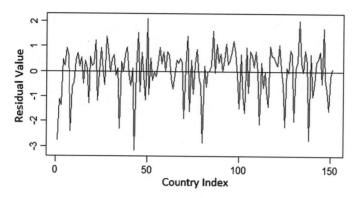

Fig. B.10 Residues

model passes the test for independence. The residue are very likely independent.
A plot of the residues is shown in Fig. B.10.

3. Randomness

The turning point test is a statistical test for evaluating the randomness of the
residues. The described model rejects the null hypothesis of the Turing point
test on a 99% significance level. Therefore, with a high likelihood (> 99%), the
residues are random.

After the residual analysis, it appears that the described multivariate regression
model has a good fit. This conclusion is additionally supported by a high coefficient of determination (0.7927).

References

1. Ackerman, S. (2014). *Snowden: NSA accidentally caused Syria's internet blackout in 2012.* The Guardian. Retrieved from http://www.theguardian.com/world/2014/aug/13/snowden-nsa-syria-internet-outage-civil-war.
2. Ackerman, E. (2013). Google gets in your face [tech to watch]. *IEEE Spectrum, 50*(1), 26–29.
3. Aldridge, I. (2013). *High-frequency trading: A practical guide to algorithmic strategies and trading systems.* Wiley.
4. Allison, J. E. (2002). *Technology, development, and democracy: International conflict and cooperation in the information age.* State University of New York. ISBN: 978-0791452134.
5. Allston, P. (2005). Labour Rights as human rights: The not so happy state of the art. In *Labour rights as human rights (Collected Courses of the Academy of European Law).* Oxford University Press. ISBN: 978-0199281060.
6. Amar, A. R. (1998). *The bill of rights: Creation and reconstruction.* Yale University Press.
7. Amiri, S. (2013). Internet penetration and its correlation to gross domestic product: An analysis of the Nordic countries. *International Journal of Business, Humanities and Technology, 3*(2), 50–60.
8. Andelfinger, V. P., & Hänisch, T. (2013). *Internet der Dinge: Technik.* Trends und Geschäftsmodelle: Verlag Versicherungswirtschaft. ISBN 978-3658067281.
9. Andres, L., Cuberes, D., Diouf, M. A., & Serebrisky, T. (2010). The diffusion of the Internet: A cross-country analysis. *Telecommunication Policy, 34,* 323–340.
10. Arab Dialogue on Internet Governance. (2012). *Conference and public consultations to establish the Arab IGF.* Beirut. Retrieved from http://css.escwa.org.lb/ictd/1759/Infonote.pdf.
11. Arsene, S. (2012). The impact of China on global Internet governance in an era of private control. In *Chinese Internet Research Conference,* Los Angeles, United States. hal-00704196v1.
12. Axelrod, R., & Hamilton, W. D. (1981). The evolution of cooperation. *Science, 211*(4489, 27), 1390–1396. https://doi.org/10.1126/science.7466396.
13. Axelrod, R. (1984). *The evolution of cooperation.* New York: Basic Books.
14. Axelrod, R. (1997). *The complexity of cooperation: Agent-based models of competition and collaboration.* Princeton, NJ: Princeton University Press.
15. Backhaus, K., Erichson, B., Plinke, W., & Weiber, R. (2006). *Multivariate Analysemethode: Eine anwendungsorientierte Einführung* (11th ed.). Berlin: Springer.

© Springer International Publishing AG 2018
H. Ünver, *Global Networking, Communication and Culture: Conflict or Convergence?*, Studies in Systems, Decision and Control 151,
https://doi.org/10.1007/978-3-319-76448-1

16. Baer, W. S. (1997). Will the global information infrastructure need transnational (or Any) governance? In B. Kahin & E. J. Wilson (Eds.), *National information infrastructure initiatives: Vision and policy design* (pp. 532–552). Cambridge: MIT Press.

17. Bagchi, K. (2005). Factors contribution to global digital divide: Some empirical results. *Journal of Global Information Technology Management, 8*, 47–65.

18. Baker, W. E., & Ronald, I. (2000). Modernization, cultural change and the persistance of traditional values. *American Sociological Review, 65*, 19–51.

19. Baliamoune, M. N. (2002). *The new economy and developing countries: Assessing the role of ICT diffusion*. WIDER Discussion Papers, World Institute for Development Economics (UNU-WIDER), No. 2002/77, ISBN: 9291902799.

20. Barabasi, A. L. (2002). *Linked*. The new science of networks. Cambridge: Perseus Publishing. ISBN 0-7382-0667-9.

21. Bartneck, N., Klaas, V., & Schönherr, H. (2008). *Prozesse optimieren mit RFID und Auto-ID: Grundlagen, Problemlösung und Anwendungsbeispiele*. Publicis Publishing, 1. Auflage. ISBN: 978-3895783197.

22. Bauer, A., & Günzel, H. (2001). *Data Warehouse Systeme - Architektur, Entwicklung, Anwendung*. dpunkt Verlag.

23. Bauer, N. (2007). *Handbuch zur industriellen Bildverarbeitung: Qualitätssicherung in der Praxis*. Fraunhofer-Allianz Vision.

24. Bauschke, M. (2010). *Die Goldene Regel: Staunen—Versehen - Handeln*. EB-Verlag, 1. Auflage, Berlin. ISBN: 978-3868930306.

25. Barzilai, G., & Barzilai-Nahon, K. (2005). Cultured technology: Internet & religious fundamentalism. *The Information Society, 21*(1), 25–40.

26. Beck, U. (1997). *Was ist Globalisierung?*. Edition Zweite Moderne, Suhrkamp, Frankfurt a.M.

27. Beck, U. (1998). *Perspektiven der Weltgesellschaft*. Edition Zweite Moderne, Suhrkamp, Frankfurt a.M.

28. Beck, U. (2007). *Weltrisikogesellschaft: Auf der Suche nach der verlorenen Sicherheit*. Suhrkamp, Frankfurt a.M. ISBN: 978-3518414255.

29. Beck, U. (2008). *Die Neuvermessung der Ungleichheit unter den Menschen*. Suhrkamp Verlag, Frankfurt a.M. ISBN: 978-3518069943.

30. Benkler, Y. (2006). *The wealth of networks: How social production transforms markets and freedom*. Yale University Press. ISBN: 0300125771.

31. Betz, J., & Kübler, H. D. (2013). *Internet governance: Wer regiert wie das Internet?* Springer VS. ISBN: 978-3531192406.

32. Blackford, R., & Broderick, D. (Eds.). (2014). *Intelligence unbound: The future of uploaded and machine minds*. Wiley-Blackwell.

33. Bloemraad, I. (2007). Unity in diversity? *Du Bois Review: Social Science and Research on Race, 4*(02), 317–336.

34. BMWi. (2014). *Dritter Monitoringbericht zur Breitbandstrategie der Bundesregierung*. Im Auftrag des Bundesministeriums für Wirtschaft und Technologie. Retrieved from http://www.bmwi.de/Dateien/BMWi/PDF/dritter-monitoringbericht-zur-breitbandstrategie, property=pdf,bereich=bmwi2012,sprache=de,rwb=true.pdf.

35. Boas, T. (2006). Weaving the authoritarian web: The control of the internet use in nondemocratic regimes. In *How revolutionary was the digital revolution* (pp. 361–378). Standford Business Books.

36. Bobkowski, P. (2008). Self-disclosure of religious identity on Facebook. *Gnovis, 9*(1),

37. Bobkowski, P., & Pearce, L. (2011). Baring their souls in online profiles or not? Religious self-disclosure in social media. *Journal for the Scientific Study of Religion, 50*(4), 744–762.

38. Bonfadelli, H. (2002). The internet and knowledge gaps: A theoretical and empirical investigation. *European Journal of Communication, 17*(1), 65–84.

39. Bostrom, N. (2006). *Welcome to a world of exponential change* (pp. 40–50). Demos.

40. Bourdieu, P. (1983). Ökonomisches Kapital - Kulturelles Kapital - Soziales Kapital. In *Die verborgenen Mechanismen der Macht* (pp. 49–80). ISBN: 978-3879756056.

41. Brinn, S., & Page, L. (1998). *The anatomy of a large-scale hypertextual web search engine.* Stanford CA, USA: Stanford University. Retrieved from http://infolab.stanford.edu/pub/papers/google.pdf.
42. Braitenberg, V., & Radermacher F. J. (2007). *Interdisciplinary approaches to a new understanding of cognition and consciousness.* Ergebnisband Villa Vigoni-Konferenz 1997 Italien. Publikationsreihe FAW/n. ISBN: 978-3-981184105.
43. Brockhaus. (2007). *Brockhaus Enzyklopaedie.*
44. Brown, P., Hugh, L., & Ashton, D. (2008). Education, Globalisation and the Future of the Knowledge Economy. *European Educational Research Journal, 7*(2), 131–156.
45. Brynjolfsson, E., & McAfee, A. (2014). *The second machine age: Work, progress, and prosperity in a time of brilliant technologies.* W. W: Norton & Company. ISBN 978-0393239355.
46. Buchner, B. K., & Ellerman, D. A. (2007). The European union emissions trading scheme: Origins, allocation, and early results. In *Review of Environmental Economics and Policy* (pp. 66–87). https://doi.org/10.1093/reep/rem003.
47. Burton, J. W. (1972). *World society.* Cambridge University Press.
48. BMBF. (2006). *Das Informatikjahr - Wissenschaftsjahr 2006.* Bundesminiterium für Bildung und Forschung. http://www.informatikjahr.de/.
49. BMBF. (2014). *Die Digitale Gesellschaft - Wissenschaftsjahr 2014.* Bundesminiterium für Bildung und Forschung. http://www.digital-ist.de/.
50. Bundesregierung. (2013). *Lebenslagen in Deutschland.* Der vierte Armuts- und Reichtumsbericht der Bundesregierung, Bundesministerium für Arbeit und Soziales (Hrsg.).
51. Bundesregierung. (2014). *Deutschlands Zukunft gestalten. Koalitionsvertrag zwischen CDU, CSU und SPD, 18. Legislaturperiode.* Retrieved from http://www.bundesregierung.de/Content/DE/_Anlagen/2013/2013-12-17-koalitionsvertrag.pdf;jsessionid=E87CD5C0034F021AD41734F9621E9758.s4t2?_blob=publicationFile&v=2.
52. Johannes Burkhardt, J. (1992). *Der Dreißigjährige Krieg.* Frankfurt a.M.: Suhrkamp Verlag. ISBN 978-3518115428.
53. Calandro, E. (2014). IGF & The future of the internet ecosystem. No. 118 Discussion on multistakeholderism in Africa. In *IGF Workshop.* Istanbul. Retrieved from http://www.intgovforum.org/cms/wks2014/index.php/proposal/view_public/118.
54. Campbell, H. (2005). Making space for religion in internet studies. *Information Society, 21*(4), 309–315. https://doi.org/10.1080/01972240591007625.
55. Castells, M., et al. (2007). *Mobile communications and society: A global perspektive.* Cambridge: MIT Press.
56. CCITT. (1968). *GAS-5 handbook: Economic studies at the national level in the field of telecommunications.* ITU, Geneva: Consultative Committee on International Telephone and Telegraph.
57. Cebrian, J. L. (1999). *Im Netz: die hypnotisierte Gesellschaft.* Der neue Bericht an den Club of Rome. Stuttgart: Deutsche Verlags-Anstalt. ISBN 3-421-05307-3.
58. Cerf, V. (2014). The internet governance ecosystem. *Communications of the ACM, 57*(4), 7.
59. Chinesische Regierung. (2010). *The Internet in China.* Information Office of the State Council of the People's Republic of China. Beijing. Retrieved from http://www.china.org.cn/government/whitepaper/node_7093508.htm.
60. Chinn, M. D., & Fairlie, R. W. (2007). The determinants of the global digital devide: A cross-country analysis of computer and Internet penetration. In *Oxford Economic Papers, 59*, 16-44. https://doi.org/10.1093/oep/gpl024.
61. Chordia, T., et al. (2013). High-frequency trading. *Journal of Financial Markets, 16*(4), 637–645.
62. Comerford, R., & Perry, T. (1998). Brooding the year 2000 millenium bug. *IEEE Spectrum, 35*(6), 68–73. https://doi.org/10.1109/6.681974.
63. Commonwealth Network. (2015). http://www.commonwealthofnations.org/commonwealth/commonwealth-membership/commonwealth-members/.
64. Club of Rome. (2015). *Club of Rome Schulen.* Retrieved from http://www.cluborfrome.de/schulen/corschulen.php.

65. Coy, W. (1987). Von QWERTY zu WYSIWYG - Texte, Tastatur, Papier. In *Sprache im technischen Zeitalter* (pp. 136–144).
66. Coy, W. (1998). Wer kontrolliert das Internet. In S. Krämer (Ed.), *Medien Computer Realität: Wirklichkeitsvorstellungen und Neue Medien*. Frankfurt/a.M.: Suhrkamp.
67. Cukier, K. N. (2010). *Data, data everywhere*. In The economist-special report: Managing information.
68. Cukier, K. N., & Mayer-Schoenberger, V. (2013a). *The rise of Big Data. How it's changing the way we think about the world*. Council on Foreign Affairs. Retrieved from http://www.foreignaffairs.com/articles/139104/kenneth-neil-cukier-and-viktor-mayer-schoenberger/the-rise-of-big-data.
69. Cukier, K. N., & Mayer-Schoenberger, V. (2013b). *Big Data: A revolution that will transform how we live, work and think*. Eamon Dolan/Houghton Mifflin Harcourt. ISBN 978-0544002692.
70. Dahrendorf, R. (1994). Das Zerbrechen der Ligaturen und die Utopie der Weltbürgergesellschaft In: *Riskante Freiheiten: Individualisierung in modernen Gesellschaften (Hrsg. Ulrich Beck und Elisabeth Beck-Gernsheim)*, Suhrkamp, Frankfurt a.M. ISBN: 3-518118161.
71. Darwin, C. (1859). *On the origin of species* (1st ed.). London: John Murray.
72. Davis, A. (2005). Should the EU have the power to set minimum standards for collective labour rights in the member states?. In *Labour Rights As Human Rights (Collected Courses of the Academy of European Law)*. Oxford University Press. ISBN: 978-0199281060.
73. Deakin, S. (2005). Social rights in a globalised economy. In *Labour rights as human rights (Collected Courses of the Academy of European Law)*. Oxford University Press. ISBN: 978-0199281060.
74. Deibert, R. (2008). *Access denied: The practice and policy of global internet filtering*. Cambridge: MIT Press.
75. Deibert, R., Palfrey, J., Rohozinski, R., Zittrain, J., & Haraszti, M. (2010). *Access controlled: The shaping of power, rights, and rule in cyberspace*. Cambridge: MIT Press.
76. Deibert, R., Zittrain, J. L., Palfrey, J., & Rohozinski, R. (2011). *Access contested: Security, identity, and resistance in Asian cyberspace*. Cambridge: MIT Press.
77. Deibert, R. (2013). Trouble at the Border: China's Internet. *Index on Censorship, 42*(2), 132–135.
78. Deitrick, J. (2008). E-jing: Using information technology to teach about Chinese religions. *Teaching Theology and Religion, 11*(3), 153–158. ISSN: 1368-4868.
79. DeNardis, L. (2009). *Protocol politics: The globalization of internet governance*. Cambridge: MIT Press.
80. Denkwerk, Z. (2014). *Das Wohlstandsquintett 2014: Zur Messung des Wohlstands in Deutschland und anderen früh industrialisierten Ländern*. Bonn: Memorandum des Denkwerks Zukunft.
81. Deutsche, B, (2014). *Hochfrequenzhandel*. Retrieved from http://deutsche-boerse.com/dbg/dispatch/de/kir/dbg_nav/about_us/15_Public_affairs/10_News/30_HFT.
82. Dickerson, M. D., & Gentry, J. W. (1983). Characteristics of adopters and non-adopters of home computer. *Journal of Consumer Research, 10*, 225–235.
83. Dierksmeyer, C., et al. (2011). *Humanistic ethics in the age of globality*. London/New York.
84. van Dijk, J. (2005). *The deepening divide, inequality in the information society* (p. 240). Thousand Oaks, London, New Delhi: Sage.
85. van Dijk, J. (2006). Digital divide research. *Achievements and Shortcomings. Poetics, 34*, 221–235.
86. van Dijk, J. (2012). *The network society* (3rd ed.). London, Thousand Oaks CA, New Delhi, Singapore: Sage Publications.
87. van Dijk, J. (2012). *The evolution of the digital devide. The digital devide turns to inequality of skills and usage*. Digital Enlightenment Yearbook 2012. J. Bus et al. (Eds.). IOS Press. https://doi.org/10.3233/978-1-61499-057-4-57.
88. Diamond, J. (2005). *Kollaps: Warum Gesellschaften überleben oder untergehen*. Frankfurt: S. Fischer. ISBN 3-10-013904-6.

89. Deutsches Institut für Normung, DIN (2008). *Technikkonvergenz, Normung und Europa - Effektive und effiziente europäische Normungs- und Standardisierungsprozesse.* Deutsches Institut für Normung. Retrieved from http://www.din.de/cmd?level=tpl-artikel& bcrumblevel=1&cmstextid=74408&languageid=de; Letzter Zugriff: 18.09.2014.

90. Dodabo. (2015). *Web-Technik und Anleitungen.* Retrieved from http://www.dodabo.de/netz/ rfcs.html

91. Dowell-Jones, M., & Kinley, D. (2011). Minding the gap: Global finance and human rights. *Ethics & International Affairs, 25*(02), 183–210.

92. Downey, A. B. (2014). *Religious affiliation, education and internet use.* arXiv preprint arXiv:1403.5534. http://arxiv.org/pdf/1403.5534v1.pdf.

93. Dreier, T., Katzenberger, P., von Lewinski, S., & Schricker, G. (1997). *Urheberrecht auf dem Weg zur Informationsgesellschaft.* Baden-Baden: Nomos Verlag.

94. Dutta, A. (2001). Telecommunications and economic development: An analysis of granger causality. *Journal of Management of Information Systems, 17*(4), 71–95.

95. Dyson, G. (2012). *Darwin among the machines: The evolution of global intelligence.* Basic Books. ISBN: 978-0465031627.

96. van Eeten, M. J., & Mueller, M. (2012). Where is the governance in Internet governance? *New Media & Society, 15*(5), 720–736.

97. Erpenbeck, J., & Sauter, W. (2007). *Kompetenzentwicklung im Netz: New Blended Learning mit Web 2.0.* Hermann Luchterhand Verlag. ISBN: 978-3472070894.

98. ESCWA Technical Paper. (2010). *Arab regional roadmap for internet governance: Framework, principles and objectives.* Economic and Social Commission for Western Asia (ESCWA). New York. Retrieved from http://www.escwa.un.org/information/publications/ edit/upload/ICTD-10-TP-5.pdf.

99. Europäischer Gerichtshof. (2014). *Judgment in Joined Cases C-293/12 and C-594/12.* Digital Rights Ireland and Seitlinger and Others. Judgment of 8 April 2014, paras. 26-27, and 37.

100. EU-Kommission. (1997). *Green paper on the convergence of the telecommunications, media, information, technology sectors, and the implications for regulation,* Brüssel.

101. EU-Kommission. (2014). *Communication on internet policy and governance.* Retrieved from http://eur-lex.europa.eu/legal-content/DE/TXT/PDF/?uri=CELEX:52014DC0072& from=EN, Brüssel.

102. EU-Kommission. (2015). *Beyond GDP: Measuring progress, true wealth, and the well-being of nations.* Retrieved from http://ec.europa.eu/environment/beyond_gdp/index_en.html

103. EU-Parlament. (2015). *Zusammenfassung der EU-Gesetzgebung.* EU Website. http://europa. eu/legislation_summaries/economic_and_monetary_affairs/institutional_and_economic_fra mework/ec0013_de.html.

104. EU-Parlament EU-Kommission, EU-Rat. (2000). Charta der Grundrechte der Europäischen Union. Amtsblatt der Europäischen Gemeinschaften C364. Retrieved from http://www. europarl.europa.eu/charter/pdf/text_de.pdf.

105. Executive Office of the President of the United States. (2014). *Big Data: seizing opportunities, preserving values.* May 2014. Retrieved from www.whitehouse.gov/sites/default/files/docs/ big_data_privacy_report_may_1_2014.pdf, p. 54.

106. Fahrmeir, L., Kneib, T., & Lang, S. (2009). *Regression - Modelle, Methoden.* Springer, Heidelberg: Anwendungen.

107. FCC Fact Sheet. (2015). *Fact sheet: Chairman wheeler proposes new rules for protecting the open internet.* Federal Communications Commission. Retrieved from http://transition. fcc.gov/Daily_Releases/Daily_Business/2015/db0204/DOC-331869A1.pdf.

108. Fischer, F. W. (2013). *Förderung der Demokratie durch Medien und Internet.* Pro Business, 1. Auflage. ISBN: 978-3863865276.

109. Fisher, W. W. (2007). When should we permit differential pricing of information. *UCLA Law Review, 55*(1).

110. Fleisch, E., & Mattern, F. (2005). *Das Internet der Dinge: Ubiquitous Computing und RFID in der Praxis:Visionen, Technologien, Anwendungen.* Handlungsanleitungen: Springer-Verlag, New York Inc, Secaucus, NJ, USA.

111. FOC. (2015). *The freedom online coaltion.* https://www.freedomonlinecoalition.com/.
112. Forum Info. (1998). *Forum Info 2000.* Forum Informationsgesellschaft der Bundesregierung: Herausforderungen 2025. Auf dem Weg in eine nachhaltige Informationsgesellschaft. Hrsg. FAW, Ulm.
113. Franois, L., & Derek, B. (2007). *Understanding national accounts.* OECD Publishing.
114. Frank, A. G. (1968). *Kapitalismus und Unterentwicklung in Lateinamerika.* 1.Auflage, Frankfurt.
115. Frank, A. G. (1978). *World Accumulation 1492–1789.* New York: Monthly Review Press.
116. Frank, A. G., & Gills, B. (1993). *The world system: Five hundred years or five thousand.* London: Routledge.
117. Freedman, D., Pisani, R., & Purves, R. (2007). *Statistics* (4th ed.). W. W. Norton & Company. ISBN: 978-0393929720.
118. Frey, C. B., & Osborne, M. A. (2013). *The future of employment: How suspectable are jobs to computerisation?.* University of Oxford. Retrieved from http://www.oxfordmartin.ox.ac.uk/downloads/academic/The_Future_of_Employment.pdf.
119. Fuchs, C. (2009). The role of income inequality in a multivariate cross-national analysis of the digital divide. *Social Science Computer Review, 27*(1), 41–58.
120. Galtung, J. (1966). Rank and social integration: A multidimensional approach. In J. Berger (Ed.), *Sociological theory in progress* (pp. 145–198). Boston: Houghton Mifflin.
121. Galtung, J. (1971a). Theorien des Friedens. In B. Meyer & F. der Konfliktregelung (Eds.), *Eine.* Leske und Budrich: Einführung mit Quellen, Opladen.
122. Galtung, J. (1971b). A structural theory of imperialism. *Journal of Peace Research, 8*, 87–117.
123. Galtung, J. (1998). *Konflikttransformation mit friedlichen Mitteln. Die Methode der Transzendenz.* Retrieved from http://homepage.univie.ac.at/silvia.michal-misak/galtung.htm.
124. Gatignon, H., & Robertson, T. S. (1991). *Innovative decission processes. In Handbook of consumer behavior.* Engelwood Cliffs, NJ: Prentice Hall.
125. Gauss, C. F. (1880). *1821.* Göttingen: Theoria combinationis observationum erroribus minimis obnoxiae. Werke IV.
126. Gauss, C. F. (1880). *1828.* Göttingen: Supplementum theoriae combinationis observationum erroribus minimis obnoxiae. Werke IV.
127. Gauss, C. F. (1964). *Abhandlungen zur Methode der kleinsten Quadrate* (Vol. 5). Physica-Verlag.
128. Gilens, M. (2012). *Affluence and influence: economic inequaltiy and politcal power in America.* Princeton University Press. ISBN: 978-0691162423.
129. Gille, L. (1986). Growth and telecommunications. *Information* (pp. 25–61). Geneva, ITU: Telecommunications and Development.
130. Goggin, G. (2006). *Cell phone culture: Mobile technology in everyday life.* London: Routledge.
131. Goldin, C. D., & Lawrence, F. K. (2009). *The race between education and technology.* Belknap Press of the Harvard University Press. ISBN: 978-0674035300.
132. Gore, A. (1992). *Wege zum Gleichgewicht.* Ein Marshallplan für die Erde: Fischer.
133. Gore, A. (2013). *The future—Six drivers of global change.* Random House Trade Paperbacks.
134. Götz, W. (2008). *Einkommen für Alle.* Bastei Lübbe Verlag. ISBN: 978-3404606078.
135. Goetze, D. (1992). Culture of Poverty—Eine Spurensuche. In S. Leibfried, & W. Voges (Hrsg.): *Armut im modernen Wohlfahrtsstaat. Kölner Zeitschrift für Soziologie und Sozialpsychologie. Sonderheft* (Vol. 32, pp. 88–103).
136. Granger, C. W. J. (1969). Investigating Causal Relations by Econometric Models and Cross-spectral Methods. *Econometrica, 37*(3), 424–438.
137. Greenberg, A. (2000). The Church and the revitalization of politics and community. *Political Science Quarterly, 115*(3), 377–394.
138. Güth, W., & Kliemt, H. (1995). Elementare spieltheoretische Modelle sozialer Kooperation, Ökonomie und Gesellschaft. In *Jahrbuch 12: Soziale Kooperation, (Hrsg.: P. Weise), Frankfurt/Main* (pp. 12–62). New York: Campus Verlag.

139. Güth, W., & Kliemnt, H. (2004). Zur ökonomischen Modellierung der Grundlagen und Wurzeln menschlicher Kulturfähigkeit. In G. Blümle, R. Klump, B. Schauenberg & H. von Senger (Eds.), *Startband 'Kulturelle Ökonomik'* (pp. 127–138).

140. Günther, S., & Luckmann, T. (2001). Asymmetries of knowledge in intercultural communication: The relevance of cultural repertoires of communcative genres. In A. Di Luzio, S. Günther, & F. Orletti (Eds.), *Culture in communication* (pp. 55–86). Amsterdam/Philadelphia: John Benjamins Publishing Company.

141. Habermas, J. (1981). *Theorie des kommunikativen Handelns* (Vol. 2, pp. 1049–1054). Suhrkamp, Frankfurt a.M.

142. Habermas, J. (1998). Jenseits des Nationalstaats? Bemerkungen zu Folgeproblemen der wirtschaftlichen Globalisierung. In Politik der Globalisierung (Ed.), *Beck* (pp. 67–84). Frankfurt a.M.: Suhrkamp.

143. Habermas, J. (2011). *Zur Verfassung Europas: Ein Essay.* Berlin: Suhrkamp.

144. Haken, H. (1982). *Synergetik: Eine Einführung 3.* Springer. ISBN: 978-3540110507.

145. Hahn, H. P., & Weiss, H. (2013). *Mobility, meaning & transformations of things.* Oxbow, Oxford: Shifting contexts of material culture through time and space. ISBN 978-1842175255.

146. Hannerz, U. (1990). Cosmopolitans and Locals in World Culture. *Theory, Culture & Society, 7,* 237–251.

147. Hardy, A. P. (1980). The role of the telephone in economic development. *Telecommunications Policy, 4*(4), 278–286.

148. Hargittai, E. (1999). Weaving the Western Web: Explaining differences in Internet connectivity among OECD countries. *Telecommunication Policy, 23,* 701–718.

149. Harvard. (2015). *Harvard Information Infrastructure Project.* Belfer Center for Science and International Affairs, John F. Kennedy School of Government, Harvard University. Retrieved from http://belfercenter.ksg.harvard.edu/project/9/harvard_information_infrastructure_project.html.

150. Hawthorne, N. (1851). *The house of the seven gables.* Ticknor & Fields.

151. Hayek, F. A. (1935). *The nature and history of the problem.*

152. Heidenbluth, N. (2012). *Datenmanagement - Beschaffung, Speicherung, Analyse, Präsentation.* Skript zur Vorlesung im Sommersemester 2012, Universität Ulm.

153. Heintz, B. (2004). Emergenz und Reduktion: Das Mikro-Makro-Problem in der Soziologie und der Philosophie des Geistes. *Kölner Zeitschrift für Soziologie und Sozialpsychologie, 56,* 1–31.

154. Heintz, B., & Werron, T. (2014). Fehlinterpretationen der Weltgesellschaftstheorie. *Kölner Zeitschrift für Soziologie und Sozialpsychologie, 66*(2), 291–302.

155. Herlyn, E. (2012). *Einkommensverteilungsbasierte Präferenz- und Koalitionsanalysen auf der Basis Selbstähnlicher Equity-Lorenzkurven: Ein Beitrag zur Quantifizierung Sozialer Nachhaltigkeit.* Springer Gabler, ISBN: 978-3834943507.

156. Hern, W. M. (1993). Has the Human Species Become a Cancer on the Planet? A Theoretical View of Population Growth as a Sign of Pathology. *Current World Leaders, 36*(6), 1089–1124.

157. Hertwich, E. G., & Peters G. P. (2009). Carbon footprint of nations: A global, trade-linked analysis. *Environmental Science & Technology, 43*(16), 6414–6420. https://doi.org/10.1021/es803496a.

158. Hettlage, R. (2004). Die Wissensgesellschaft im Verzauberungs- und Entzauberungszirkel. *Soziologisches Revue, 27,* 407–424.

159. Heylighen, F., & Campbell, D. T. (1995). Selection of organisation at the social level. *World Futures, 45,* 181–212.

160. Heylighen, F., & Bernheim, J. (2000a). Global Progress I: Empirical Evidence for Increasing Quality of Life. *Journal of Happiness Studies, 1*(3), 323–349.

161. Heylighen, F., & Bernheim, J. (2000b). Global Progress II: Evolutionary Mechanisms and their Side-effects. *Journal of Happiness Studies, 1*(3), 351–374.

162. Heylighen, F. (2007). The global superorganism: An evolutionary-cybernetic model of the Emerging Network Society. *Social Evolution & History, 6*(1), 57–117.

163. Heylighen, F. (2011). Conceptions of a global brain: An historical review. In L. E. Grinin, R. L. Carneiro, A. V. Korotayev & F.Spier (Eds.), *Evolution: Cosmic, biological, and social* (pp. 274-289). Uchitel Publishing.

164. Heylighen, F. (2013). Return to Eden? Promises and Perils on the road to a global superintelligence. In B. Goertzel & T. Goertzel (Eds.), *The end of the beginning: Life, society and economy on the brink of the singularity.* Retrieved from http://pespmc1.vub.ac.be/Papers/BrinkofSingularity.pdf.

165. Hillebrand, A., & Büllingen, F. (2001). *Internet-Governance: Politiken und Folgen der institutionellen Neuordnung der Domainverwaltung durch ICANN.* Wissenschaftliche Institut für Kommunikationsdienste, Diskussionsbeitrag Nr. 218, Bad Honnef.

166. Hirsch, J. (1992). *Einkommen und Kinderzahl, Wirtschaftswachstum und Bevölkerungsentwicklung: Eine mikroökonomisch-statische und makroökonomisch-dynamische Analyse nichtlinearer Systeme.* Tübingen: Franke Verlag.

167. Hoftede, G., Hofstede, G. J., & Minkov, M. (2010). *Cultures and organizations: software of the mind: intercultural cooperation and its importance for survival.* McGraw-Hill.

168. Horsfield, P., & Teusner, P. (2007). A mediated religion: historical perspectives on Christianity and the internet. *Studies In World Christianity, 13*(3), 278–295.

169. Hoover, S. M., Clark, L. S., & Rainie, L. (2004). *Faith Online: 64% of Wired Americans have used the Internet for spiritual or religious Information.* Pew Internet and American Life Project. Retrieved from http://www.pewinternet.org/reports/toc.asp?Report=119.

170. Hu, H. L. (2011). The political economy of governing ISPs in China: Perspectives of net neutrality and vertical integration. *The China Quarterly, 207,* 523–540.

171. Hubertus, P. (1991). *Alphabetisierung und Analphabetismus - Eine Bibliographie.* Bundesverband Alphabetisierung. ISBN: 978-3-929800-00-5.

172. Hufford, K. D., & Zeleny, M. (1991). All autopoietic systems must be social systems. *Journal of Social and Biological Structures, 14*(3), 311–332.

173. Huntington, S. P. (1993). Clash of Civilizations? *Foreign Affairs, 72*(3), 22–49.

174. Huntington, S. P. (1996a). *The clash of civilizations and the remaking of world order.* New York: Simon & Schuster. ISBN 978-0743231497.

175. Huntington, S. P. (1996b). *The clash of civilizations? The Debate. Foreign Affairs.* ISBN: 978-0140267310.

176. IAB. (2014). *Internet Architecture Board.* Retrieved from http://www.iab.org/.

177. IAB Statement. (2014). *IAB statement on internet confidentiality.* Retrieved from https://www.iab.org/2014/11/14/iab-statement-on-internet-confidentiality/.

178. IANA. (2014). *Internet Assigned Numbers Authority.* Retrieved from http://www.internetassignednumbersauthority.org/.

179. IANA Root. (2014). *IANA Root Zone Database.* Retrieved from http://www.iana.org/domains/root/db.

180. ICANN. (2014). *Internet Corporation for Assigned Names and Numbers.* Retrieved from https://www.icann.org/.

181. ICANN Functions. (2014). *ICANN Functions Fact Sheet.* Retrieved from https://www.icann.org/en/system/files/files/iana-factsheet-24mar14-en.pdf.

182. IEEE Standars Association. (2015). *Institute of Electrical and Electronics Engineers Standards Association.* Retrieved from http://standards.ieee.org/index.html.

183. Bradner, S. (2013). *IETF structure and internet standards process.* Vancouver, Kanada. Retrieved from http://www.ietf.org/edu/process-oriented-tutorials.html#newcomers.

184. IEFT 88. (2013). *Internet Engineering Task Force Meeting 88.* Vancouver. Kanada. Retrieved from http://www.ietf.org/meeting/88/.

185. IGF. (2009). *Internet Governance Forum - Transcripts.* Sharm-El-Sheikh. Retrieved from http://www.intgovforum.org/cms/2009/sharm_el_Sheikh/Transcripts/Sharm%20El%20Sheikh%2018%20November%20Stock%20Taking%20Part%20I.pdf.

186. IGF. (2014). *Internet Governance Forum.* Istanbul. Retrieved from http://www.intgovforum.org/cms/igf-2014.

187. IGF. (2015). *Internet Governance Forum.* Jeao Pessoa. Brasilien. Retrieved from http://www.intgovforum.org/cms/igf-2015-website.
188. ISOC. (2014). *Internet Society.* Retrieved from http://www.internetsociety.org/.
189. ITU. (1994). *Open Systems Interconnection - Model and Notation.* ISO/IEC 7498-1:1994. Retrieved from http://handle.itu.int/11.1002/1000/2820.
190. ITU. (2010). *Definitions of World Telecommunication/ICT Indicators.* Retrieved from https://www.itu.int/ITU-D/ict/material/TelecomICT_Indicators_Definition_March2010_for_web.pdf.
191. ITU. (2012). *Measuring the Information Society.* ISBN: 978-9261140717, p. 2.
192. ITU. (2013). *Statistics.* http://www.itu.int/en/ITU-D/Statistics/Pages/stat/default.aspx.
193. ITU. (2014a). *The State of Broadband 2014: Broadband for All.* A Report by the Broadband Commission. ITU / UNESCO. Retrieved from http://www.broadbandcommission.org/documents/reports/bb-annualreport2014.pdf.
194. ITU. (2014b). *Measuring the Information Society.* ISBN: 978-9261152918.
195. ITU-T (2015). *ITU Telecommunication Standardization Sector.* Retrieved from http://www.itu.int/en/ITU-T/Pages/default.aspx.
196. Jasper, W. (2014). Milestones in the UN's march for control of the internet. *New American, 30*(21), 19–20.
197. Johnson, D., Crawford, S., & Palfrey, J. (2004). The accountable net. *Virginia Journal of Law & Technology, 9,* 6–33.
198. Joorabchi, T. N., Osman, M. N., & Hassan, M. S. (2012). The relationship between person factors and its effect on internet usage. *Journal of Mass Communication and Journalism, 2,* 139.
199. Jiang, M.(2012). Authoritarian informationalism: China's approach to internet sovereignty. In P. O'Neil & R. Rogowski (Eds.), *Essential readings of comparative politics* (4th Ed.). WW Norton & Company, New York.
200. Jipp, A. (1963). Wealth of nations and telephone density. In *Telecommunications Journal,* 199–201.
201. Kämpke, T., Pestel, R., & Radermacher, F. J. (2003). A computational concept for normative equity. *European Journal of Law and Economics, 15,* 129–163.
202. Kämpke, T., & Radermacher, F. J. (2005). Equity analysis by functional approach. *Data analysis and decision support* (pp. 241–248). Heidelberg: Springer.
203. Kämpke, T. (2010). The use of mean values vs. medians in inequality analysis. *Journal of Economic and Social Measurement, 35,* 34–62.
204. Kämpke, T., & Radermacher, F. J. (2011). *Analytische Eigenschaften von Equity. Lorenzkurven,* FAW working paper, Ulm.
205. Kämpke, T. (2012). Income distribution and majority patterns. *International Journal of Computational Economics and Econometrics, 2,* 155–178.
206. Kämpke, T., & Radermacher, F. J. (2014). *Income Modeling and Balancing-A rigorous treatment of distribution patterns.* Lecture Notes in Economics and Mathematical Systems: Springer. ISBN 978-3319132235.
207. Kahin, B. J., & Keller, H. (1997). *Coordinating the internet.* Cambridge, MA: MIT Press.
208. Kahn, R. E. (1994). The role of government in the evolution of the internet. *Communications of the ACM, 37,* 15–19.
209. Kant, I. (1788). *Kritik der praktischen Vernunft.* Oldenbourg Akademieverlag. ISBN: 978-3050035765.
210. Kapitza, S. (2006). *Global population blow up and after.* Report to the Club of Rome and Global Marshall Plan Initiative: The Demographic Recolution and Information Society.
211. Kapitza, S. (2009). Global population blow-up and after: The demographic revolution and sustainable development. *Bulletin of the Georgian National Academy of Sciences, 3*(1).
212. Kauffmann, R. J., et al. (2005). Is there a global digital divide for digital wireless phone technologies? *Journal of Assosiation for Information Systems, 6,* 338–382.
213. Keil, S., & Thaidigsmann, S. I. (2012). *Zivile Bürgergesellschaft und Demokratie: Aktuelle Ergebnisse der empirischen Politikforschung.* Springer VS. ISBN: 978-3658008741.

214. Keeney, R. L., & Raiffa, H. (1993). *Decisions with multiple objectives: Preferences and value tradeoffs*, New York.
215. Kernighan, B. W., & Ritchie, D. M. (1988). *The C Programming Language*. Prentice Hall.
216. Kesseler. (2004).
217. Kiiski, S., & Pohjola, M. (2002). Cross-country diffusion of the Internet. *Information Economics and Policy*, *14*, 297–310.
218. Kinross, L. (1979). *The Ottoman centuries: The rise and the fall of the Turkish Empire*. New York: Harper Perennial. ISBN 978-0688080938.
219. Kleinsteuber, H. J. (1996). *Der "'Information Superhighway"': Amerikanische Visionen und Erfahrungen*. Opladen: Westdeutscher Verlag.
220. Klein, H. (2004). Understanding WSIS: An institutional analysis of the UN world summit on the information society. *Information Technology & International Development*, *1*(3–4), 3–14.
221. Kleinwächter, W. (2004a). Beyond ICANN vs ITU? How WSIS tries to enter the new territory of Internet governance. *Gazette*, *66*(3–4), 233–251.
222. Kleinwächter, W. (2004b). WSIS: a new diplomacy? Multistakeholder approach and bottom up policy in global ICT governance. *Information Technology & International Development*, *1*(3–4), 3–13.
223. Kleinwächter, W. (2000). ICANN between Technical Mandate and Political Challenges. *Telecommunications Policy*, *24*, 553–563.
224. Kleinwächter, W. (2015). *Gibt die US-Regierung die Aufsicht über den Internet Root ab?*. Retrieved from http://www.heise.de/tp/artikel/43/43887/1.html.
225. Kolkman, O. (2014). *A framework for describing the Internet Assigned Numbers Authority (IANA)*. IETF Internet draft, work in progress.
226. Kooiman, J. (1993). *Modern governance: New government-society interactions*. London: Sage.
227. Kornhuber, H. H. (1992). Gehirn, Wille, Freiheit. In *Revue de metaphysique et de morale* (Bd. 97, H. 2, pp. 203–223).
228. Kornhuber, H. H., & Deecke, L. (2009). *Wille und Gehirn*. Edition Sirius im Aisthesis-Verlag, Bielefeld/Locarno 2007, 2. überarbeitete Auflage.
229. Köhler, K. H., & Salz, A. (2007). *Migration als Herausforderung*. Deutsche Unesco-Kommission, Bonn: Praxisbeispiele aus den UNESCO-Projektschulen. ISBN 978-3927907973.
230. Krause, C. (2005). Interdependenzen zwischen Staat und Buddhismus in der Volksrepublik China. *China Heute XXIV, Nr. 6*(142), 222–233. http://www.china-zentrum.de/fileadmin/PDFs/Religionen/Chh-142-Krause-Buddhismus.pdf.
231. Kruse, H. J., & Ritter, H. (1998). *Die Entdeckung der Intelligenz oder können Ameisen denken*. Beck, München: Verlag Ch.
232. Küng, H., & Kuschel, K. J. (1993a). *Weltfrieden durch Religionsfrieden*. Piper Verlag. ISBN: 978-3492118620.
233. Küng, H., & Kuschel, K. J. (1993b). *Erklärung zum Weltethos*. Piper Verlag. ISBN: 978-3492219587.
234. Küng, H. (1995). *Ja zum Weltethos: Persepektiven für die Suche nach Orientierung*. Piper Verlag. ISBN: 978-3492038171.
235. Küng, H. (1996). *Weltethos und Erziehung*. Luzern: Hans-Erni-Stiftung.
236. Küng, H. (2000). *Weltethos für Weltpolitik und Weltwirtschaft*. Piper Verlag. ISBN: 978-3492230803.
237. Küng, H. (2004). *Spurensuche: Die Weltreligionen auf dem Weg*. Piper Verlag, München, 6. Auflage. ISBN: 978-3492041034.
238. Küng, H., Leisinger, K. M., & Wieland, J. (2010). *Manifest Globales Wirtschaftsethos: Konsequenzen und Herausforderungen für die Weltwirtschaft*. München: Deutscher Taschenbuch Verlag. ISBN 978-423346283.
239. Küng, H. (2012). *Handbuch Weltethos: Eine Vision und ihre Umsetzung*. München: Piper Verlag. ISBN 978-3492300599.

240. Kurzweil, R. (2001). *The law of accelerating returns*. KurzweilAI. http://www.kurzweilai. net/the-law-ofaccelerating-returns.
241. Kurzweil, R. (2005). *The singularity is near: When humans transcend biology*. Penguin.
242. Kurzweil, R. (2010). *How my predictions are faring* (pp. 1–146). KurzweilAI.
243. Kuschel, K. J. (2001). *Streit um Abraham: Was Juden*. Christen und Muslime trennt - und was sie eint: Patmos Verlag, Düsseldorf. ISBN 978-3491690301.
244. Kuschel, K. J. (2007). *Juden-Christen-Muslime: Herkunft und Zukunft*. Düsseldorf: Patmos Verlag. ISBN 978-3491725003.
245. LAC IGF (2014). *Latin Amrica and Carribian Internet Governance Forum*. San Salvador. Retrieved from http://www.lacigf.org/en/lacigf7/agenda.html.
246. Landgericht Berlin. (2014): *Versäumnisurteil: Geschäftsnummer 15 O 4413*. Retrieved from http://zap.vzbv.de/60ef0191-455d-4b5c-be58-5157c5c7ea55/WhatsApp-LG-Berlin-15_0_44_13.pdf.
247. Laszlo, E. (2000). *Das fünfte Feld*. Materie, Geist und Leben - Vision der neuen Wissenschaften: Bastei Lübbe.
248. Lear, E. (2014). The internet assigned numbers authority transition. *IEEE Internet Computing, 4*, 62–65.
249. O'Leary, S., & Brasher, B. (1996). The unknown God of the internet. Religious communications from the ancient agora to the virtual forum. In C. Ess (Ed.), *Philosophical perspectives on computer-mediated communication* (pp. 233–269). Albany: State University of New York Press.
250. Leib, V. (2001). Das Doppelgesicht ICANNs: Koordination und Regulierung des Internet. In H. Kubicek et al. (Hrsg.), *Internet@Future (Jahrbuch Telekommunikation und Gesellschaft 2001)*. Hüthig Heidelberg, pp. 124-126.
251. Lessing, G. E. (1987). *Nathan der Weise*. Philipp Reclam jun: Verlag GmbH, Ditzingen. ISBN 978-3150000038.
252. Leubolt, B. (2011). *Staat und politische Ökonomie in Brasilien: Die Regierung Lula im Spiegelbild der Geschichte*. SRE-Discussion Papers, WU Vienna University of Economics and Business, Vienna. Retrieved from http://epub.wu.ac.at/3152/1/sre-disc-2011_01.pdf.
253. Leuphana Universität. (2014). *MedienKulturWiki*. Retrieved from http://www2.leuphana. de/medienkulturwiki/medienkulturwiki2/index.php/Kommunikationl; Letzter Zugriff: 08.08.2014.
254. Levinson, N. (2008). The internet governance ecosystem: assessing multistakeholderism and change. In *Conference Papers American Political Science Association* (pp. 1–22).
255. Levinson, N. (2010). Unexpected allies in global governance arenas? Cross-cultural collaborative knowledge processes and the internet governance forum. In *Conference Papers International Studies Association, Annual Meeting* (pp. 1–27).
256. Lewis, O. (1966). *La vida: A Puerto Rican family in the culture of poverty - San Juan and New York*. New York Random House.
257. Lienkamp, M., Lennart, S. L., & Tang, T. (2012). *Analyse der rechtlichen Situation von teleoperierten (und autonomen) Fahrzeugen*. Lehrstuhl für Fahrzeugtechnik. TU München. Retrieved from http://www.ftm.mw.tum.de/uploads/media/07_Lutz.pdf.
258. Limage, L. J. (2001). *Democratizing education and educating democratic citizens: International and historical perspectives*. Routledge. ISBN: 978-0815335702.
259. Lind, N. (2014). *Encyclopedia of quality of life and well-being research*. Netherlands: Springer. ISBN 978-9400707528.
260. Lindner, J. (2005). *Informationsübertragung: Grundlagen der Kommunikationstechnik*. Springer.
261. Lindner, R. (1999). Was ist 'Kultur der Armut'? Anmerkungen zu Oscar Lewis. In S. Herkommer (Hrsg.) *Soziale Ausgrenzungen* (pp. 171–178).
262. Lovelock, J. E., & Watson, A. J. (1983). Biological homeostasis of the global environment: the parable of Daisyworld. *Tellus B, 35*(4), 284–289.
263. Lovelock, J. E. (2000). *Gaia: A new look at life on Earth*. Oxford University Press. ISBN: 0-192862189.

264. Luhmann, N. (1975). Die Weltgesellschaft. In *Soziologische Aufklärung 2, Aufsätze zur Theorie der Gesellschaft* (pp. 51–71). Opladen.
265. Luhmann, N. (1995a). *Social systems.* Stanford University Press.
266. Luhmann, N. (1995b). Inklusion und Exklusion. In *Soziologische Aufklärung* (Vol. 6, pp. 237–264). Opladen.
267. Luhmann, N. (1997). *Die Gesellschaft der Gesellschaft.* Frankfurt/a.M: Suhrkamp Verlag. ISBN 3-518-58240-2.
268. Lohse, S. (2011). Zur Emergenz des Sozialen bei Niklas Luhmann. Zeitschrift für Soziologie, Jg. 40. *Heft, 3*, 190–207.
269. Lovelock, J. E. (1995). *Das Gaia-Prinzip: Die Biographie unseres Planeten.* Aus dem Englischen von P: Gillhofer und B. Müller, Insel Taschenbuch, Frankfurt/a.M. ISBN 3-458-33242-1.
270. Machill, M. (2001). Wer regiert das Internet? Empfehlungen der Bertelsmann Stiftung zu Internet Governance. *Wer regiert das Internet?* (pp. 17–49). Bertelsmann Stiftung: ICANN als Fallbeispiel für Global Internet Governance.
271. MacKinnon, R. (2012). *Consent of networked: The world-wide struggle for internet freedom.* New York: Basic Books.
272. Magnis, C. (2015). *Huntingtons Spur.* Cicero: Magazin für politische Kultur.
273. Majerova, I. (2012). Comparison of old and new methodology in human development and poverty indexes: A case of the least developed countries. International Business Information Management Association (IBIMA). *Journal of Economic Studies and Research, 2012.* https://doi.org/10.5171/2012.290025.
274. Mankiw, G. N. (2004). *Grundzüge der Volkswirtschaftslehre* (3rd ed.). Stuttgart: Schäffer-Poeschel.
275. Marsh, C. (2011). *Religion and the state in Russia and China: Suppression, survival, and revival* (1st ed.). Bloomsbury Academic. ISBN: 978-1441112477.
276. Martiny, A. (2001). *Korruption: Wuchernder Krebsschaden in der Gesellschaft.* Aus Politik und Zeitgeschichte. Beilage zur Wochenzeitung Das Parlament, B, pp. 32–33. Retrieved from http://www.bpb.de/system/files/pdf/CQIOO5.pdf.
277. Mattern, F., & Flörkemeier, C. (2010). Vom Internet der Computer zum Internet der Dinge. *Informatik-Spektrum, 33*(2), 107–121.
278. Maturana, H. R., Varela, F. J., & Uribe, R. (1974). Autopoiesis: The organization of living systems, its characterization and a model. *Biosystems, 5*, 187–196. https://doi.org/10.1016/0303 2647(74)90031 8.
279. Maturana, H., & Varela, F. J. (1980). *Autopoiesis and cognition: The realization of the living.* Boston: D. Reidel. ISBN 978-90-277-1016-1.
280. Maturana, H., & Varlela, F. J. (1984). *Der Baum der Erkenntnis.* Die biologischen Wurzeln des menschlichen Erkennens: Deutschsprachige Ausgabe, Scherz Verlag, Bern und München.
281. Mayntz, R. (1998). *New challenges to governance theory.* Jean Monnet Chair Papers 50, The Robert Schuman Centre at the European University Institute, Florenz. Retrieved from http://www.iue.it/RSC/Mayntz.htm.
282. Mbarika, V., Kah, M., Musa, P., & Meso, P. (2003). Predictors of growth of teledensity in developing countries: A focus on middle and low-income countries *Electronic Journal on Information Society in Developing Countries, 1*(12), 1–16.
283. Mayer-Schönberger, V. (2013). *Big Data: A revolution that will transform how we live, work and think.* London: John Murray. ISBN 978-0544002692.
284. Meadows, D., Meadows, D. H., Milling, P., & Zahn, E. (1972). *Grenzen des Wachstums.* Bericht des Club of Rome zur Lage der Menschheit: Deutsche Verlags-Anstalt. ISBN 978-3421026330.
285. Mersch, P. (2012). *Systemische Evolutionstheorie: Eine systemtheoretische Verallgemeinerung der Darwinschen Evolutionstheorie.* Norderstedt: Books on Demand. ISBN 978-3-8482-2738-9.
286. Meyer, J. W., Boli-Bennett, J., & Chase-Dunn, C. (1975). Convergence and divergence in development. *Annual Review of Sociology*, 223–246.

287. Meyer, J. W., & Hannan, M. T. (1979). *National development and the world system: Educational, economic, and political change 1950–1970*. University of Chicago Press.
288. Meyer, J. W. (1980). The World polity and the authority of the nation-state. *Studies of the Modern World System* (pp. 109–137). New York: Academic Press.
289. Meyer, J., Boli, J., Thomas, G. M., & Ramirez, F. O. (1997). World society and the nation-state. *The American Journal of Sociology, 103*(1), 144–181. Retrieved from https://webfiles.uci.edu/schofer/classes/2012soc219IT/readings/5h%20Meyer%20et%20al.%201997%20World%20Society%20Nation%20State.pdf.
290. Meyer, J. (2005). *Weltkultur: Wie die westlichen Prinzipien die Welt durchdringen*. Suhrkamp Verlag, Frankfurt a.M., ISBN: 978-3518416518.
291. Miller, J. G. (1978). *Living systems*. New York: McGraw Hill.
292. Miller, S. J. (2006). *The method of least squares*. Mathematics Department Brown University (pp. 1–7). Retrieved from http://web.williams.edu/Mathematics/sjmiller/public_html/105Sp10/handouts/MethodLeastSquares.pdf.
293. Miller, J. H., Page, S. E., & LeBaron, B. (2008). *Complex adaptive systems: An introduction to computational models of social life*. Princeton University Press.
294. Miller, B. J., Mundey, P., & Hill, J. P. (2013). Faith in the age of Facebook: Exploring the links between religion and social network site membership and use. *Sociology of Religion, 74*(2), 227–253.
295. Mises, v. L. (1920). *Economic calculation problem in the socialist commonwealth*.
296. Mittelstraß, J. (1996). *Leonardo-Welt: Über Wissenschaft. Forschung und Verantwortung*: Suhrkamp Verlag, Frankfurt a.M.
297. Moore, G. E. (1965). Cramming more components onto integrated circuits. *Electronics, 38*(8).
298. Moore, G. (1975). Progress in digital integrated electronics. In *IEEE International Electron Devices Meeting* (pp. 11–13).
299. Moré, J. J. (1978). The Levenberg-Marquardt algorithm: implementation and theory. *Numerical analysis* (pp. 105–116). Berlin Heidelberg: Springer.
300. Murray, C., Hoane, A. J., & Hsu, F. (2002). Deep Blue. *Artificial Intelligence, 134*(1–2), 57–83.
301. Müller, M., Kürbis, B., & Page, C. (2007). Democratizing global communication? Global civil society and the campaign for communication rights in the information society. *International Journal of Communication, 1*, 267–296.
302. Müller, M., & Kürbis, B. (2014). *Roadmap for globalizing IANA: Four principles and a proposal for reform*. Technical Report, Internet Governance Project. Retrieved from www.internetgovernance.org/wordpress/wp-content/uploads/ICANNreformglobalizingIANAfinal.pdf.
303. Nachhaltigkeitsbeirat, B.-W. (2012). *Gutachten Energiewende - Implikationen für Baden-Württemberg*, Stuttgart. Retrieved from http://www.nachhaltigkeitsbeirat-bw.de/mainDaten/dokumente/dokumente.htm; Letzter Zugriff 16.09.2014.
304. Nachtigall, W. (2001). *Natur macht erfinderisch*. Ravensburger Buchverlag, 2. Auflage. ISBN: 978-3473358908.
305. Nachtigall, W., & Pohl, G. (2013). *Bau-Bionik: Natur-Analogien-Technik*. Springer Vieweg, 2. Auflage. ISBN: 978-3540889946.
306. Nagy, B., Farmer, J. D., Trancik, J. E., & Gonzalez, J. P. (2011). Superexponential long-term trends in information technology. *Technological Forecasting and Social Change, 78*, 1356–1364.
307. Neill, S. (1991). *A History of Christian Missions (Penguin History of the Church)* (Vol. 2). Auflage, London: Penguin Books. ISBN 978-0140137637.
308. NETmundial. (2014). *Global Multistakeholder Meeting on the Future of Internet Governance*. Sau Paulo. Retrieved from http://netmundial.br/.
309. Neirynck, J. (2007). *Der göttliche Ingenieur: Die Evolution der Technik*. expert; 7. Auflage. ISBN: 978-3816927747.
310. Neuberger, B. (2008). *Die Bedeutung der Religion im Staat Israel*. Bundeszentrale für politische Bildung. Retrieved from http://www.bpb.de/internationales/asien/israel/45108/staat-und-religion.

311. Neuhoff, K. (2007). *Bildung als Voraussetzung für eine Kultur der Verständigung.* Vortrag auf der 7. Pädagogischen Konferenz: "'Verständigung der Kulturen - Kultur der Verständigung in der Schule'", Gewerkschaft für Erziehung und Wissenschaft (GEW), Zentrum Kloster Drübeck. Retrieved from www.rpi-virtuell.net/workspace/6AD6A4BD-9407-434B-A79D-BAB7E44E15DD/2007/druebeck/neuhoff_bildung.pdf.

312. Nocetti, J. (2015). Contest and conquest: Russia and global internet governance. *International Affairs, 91*(1), 111–130.

313. Norris, P. (2001). *Digital divide? Civic engagement, information poverty, and the Internet worldwide.* Cambridge: Cambridge University Press.

314. Noveck, B. S. (2009). *Wiki government: How technology can make government better, democracy stronger, and citizens more powerful.* Washington DC: Brookings Institution Press.

315. Novitz, T. (2005). The European Union and International Labour Standards: The dynamics of dialogue between the EU and the ILO. In *Labour rights as human rights (Collected Courses of the Academy of European Law).* Oxford University Press. ISBN: 978-0199281060.

316. NTIA. (2014). *NTIA Announces Intent to Transition Key Internet Domain Name Functions.* National Telecommunications & Information Administration. US Department of Commerce. Retrieved from http://www.ntia.doc.gov/press-release/2014/ntia-announces-intent-transition-key-internet-domain-name-functions.

317. OECD. (2000). *Learning to brigde the Digital Divide.* Paris: Organisation for Economic Co-operation and Developement.

318. OECD. (2010). *The economic and social role of internet intermediaries.* OECD Directorate for Science Technology and Industry. Retrieved from http://www.oecd.org/dataoecd/49/4/44949023.pdf.

319. OECD. (2011). *Public spending on Education. Directorate of Employment, Labour and Social Affairs.* Social Policy Division. Retrieved from http://www.oecd.org/social/family/PF1.2%20Public%20spending%20on%20education%20-%20updated%20181012.pdf.

320. OECD. (2014a). *Data-driven Innovation for Growth an Well-being.* Interim Synthesis Report. Retrieved from http://www.oecd.org/sti/inno/data-driven-innovation-interim-synthesis.pdf.

321. OECD. (2014b). *Standard für den automatischen Informationsaustausch über Finanzkonten: Gemeinsamer Meldestandard.* Ausschuss für Steuerfragen (CFA) der OECD in Zusammenarbeit mit G20 und EU. Retrieved from http://www.oecd.org/ctp/exchange-of-tax-information/standard-fur-den-automatischen-informationsaustausch-von-finanzkonten.pdf.

322. OECD. (2015). *Better life index.* Retrieved from http://www.oecdbetterlifeindex.org/de/.

323. Olawuji, J. O., & Mgbole, F. (2012). Technological convergence. *Science Journal of Physics, 2012,* Article ID sjp-221, 5 pp. https://doi.org/10.7237/sjp/221.

324. Owen, B. (2002). A Novel conference: The origins of TPRC. In S. Braman (Ed.), *Communication Research & Policy: A Sourcebook* (pp. 347–356). Cambridge: MIT Press.

325. Oxford Dictionary. (2014). *Oxford Dictionary online.* Retrieved from http://www.oxforddictionaries.com; Letzter Zugriff: 08.08.2014.

326. Palm, G. (1994). *Assoziatives Gedächtnis und Gehirntheorie.* In Evolution und Intelligenz: Universitätsverlag Ulm. ISBN 3-89559-021-5.

327. Panko, R. R. (2008). IT employment prospects: Beyond the dotcom bubble. *European Journal of Information Systems (2008) 17,* 182–197. https://doi.org/10.1057/ejis.2008.19.

328. Parsons, T. (1969). Full citizenship for the Negro American? In T. Parsons (Ed.), *Politics and social structure* (pp. 252–291). New York: The Free Press.

329. PCAST. (2014). *Big Data and privacy: A technological perspecitve.* A Report to the President. Executive Office of the President, President's Council of Advisors on Science and Technology. Retrieved from https://www.whitehouse.gov/sites/default/files/microsites/ostp/PCAST/pcast_big_data_and_privacy_-_may_2014.pdf.

330. Pearson, K. (1896). Mathematical contributions to the theory of evolution. III. Regressions, heredity, and panmixia. *Philosophical Transactions ot the Royal Society Ser., A187,* 253–318.

331. Pearson, K. (1920). Notes on history of correlation. *Biometrika, 13,* 25–45.

332. Pennachin, C., & Goertzel, B. (2007). *Contemporary approaches to artificial general intelligence* (pp. 1–30). Springer, Berlin: Artificial General Intelligence.

333. PEW Research Center. (2010a). *Pew-templeton global religious futures*. Pew ResearchReligion & Public Life Project, Washington D.C. Retrieved from http://www.globalreligiousfutures.org/; Letzter Zugriff: 18.09.2014.
334. PEW Research Center. (2010b). *The global religious landscape: A report on the size and distribution of the World's major religious groups as of 2010*. Retrieved from http://www.pewforum.org/2012/12/18/global-religious-landscape-exec/.
335. PEW Research Center. (2011). *The civic and community engagement of religiously active Americans: Part 3: Technology and religious group members*. Pew Research Center, Internet, Science & Tech. Retrieved from http://www.pewinternet.org/2011/12/23/part-3-technology-and-religious-group-members/.
336. PEW Research Center (2012). *The global religious landscape, Appendix A: Methodology*. Retrieved from http://www.pewforum.org/2014/04/04/methodology-2/
337. PEW Research Center. (2012). *The global religious landscape, Appendix B: Data Sources by Country*. Retrieved from http://www.pewforum.org/files/2012/12/globalReligion-appB.pdf.
338. PEW Research Center. (2014). *Table: Religious diversity index scores by country*. Retrieved from http://www.pewforum.org/2014/04/04/religious-diversity-index-scores-by-country/
339. PEW Research Center. (2015). *Internet seen as positive influence on education but negative on morality in emerging and developing nations*. Pew Research Center, Global Attitudes & Trends. Retrieved from http://www.pewglobal.org/2015/03/19/.
340. PEW Research Center. (2017). Europe's growing Muslim population. *Religion & Public Life*. Retrieved from http://www.pewforum.org/2017/11/29/europes-growing-muslim-population/.
341. Picketty, T. (2014). *CAPITAL in the twenty-first century*. Cambridge, MA: The Belknap Press of Harvard Univ. Press.
342. Pogge, T. (2007). *World poverty and human rights: Cosmopolitan responsibilities and reforms* (2nd ed.). Wiley. ISBN: 978-0745641447.
343. Porter, C. E., & Donthu, N. (2006). Using the technology acceptance model to explain how attitudes determine Internet usage: The role of perceived access barriers and demographics. *Journal of Business Research, 59*.
344. Poundstone, W. (1995). *Prisoner's dilemma: John von Neumann, game theory, and the puzzle of the bomb*. Anchor/Random House. ISBN: 978-0385415804.
345. Powers, W. T. (1973). *Behavior: The control of perception*. Chicago: Aldine.
346. Powers, W. T. (1989). *Living control systems*. New Canaan, CT: Benchmark Publications.
347. Pratama, A. R., & Al-Shaikh, M. (2012). *Relation and growth of internet penetration rate with human development level from 2000 to 2010*. International Business Information Management Association (IBIMA). Communications of IBIMA Vol. 2012. https://doi.org/10.5171/2012.778309.
348. Pronk, J. (2007). Die Globale Apartheid Überwinden. *Eins. Entwicklungspolitik Information Nord-Süd Nr., 8–9*, 26–31.
349. Prutz, H. (1964). *Kulturgeschichte der Kreuzzüge*. Hildesheim: Georg Olms Verlag. ISBN 978-3487004891.
350. Purkayastha, P., & Bailey, R. (2014). U.S. Control of the Internet. *Monthly Review: An Independent Socialist Magazine, 66*(3), 103–127.
351. Quibria, M. G., Shamsun, N. A., Tschanh, T., & Reyes-Macasaquit, M. L. (2003). Digital Divide: Determinants and policies with special reference to Asia. *Journal of Asian Economics, 13*, 811–825.
352. Radermacher, F. J. (1994). *Das Bild des Menschen aus der Sicht der evolutionären Erkenntnistheorie*. In: Evolution und Intelligenz. Universitätsverlag Ulm. ISBN: 3-89559-021-5.
353. Radermacher, F. J. (1997a). *Das digitale Bewusstsein*. Bild der Wissenschaft, 7, 70–71.
354. Radermacher, F. J. (1997b). *Globalisierung und weltweite Herausforderungen: Die Rolle der Pädagogik*. Beitrag zur Internationalen Pädagogischen Werktagung, Salzburg. Retrieved from http://www.ziel.org/nachklang/NutzenNachkl_Radermacher.pdf.
355. Radermacher F. J. (1998a). Globalisation and information technology. In U. Bartosch & J. Wagner (Eds.), *International Conference in celebration of the 85th birthday of Carl-Friedrich von Weizsäcker* (pp. 105–107). Evangelical Academy Tutzing.

356. Radermacher, F. J. (1998b). Intelligenz - Kognition - Bewusstsein: Systemtheoretische Über-
     legungen, technische Möglichkeiten, philosophische Fragen. In C. Stadelhofer (Ed.), *Inter-
     disziplinäre Beiträge zur Kommunikation und zum Mensch-Technik-Verhältnis* (Vol. 6, pp.
     146–193). Bielefeld: Kleine Verlag GmbH.
357. Radermacher, F. J., van Dijk, J. A. G. M., & Pestel, R. (1999). The European Way to the
     Global Information Society. *IPTS Report, No., 32*, 10–16.
358. Radermacher, F. J. (2000). Wissensmanagement in Superorganismen. In C. Hubig (Ed.),
     Unterwegs zur Wissensgesellschaft (pp. 62–81). Berlin: Edition Sigma.
359. Radermacher, F. J. (2002). *Balance oder Zerstörung: Ökosoziale Marktwirtschaft als Schlüs-
     sel zu einer weltweiten nachhaltigen Entwicklung.* Wien: Ökosoziales Forum Europa.
360. Radermacher, F. J. (2003). *Mit- und Gegeneinander der Kulturen in der globalen Informa-
     tionsgesellschaft : ein 'Balanced Way' als Zukunftsentwurf.* In K. R. Kegler & M. Kerner
     (Eds.), *Köln Böhlau Technik, Welt, Kultur: technische Zivilisation und kulturelle Identitäten
     im Zeitalter der Globalisierung* (pp. 49–73). ISBN: 3-412112038.
361. Radermacher, F. J. (2004a). *Global Marshall Plan/Ein Planetary Contract. Für eine weltweite
     Ökosoziale Marktwirtschaft.* Ökosoziales Forum Europa (ed.), Wien.
362. Radermacher, F. J. (2004b). *Perspektiven für den Globus - Welche Zukunft liegt vor uns?* zfv -
     Zeitschrift für Geodäsie, Geodateninformation und Landemanagement, Teil 1 in Heft 3/2004,
     129 Jg., Teil 2 in Heft 4, pp. 242–248.
363. Radermacher, F. J. (2005a). *Was macht Gesellschaften reich? - Die Infrastruktur als
     wesentlicher Baustein*, Metropolis-Verlag.
364. Radermacher, F. J. (2005b). *Die Zukunft der Infrastrukturen - Intelligente Netzwerke für eine
     Nachhaltige Entwicklung.* in: Metropolis-Verl., ISBN: 3895185027, pp. 97-112.
365. Radermacher, F. J. (2006a). *Globalisierung gestalten: Die neue zentrale Aufgabe der Poli-
     tik. Das Wirken des Bundesverbands für Wirtschaftsförderung und Aussenwirtschaft für eine
     globale Rahmenordnung einer Ökosozialen Marktwirtschaft.* Horizonte Verlag GmbH.
366. Radermacher, F. J. (2007a). *Bewusstsein, Ressourcenknappheit, Sprache: Überlegungen zur
     Evolution leistungsfähiger Systeme in Superorganismen.* Ulm: FAW/n Report.
367. Radermacher, F. J. (2007b). *Welt mit Zukunft, Ueberleben im 21.* Jahrhundert: Murmann
     Verlag, Hamburg.
368. Radermacher, F. J., Obermüller, M., & Spiegel, P. (2009). *Global Impact: Der neue Weg zur
     globalen Verantwortung.* München: Carl Hanser Verlag. ISBN 978-3446417304.
369. Radermacher, F. J. (2010a). *Die Zukunft unserer Welt.* Edition Stifterverband: Navigieren in
     schwierigem Gelaende.
370. Radermacher, F. J. (2010b). *Weltklimapolitik nach Kopenhagen - Umsetzung neuer Poten-
     ziale.* FAW/n Report, Forschungsinstitut für Anwendungsorientierte Wissensverarbeitung/n.
     Retrieved from http://www.faw-neu-ulm.de/weltklimapolitik-nach-kopenhagen.
371. Radermacher, F. J. (2011). *Welt mit Zukunft.* Die Ökosoziale Perspektive: Murmann Verlag,
     Hamburg.
372. Radermacher, F. J., Riegler, J., & Weiger, H. (2011). *Ökosoziale Marktwirtschaft.* Historie,
     Programm, Perspektive eines zukunftsfähigen globalen Wirtschaftssystems: Oekom Verlag.
     ISBN 978-3865812599.
373. Radermacher, F. J. (2013). *Die Ressourcen der Erde setzen uns Grenzen - vom sächsischen
     Bergmann Hans Carl von Carlowitz 1713 bis zum neuen Report an den Club of Rome 2052.* In:
     Die Erfindung der Nachhaltigkeit - Leben, Werk und Wirkung des Hans Carl von Carlowitz.
     Sächsische Hans-Carl-von-Carlowitz-Gesellschaft e. V. (Hrsg.), pp. 141–155, oekom Verlag.
374. Radermacher, F. J. (2014a). *Chancen und Risiken durch Robotik.* Computerwoche
     29.09.2014. Retrieved from http://www.computerwoche.de/a/franz-josef-radermacher-und-
     gunter-dueck-im-gespraech,3068463.
375. Radermacher, F. J. (2014b). *Kann die 2-Grad-Obergrenze noch eingehalten werden?*
     Forschungsinstitut für Anwendungsorientierte Wissensverarbeitung/n (FAW/n-Bericht), Ulm.
376. Radermacher, F. J. (2014c). Can we still comply with the maximum limit of 2o C? Approaches
     to a new climate contract. *CADMUS, 2*(3), 152–161.

377. Radermacher, F. J. (2015a). *Algorithmen, Maschinelle Intelligenz, Big Data: Einige Grund-satzüberlegungen*. Schwerpunktheft "'Big Data contra große Datensammlungen. Chancen und Risiken für die Gesundheitsforschung'" des Bundesgesundheitsblattes.
378. Radermacher, F. J. (2015b). *Überlegungen zur Ausgestaltung von Club of Rome Schulen*. Forschungsinstitut für Anwendungsorientierte Wissensverarbeitung/n (FAW/n) Working Paper.
379. Ramirez, F. O., & Meyer, J. W. (1980). Comparative education: The social construction of the modern world system. *Annual Review of Sociology*, 369–399.
380. Randers, J. (2012). *2052—A Global Forecast fort he Next Forty Years*. Chelsea Green Publishing.
381. Rane, H., Ewart, J., & Martinkus, J. (2014). *Media framing of the Muslim World: Conflicts*. Palgrave Macmillan, Hampshire: Crises and Contexts. ISBN 978-1137334824.
382. Rapoport, A., & Chammah, A. M. (1965). *Prisoner's dilemma: a study in conflict and cooperation*. University of Michigan Press.
383. Reichholf, J. R. (2002). Der Mensch als Produkt der Evolution. In A. Beyer (Ed.), *Fit für Nachhaltigkeit?*. Opladen.
384. Reinefeld, A. (2006). *Entwicklung der Spielbaum-Suchverfahren: Von Zuses Schachhirn zum modernen Schachcomputer*. Berlin, Heidelberg: Springer.
385. Reisig, W. (2010). *Petrinetze*. Springer.
386. Reisz, R. D., & Stock, M. (2007). Theorie der Weltgesellschaft und statistische Modelle im soziologischen Neoinstitutionalismus. Zeitschrift für Soziologie, Jg. 36. *Heft, 2*, 82–99.
387. Renz, A., Schmid, H., Takim, A., & Ucar, B. (2008). *Verantwortung für das Leben: Ethik in Christentum und Islam*. Theologisches Forum Christentum - Islam: Verlag Friedrich Pustet, Regensburg. ISBN 978-3791721866.
388. Research ICT Africa. (2014). *Mapping Multistakeholder Participation in Internet Governance from an African Perspective*. Results of a survey on African Internet Governance. Retrieved from http://www.researchictafrica.net/presentations/Presentations/RIA_2014_-_Survey_results_on_Mapping_Multistakeholderism_in_Internet_Governance_from_Africa. pdf.
389. Riegler, J. (2009). *Den Blick nach vorn - Ökosozial leben und wirtschaften*. Club Niederösterreich in Kooperation mit dem Ökosozialen Forum Steiermark und dem Ökosozialen Forum Österreich.
390. Robb, F. (1989). Cybernetics and suprahuman autopoietic systems. *Systems Practice, 2*(1), 47–74.
391. Rogall, H. (2008). *Ökologische Ökonomie: Eine Einführung*. VS Verlag für Sozialwissenschaften. ISBN: 978-3531160580.
392. Rogall, H. (2009). *Vom Homo oeconomicus zum Homo cooperativus*. In Vorgänge, Zeitschrift für Bürgerrechte und Gesellschaftspolitik, 48. Jg.
393. Rogall, H. (2011). *Grundlagen einer nachhaltigen Wirtschaftslehre: Volkswirtschaftslehre für Studierende des 21. Jahrhunderts*. Metropolis, 1. Auflage. ISBN: 978-3895188602.
394. Rogers, E. (2003). *Diffusion of innovations*. New York: Free Press. ISBN 978-0743222099.
395. Root Server. (2014). *Root Servers Homepage*. 2014. http://www.root-servers.org/.
396. de Rosnay, J. (1979). *The macroscope*. New York: Harper & Row.
397. de Rosney, J. (2002). *Superorganismus Erde: Gedanken aus der Sicht eines utopischen Biologen*. Tec21, Band 128, Nr.27/28 Natur und Natürlichkeit, pp. 22–23. Retrieved from http://retro.seals.ch/cntmng?type=pdf&rid=sbz-004:2002:128::798&gathStatIcon=true.
398. Rouvinen, P. (2006). Diffusion of digital mobile telephony: Are developing countries different? *Telecommunications Policy, 30*, 46–63.
399. Russell, P. (1982). *The awakening Earth: The global brain*. London: Routledge & Kegan Paul.
400. Samir, S (2014). *The ITU and unbundling internet governance*. Council on Foreign Relations. Retrieved from http://www.cfr.org/internet-policy/itu-unbundling-internet-governance/p33656.

401. Sanders, A. (2011). *Bildung für das Leben in der Weltgesellschaft-Eine dokumenten-und fallanalytisch gestützte Untersuchung des Bildungskonzepts der UNESCO-Projektschulen Deutschlands.* Dissertation/ Universitätsbibliothek, Leuphana Universität Lüneburg. Retrieved from http://opus.uni-lueneburg.de/opus/volltexte/2013/14259/pdf/DissAnjaSanders.pdf.

402. Saunders, R. J., Warford, J. I., & Wellenius, B. (1994). *Telecommunications and economic development.* Baltimore: John Hopkins University Press.

403. Schauer, T., & Radermacher, F. J. (2001). *The challenge of the digital divide: Promoting a global society dialogue.* Ulm: Universitäts-Verlag.

404. Schlensog, S. (2006). *Der Hinduismus.* Glaube, Geschichte, Ethos: Piper Verlag, München. ISBN 3-492-04850-1.

405. Schiller, B. (2014). *First degree price discrimination using Big Data.* Brandeis University. Retrieved from http://www.benjaminshiller.com/images/First_Degree_PD_Using_Big_Data_Jan_27,_2014.pdf.

406. Schmidt, H. (1998). *Allgemeine Erklärung der Menschenpflichten - Ein Vorschlag.* Piper Verlag. ISBN: 978-3492226646.

407. Schmidhuber, J. (2012). Philosophers and futurists, catch up! response to the singularity. *Journal of Consciousness Studies, 19,* 173–182.

408. Scholl-Latour, P. (2011). *Arabiens Stunde der Wahrheit: Aufruhr an der Schwelle Europas.* Propyl"aen Verlag. ISBN: 978-3549073667.

409. Schöning, U. (2008). *Ideen der Informatik: Grundlegende Modelle und Konzepte der theoretischen Informatik.* Oldenbourg Verlag.

410. Schubertus, K., & Klein, M. (2011). *Das Politiklexikon* (5th ed.). Bonn: Dietz.

411. Schwinn, T. (2004). Von der historischen Entstehung zur aktuellen Ausbreitung der Moderne. Max Webers Soziologie im 21. Jahrhundert. *Berliner Journal für Soziologie, 14,* 527–544.

412. Schwinn, T. (2006a). Konvergenz, Divergenz Oder Hybridisierung? *Kölner Zeitschrift für Soziologie und Sozialpsychologie, 58*(2), 201–232.

413. Schwinn, T. (2006b). *Die Vielfalt und Einheit der Moderne: Kultur- und strukturvergleichende Analysen.* VS Verlag für Sozialwissenschaften. ISBN: 978-3531144276.

414. Sen, A. (2002). *Ökonomie für den Menschen: Wege zu Gerechtigkeit und Solidarität in der Marktwirtschaft.* Deutscher Taschenbuch Verlag. ISBN: 978-3423362641.

415. Sen, A. (2009). *The idea of justice.* Harvard University Press. ISBN: 978-0674036130.

416. Sen, A. (2010). *Die Identitätsfalle. Warum es keinen Krieg der Kulturen gibt.* dtv, München 2010 (Originaltitel: Identity and Violence. The Illusion of Destiny, übersetzt von Friedrich Griese), ISBN: 978-3-423-34601-6.

417. Sherkat, D. E., & Christopher, G. E. (1999). Recent developments and current controversies in the sociology of religion. *Annual Review of Sociology, 25*(1), 363–394.

418. Sesink, W. (1993). *Menschliche und künstliche Intelligenz: Der kleine Unterschied.* Stuttgart: Klett-Cotta Verlag. ISBN 9783608954982.

419. Shalbak, M. (2012). *Mauern - Israelische Trennungsstrategien und ihre Folgen für Raumentwicklung in den palästinensischen Gebieten.* Dissertation, Fakultät für Architektur, Karlsruhe Institut für Technologie.

420. Shannon, C. E. (1948). A mathematical theory of communication. *Bell Systems Technical Journal, 27*(379–423), 623–656.

421. Shiu, A., & Lam, P. L. (2008). *Causal relationship between telecommunications and economic growth. A Study of 105 countries.* 17th Biennial Conference of the International Telecommunications Society, Montreal. Retrieved from http://www.imaginar.org/taller/its2008/192.pdf.

422. Siemens. (2014). *Internet der Dinge. Fakten und Prognosen: Die Vernetzung der Welt.* Retrieved from http://www.siemens.com/innovation/de/home/pictures-of-the-future/digitalisierung-und-software/internet-of-things-fakten-und-prognosen.html.

423. Sigalas, E. (2010). Cross-border mobility and European identity: The effectiveness of intergroup contact during the ERASMUS year abroad. *European Union Politics, 11*(2), 241–265. https://doi.org/10.1177/1465116510363656.

424. Smith, A.(1776). *An inquiry into the nature and causes of the wealth of nations*. Band 1. Nachdruck von 1981. Indianapolis Indiana USA. ISBN: 0-86597-006-8.
425. Smith, C., & Snell, P. (2009). *Souls in transition: The religious and spiritual lives of emerging adults*. New York: Oxford University Press.
426. Solte, D. (2008). *Weltfinanzsystem am Limit: Einblicke in den "'Heiligen Gral"' der Globalisierung*. Terra Media Verlag, ISBN: 978-3981171525.
427. Solte, D. (2015). *Wann haben wir genug?: Europas Ideale im Fadenkreuz elitärer Macht*. Goldegg Verlag, 1. Auflage, ISBN: 978-3902991386.
428. Soeffner, H. G. (2005). Methodological cosmopolitanism-How to maintain cultural diversity despite economic and cultural globalization. In H.-J. Aretz & C. Lahusen (Eds.), *Die ORdnung der Gesellschaft* (pp. 413–427). Frankfurt a.M.: Peter Lang.
429. Sorrentino, M., & Niehaves, B. (2010). Intermediaries in E-Inclusion: A literature review. In *Proceedings of the 43rd Hawaii International Conference on System Sciences* (pp. 1–10).
430. Spencer, H. (1895). *The principles of sociology* (Vol. 6). Appleton.
431. Stagl, J. (1998). Homo Collector: Zur Anthropologie und Soziologie des Sammelns. In *Aleida Assmann und andere (Herausgeber): Sammler - Bibliophile - Exzentriker* (p. 50). Gunter Narr Verlag, Tübingen, ISBN: 3-8233-5700-X.
432. Statistisches Bundesamt. (2005). *Bruttoinlandsprodukt (BIP)*. Retrieved from https://www.destatis.de/DE/ZahlenFakten/GesamtwirtschaftUmwelt/VGR/Methoden/BIP.html
433. Steinbuch, K. (1961). *Automat und Mensch*. Springer, Berlin: Über menschliche und maschinelle Intelligenz.
434. Steinbuch, K. (1966). *Die informierte Gesellschaft*. Geschichte und Zukunft der Nachrichtentechnik: Stuttgart, dva.
435. Steiner, F. (2005). Einsteins kosmische Religiosität. *Albert Einstein* (pp. 191–217). Berlin Heidelberg: Springer.
436. Steinmüller, W. E. (2001). ICT and the possibilities of leapfrogging by developing countries. *International Labour Review, 140*, 193–210.
437. Stelter, D. (2014). Die Schulden im 21. Jahrhundert. *Frankfurter Societäts-Medien GmbH* (Vol. 1). Auflage. ISBN: 978-3956010774.
438. Stern, N. (2008). The economics of climate change. *American Economic Review, 98*(2), 1–37. https://doi.org/10.1257/aer.98.2.1.
439. Stichweh, R. (1994). Nation und Weltgesellschaft. In *Das Prinzip Nation in modernen Gesellschaften, VS Verlag für Sozialwissenschaften* (pp. 83–96).
440. Stichweh, R. (1995). *Zur Theorie der Weltgesellschaft. Soziale Systeme, 1*(1), 29–45.
441. Stichweh, R. (1997). *Inklusion/Exklusion und die Theorie der Weltgesellsehaft* VS Verlag für Sozialwissenschaften (pp. 601–607).
442. Stichweh, R. (1999). Globalisierung von Wirtschaft und Wissenschaft. Produktion und Transfer wissenschaftlichen Wissens in zwei Funktionssystemen der modernen Gesellschaft. *Soziale Systeme, 5*(1), 27–39.
443. Stichweh, R. (2000). *Die Weltgesellschaft*. Soziologische Analysen, Frankfurt a.M.
444. Stiglitz, J., Sen, A., & Fitoussi, J. P. (2008). *Issues paper, commission on the measurement of economic performance and social progress*. Retrieved from http://www.stiglitz-sen-fitoussi.fr/documents/Issues_paper.pdf.
445. Stiglitz, J. (2012). *Der Preis der Ungleichheit: Wie die Spaltung der Gesellschaft unsere Zukunft bedroht*. Siedler Verlag, 4. Auflage. ISBN: 978-3827500199.
446. Stock, G. (1993). *Metaman: The merging of humans and machines into a global superorganism*. New York: Simon & Schuster. ISBN 978-0671707231.
447. Tagesspiegel. (2014). *Studie: Allgemeine Geschäftsbedingungen werden im Netz kaum gelesen*. Retrieved from http://www.tagesspiegel.de/medien/digitale-welt/agbs-im-internet-studie-allgemeine-geschaeftsbedingungen-werden-im-netz-kaum-gelesen/10973008.html.
448. Tambini, D., Leonardi, D., & Marsden, C. (2007). *Codifying Cyberspace: Communications self-regulation in the age of internet convergence*. London: Routledge.
449. Tayler, E. B. (1871). *Primitive culture—Researches into the development of mythology, philosophy, religion, language, art, custom* (Vol. 2). Cambridge University Press. Original veröffentlicht 1871. ISBN: 9781108017510.

450. Taylor, D. (2014). NETmundial: Global multistakeholder meeting on the future of internet governance. *Journal of Internet Law, 17*(12), 25–27.
451. Töpfer, K. (2006). *Atlas der Globalisierung: Die neuen Daten und Fakten zur Lage der Welt.* TAZ, Auflage 2., ISBN: 978-3937683072.
452. Torero, M., & von Braun, J. (Eds.). (2006). *Information and communication technologies for development and poverty reduction: The potential of telecommunications.* Baltimore: International Food Policy Research Institute, John Hopkins University Press.
453. Tsatsou, P. (2009). Reconceptualising time and space in the era of electronic media and communications. *Journal of Media and Communications.* ISSN: 1836-5132.
454. Tsatsou, P. (2011). Digital divides revisited: What is new about divides and their research? *Media, Culture & Society, Jg., 33*(2), 317–331.
455. Turchin, V. (1977). *The phenomenon of science. A cybernetic approach to human evolution.* New York: Columbia University Press.
456. Turchin, V. (1981). *The inertia of fear and the scientific worldview.* New York: Columbia University Press.
457. Turing, A. M. (1950). Computing machinery and intelligence. *Mind, 59,* 433–460.
458. UN Economic Commission for Africa. (2014). *African Internet Governance Forum.* Garki-Abuja. Retrieved from http://www.uneca.org/afigf.
459. United Nations. (1948). *Allgemeine Erklärung der Menschenrechte.* Resolution der Generalversammlung. Retrieved from http://www.un.org/depts/german/menschenrechte/aemr.pdf.
460. United Nations. (2012). *Realizing the future we want for all.* Report to the Sectretary-General. UN System Task Team on the Post-2015 UN Development Agenda. New York. Retrieved from http://www.un.org/millenniumgoals/pdf/Post_2015_UNTTreport.pdf.
461. United Nations. (2013). *World population prospects: The 2012 revision.* Department of Economic and Social Affairs, Population Division, New York. Retrieved from http://esa.un.org/wpp/Documentation/pdf/WPP2012_Volume-I_Comprehensive-Tables.pdf.
462. United Nations. (2015). *Millenium Development Goals.* Retrieved from http://www.un.org/millenniumgoals/bkgd.shtml.
463. UNCTAD. (2014). *United Nations Conference on Trade and Development.* Retrieved from http://unctad.org/en/Pages/Statistics.aspx
464. UNDP. (2013a). *Human Development Report.* United Nations Development Programme: The Rise of the South. Human Progress in a Diverse World. ISBN 978-9211263404.
465. UNDP. (2013b). *Human Development Report. Technical Notes.* United Nations Development Programme. Retrieved from http://hdr.undp.org/sites/default/files/hdr_2013_en_technotes.pdf
466. UNDP. (2014): *Education Index.* Human Development Reports. Retrieved from http://hdr.undp.org/en/content/education-index
467. UNESCO. (1983): *Weltkonferenz über Kulturpolitik.* Schlussbericht UNESCO-Konferenz vom 26. Juli bis 6. August 1982 in Mexiko-Stadt. Hrsg. Deutsche UNESCO-Kommission. München: K. G. Saur. (UNESCO-Konferenzberichte, Nr. 5), p. 121.
468. UNESCO. (2000). *World Education Forum.* Retrieved from http://www.unesco.org/education/efa/wef_2000/.
469. UNESCO. (2002). *Allgemeine Erklärung zur kulturellen Vielfalt.* UNESCO heute, Zeitschrift der Deutschen UNESCO-Kommission, Ausgabe 1-2.
470. UNESCO. (2003). *United Nations Literacy Decade.* Retrieved from http://www.unesco.org/new/en/education/themes/education-building-blocks/literacy/un-literacy-decade/.
471. UNESCO. (2012). *Adult and Youth Literacy, 1990–2015. Analysis of data for 41 selected countries.* UNESCO Institute for Statistics. ISBN: 978-92-9189-117-7.
472. UNESCO. (2013). *Technology.* A Report by the Broadband Commission Working Group on Education: Broadband and Education. Advancing the Education for All Agenda.
473. UNESCO. (2013). *UNESCO Institute for Statistics.* Retrieved from http://www.uis.unesco.org/Pages/default.aspx.
474. UNESCO. (2014). *UNESCO International Literacy Day 2014.* Retrieved from http://www.unesco.org/new/en/unesco/events/prizes-and-celebrations/celebrations/international-days/literacy-day/.

475. UNESCO. (2014). *UNESCO Institute for Statistics. Education: Adult Literacy Rate.* Retrieved from http://data.uis.unesco.org/Index.aspx?DataSetCode=EDULIT_DS& popupcustomise=true&lang=en.

476. UNESCO. (2015). *Education for all global monitoring report.* Retrieved from http://www. unesco.org/new/en/education/themes/leading-the-international-agenda/efareport/.

477. Ünver, H., Manske, J., Dokter, N. Hofman, K., & Ramsden, R. (2013). Getting Prepared for a globalised digital world. In European Institute of Innovation and Technology (Ed.), *Annual Innovation Forum, Brussels* (pp. 20–27). Retrieved from http://eit.europa.eu/sites/default/ files/EITFoundation-YoungLeadersGroup.pdf.

478. Ünver, H. (2013). *Bildungsniveau und Internetzugang - Weltweite Empirische Analysen.* Bachelorarbeit Wirtschaftswissenschaften: Universität Ulm.

479. Ünver, H. (2014). Explaining education level and internet penetration by economic reasoning - Worldwide analysis from 2000 through 2010. *International Journal for Infonomics (IJI)*, 7(1/2), 898–912.

480. Ünver, H. (2015). *Globale Vernetzung, Kommunikation und Kultur - Konflikt oder Konvergenz?* Doctoral dissertation, Ulm University. Retrieved from https://www.deutsche-digitale-bibliothek.de/binary/SOHVD7ES3CC34KV43AI6PJQUB6ZY3JCY/full/1.pdf.

481. Ünver, H. (2017). Measuring the Global Information Society ? Explaining digital inequality by economic level and education standard. In *IOP Conference Series: Materials Science and Engineering* (Vol. 173, Issue 1, pp. 1–20). Retrieved from http://iopscience.iop.org/article/ 10.1088/1757-899X/173/1/012021/pdf.

482. Viertel, K. (2014). *Geschichte der gleichmäßigen Konvergenz: Ursprünge und Entwicklungen des Begriffs in der Analysis des 19.* Jahrhunderts: Springer-Verlag.

483. Vinge, V. (1993). The Coming Technological Singularity. *Whole Earth Review*, 81, 88–95.

484. Vodafone. (2013). *Mobile Money Webinar.* Retrieved from http://www.vodafone.com/ content/dam/vodafone/investors/company_presentations/2013/Mobile_Money_Webinar_ transcript.pdf.

485. Vogt, O. A. (1997). *Analyse des makroökonomischen Schadens von Korruption.* In Korruption im Wirtschaftsleben: Deutscher Universitätsverlag. ISBN 978-382446447.

486. Voss, K. (2014). *Internet und Partizipation: Bottom-Up oder Top-Down?.* Springer VS: Politische Beteiligunsmöglichkeiten im Internet. Bürgergesellschaft und Demokratie. ISBN 978-3658010270.

487. Wagner, B. (2001). Kulturelle Globalisierung: Weltkultur, Glokalität und Hybridisierung. In B. Wagner (Ed.), *Kulturelle Globalisierung - Zwischen Weltkultur und kultureller Fragmentierung* (pp. 9–38). Essen: Klartext.

488. Wahl, P. (2000). *Zwischen Hegemonialinteressen, Global Governance und Demokratie - Zur Krise der WTO.* Internationale Politik und Gesellschaft. 2000. Retrieved from http://library. fes.de/pdf-files/ipg/ipg-2000-3/artwahl.pdf.

489. Wallerstein, I. (1974). *The modern world-system I: Capitalist agriculture and the origins of the European world-economy in the sixteenth century.* New York: Academic.

490. Wallerstein, I. (1987). World-systems analysis. *Social Theory Today*, 309–324.

491. Wallerstein, I. (1999). *The end of the world as we know it.* Minneapolis.

492. Wareham, J., & Levy, A. (2002). Who will be the adopters of 3G mobile computing devices? A probit estimation of mobile telecom diffusion. *Journal of Organizational Computing and Electronic Commerce*, 12(2), 161–174.

493. Warschauer, M. (2004). *Technology and social inclusion: Rethinking the digital divide.* MIT Press. ISBN: 978-0262731737.

494. World Culture Forum. (2009). Retrieved from http://www.wcf-dresden.com/.

495. Weingardt, M. A. (2007). *Religion macht Frieden: Das Friedenspotential von Religionen in politischen Gewaltkonflikten.* Stuttgart: Kohlhammer Verlag. ISBN 978-3170198814.

496. Wenger, A. (2008). *Bildung in einer sich formierenden Weltgesellschaft.* Forschungsinstitut für Anwendungsorientierte Wissensverarbeitung/n (FAW/n): Ein Schlüssel zur Förderung einer Nachhaltigen Entwicklung. ISBN 978-3981184112.

497. Wells, H. G. (1938). *World Brain.* Limited, London: Metheum & Co.

498. Wikipedia, Die freie Enzyklopädie. *Liste von Religionen und Weltanschauungen*, erstellt von Benutzer:Neitram basierend auf der leeren Weltkarte en:Image:BlankMap-World.png. Retrieved from http://upload.wikimedia.org/wikipedia/de/2/2b/Weltreligionen.png.

499. Wieland, R., & Scherrer, K. (2000). *Arbeitswelten von morgen: Neue Technologien und Organisationsformen, Gesundheit und Arbeitsgestaltung, flexible Arbeitszeit- und Beschäftigungsmodelle*. VS Verlag für Sozialwissenschaften. ISBN: 978-3531135670.

500. Wiener, N. (1948). *Cybernetics: Or control and communication in the animal and machine.* MIT Press.

501. Wiener, N. (1950). *The human use of human beings: Cybernetics and society.* London: Free Association Books.

502. Wilson, W. P. (2000). *The Internet Church.* Word Pub. ISBN: 978-0849916397.

503. Winner, E., Goldstein, T., & Vincent-Lancrin, S. (2013). *Art for Art's Sake? The impact of arts education.* Educational Research and Innovation, OECD Publishing. Retrieved from http://dx.doi.org/10.1787/9789264180789-en.

504. Wireless Local Loop: *Research Assignment 48740 Communications Networks*, University of Technology, Sydney. Retrieved from http://services.eng.uts.edu.au/~kumbes/ra/Access-Networks/wll/Kent/Default.htm#Introduction; Letzter Zugriff: 30.09.2014.

505. Wood, R. (2002). *Faith in action.* University of Chicago Press.

506. Wohlmeyer, H. (2006). *Globales Schafe Scheren: Gegen die Politik des Niedergangs.*EDITION VA bENE, Klosterneuburg, Auflage: 1. ISBN: 978-3851671834.

507. World Bank. (2012). *Inclusive green growth: The pathway to sustainable development.* Washington: World Bank. ISBN 978-0821395516. Retrieved from http://siteresources.worldbank.org/EXTSDNET/Resources/Inclusive_Green_Growth_May_2012.pdf.

508. World Bank. (2013). *World Bank. World DataBank.* Retrieved from http://databank.worldbank.org/data/home.aspx.

509. WGIG. (2005). *Report of the working group on internet governance*, Chateau de Bossey. Retrieved from http://www.wgig.org/docs/WGIGREPORT.pdf.

510. Wilkinson, R., & Pickett, K. (2011). *The spirit level: Why more equal societies almost always do better.* Bloomsbury Press. ISBN: 978-1608193417.

511. Wikipedia. (2017). *Vernetzung.* Retrieved from https://de.wikipedia.org/wiki/Vernetzung.

512. Wise, K. D., Anderson, D. J., Hetke, J. F., Kipke, D. R., & Najafi, K. (2004). Wireless implantable microsystems: High-density electronic interfaces to the nervous system. *Proceedings of the IEEE, 92*(1), 76–97.

513. Wöhlcke, M. (2003). *Das Ende der Zivilisation. Über soziale Entropie und kollektive Selbstzerstörung.*

514. World Summit on the Information Society. (2003). *World Summit on Sustainable Development.* United Nations. Johannesburg. Retrieved from http://www.un.org/events/wssd/.

515. World Summit on the Information Society. (2003a). *World Summit on the Information Society.* United Nations and International Telecommunication Union. Geneva. Retrieved from http://www.itu.int/dms_pub/itu-s/md/03/wsis/doc/S03-WSIS-DOC-0004!!PDF-E.pdf.

516. World Summit on the Information Society. (2003b). *Declaration of principles—Building the Information Society: A global challenge in the new Millennium.* Geneva. Retrieved from http://www.itu.int/wsis/docs/geneva/official/dop.html.

517. World Summit on the Information Society. (2005). *World Summit on the Information Society.* United Nations and International Telecommunication Union. Tunis. Retrieved from http://www.itu.int/wsis/docs2/tunis/off/6rev1.html.

518. W3C. (2015). *World Wide Web Consortium.* Retrieved from http://www.w3.org/.

519. Yitzhaki, S., & Schechtman, E. (2012). *The Gini methodology: A primer on a statistical methodology* (Vol. 272). Springer Science & Business Media.

520. Yunus, M. (2008). *Die Armut besiegen.* Carl Hanser Verlag GmbH & Co. KG, 1. Auflage. ISBN-13: 978-3446412361.

521. Yusof, S. H., et al. (2013). The framing of International Media on Islam and terrorism. *European Scientific Journal, 9*(8).

522. Zhang, X. (2013). Income disparity and digital divide: The Internet Consumption Model and cross-country empirical research. *Telecommunication Policy, 37*, 515–529.
523. Ziegler, J. (2005). *Das Imperium der Schande - Der Kampf gegen die Armut und Unterdrückung*. C. Bertelsmann Verlag. ISBN: 978-3570008782.
524. Ziegler, J. (2006). *Alle Menschen können satt werden, doch der Hunger wächst.* Tischgespräch: Wer verhungert wird ermordet. In: Greenpeace Magazin1.07, p.45.
525. Zillien, N. (2009). *Digitale Ungleichheit: Neue Technologien und alte Ungleichheiten in der Informations- und Wissensgesellschaft.* VS Verlag für Sozialwissenschaften, 2. Auflage. ISBN: 978-3531166735.
526. Zucman, G. (2014). *Steueroasen: Wo der Wohlstand der Nationen versteckt* (wird ed.). Deutsche Erstausgabe: Suhrkamp Sonderdruck. ISBN 978-3518060735.
527. Zuse, K. (1970). *Der Computer: Mein Lebenswerk* (Vol. 5). Auflage, Berlin: Springer (Erstveröffentlichung Verlag Moderne Industrie). ISBN: 978-3642120954.